Parrots of the Wild

The publisher gratefully acknowledges the generous support of the General Endowment Fund of the University of California Press Foundation.

Parrots of the Wild

*A Natural History of the World's
Most Captivating Birds*

CATHERINE A. TOFT AND TIMOTHY F. WRIGHT

Foreword by   James D. Gilardi

IN COLLABORATION WITH THE WORLD PARROT TRUST

UNIVERSITY OF CALIFORNIA PRESS

University of California Press, one of the most distinguished university presses in the United States, enriches lives around the world by advancing scholarship in the humanities, social sciences, and natural sciences. Its activities are supported by the UC Press Foundation and by philanthropic contributions from individuals and institutions. For more information, visit www.ucpress.edu.

University of California Press
Oakland, California

Library of Congress Cataloging-in-Publication Data

Toft, Catherine Ann, 1950–2011, author.
    Parrots of the wild : a natural history of the world's most captivating birds / Catherine A. Toft
and Timothy F. Wright in collaboration with the World Parrot Trust ; foreword by James
D. Gilardi. — First edition.
        pages cm
    Includes bibliographical references and index.
    ISBN 978-0-520-23925-8 (cloth : alk. paper) — ISBN 0-520-23925-3 (cloth : alk. paper) —
    ISBN 978-0-520-96264-4 (ebook) — ISBN 0-520-96264-8 (ebook)
    1. Parrots.   I. Wright, Timothy F., 1967- author.   II. World Parrot Trust.   III. Title.
    QL696.P7T64 2015
    598.7'1—dc23

                                                                                        2015001989

Printed in China

24 23 22 21 20 19 18 17 16 15
10 9 8 7 6 5 4 3 2 1

The paper used in this publication meets the minimum requirements of ANSI/NISO Z39.48–1992 (R 2002)
(*Permanence of Paper*).

To Dr. Don Merton (1939–2011)
For his tireless devotion to the conservation of parrots;

To Keiki, Mickey, and Freestone
Cockatoos all
For starting me on this path.

All royalties from the sales of this book will be donated to the World Parrot Trust to support research on and conservation of wild parrots.

CONTENTS

Foreword by James D. Gilardi   ix

Preface and Acknowledgments   xiii

PART I   **The World of Parrots: Introducing the Psittaciformes**

1   What Are the Parrots and Where Did They Come From? The Evolutionary
History of the Parrots   3
Phylogeny

The Marvelous Diversity of Parrots   3

Reconstructing Evolutionary History   5

The Evolution of Parrots   8

Some Parrot Enigmas   29

PART II   **The Functional Parrot: Physiology, Morphology, and Behavior**

2   The Thriving Parrot: The Foods and Beaks of Parrots   39
Foraging Ecology

Introduction to the Diets of Parrots   39

The Seed Eaters   41

Other Diets   62

3   The Sensible Parrot: How Parrots Perceive and Use Information   81
Sensory Biology and Communication

Introduction to Sensory Biology   81

Visual Communication   82

Vocal Communication   94

4   The Thinking Parrot: The Brains of Parrots and How They Use Them   117
Cognitive Ethology

      Introduction to the Study of Learning and Cognition   117

      The Brains of Birds   118

      Cognition in Parrots   128

      Mental Distress in Captive Parrots   138

     PART III   **The Lives of Parrots: Mating, Life History, and Populations**

5   Sex and Marriage: The Mating Systems of Parrots   147
Behavioral Ecology

      Introduction to Mating Systems and How They Evolve   147

      Monogamy   148

      Other Mating Systems   162

6   From the Cradle to the Grave: The Life Histories of Parrots   177
Evolutionary Ecology

      Introduction to the Study of Life Histories   177

      Life in the Nest Hollow   178

      Life out in the World   200

7   Populations of Parrots: Conservation and Invasion Biology   217
Population Ecology

      Introduction to Conservation Biology   217

      Parrots in Peril   219

      Parrots as Invasive Species   246

    Epilogue: Themes and Threads Uniting the Chapters of This Book   261

      Parrots as the Most Human of Birds   261

      The Future of Wild Parrots   264

    Notes   267

    Bibliography   277

    Index   339

We are all familiar with orchids, the exquisite flowers that come in a bewildering variety of shapes and colors. It wasn't until I first climbed into the top of the Guatemalan forest for the first time, however, that I saw my first real, wild orchid. Only there, seeing them grow in the topical canopy along a limb festooned with bromeliads, vines, and other epiphytes, could I begin to understand where and how these plants really live, and why their unusual features evolved. It was also from this perch, high in the rainforest canopy, that I saw my first wild parrots. The birds were immediately familiar, just like the orchids, as I'd grown up with pet parrots, but the lives of these wild birds differed from that of a caged bird in nearly every way imaginable.

It had never dawned on me that the parrots I raised as a kid were really wild animals that had made a massive transition from the tropical forest to life in a cage in Southern California. Yet here they were, very much the same creatures, flying rapidly through the forest, socializing in fluid flocks, and feeding on all sorts of seeds, fruits, and flowers. All day and night, these birds were using their highly evolved bodies and years of learning to survive and thrive in the tropical forest. They were making life-and-death decisions about partners, food, and predators, with no knowledge whatsoever of humankind, cages or aviaries.

Jumping forward twenty-five years, I now have the privilege of directing an international organization devoted to all parrots, with aims to save rare parrots in the wild and to help ensure that the millions of parrots living with humans receive the best possible care. At the World Parrot Trust, we are forever with a foot in each camp. We use our understanding of wild parrots both to help save them from extinction and to help inform our best practices in captive bird care. On the flip side, living with parrots in and around our homes gives us an intimate window into their lives and needs, and often inspires questions to explore in their wild brethren.

A great deal of our work at the Trust involves applying what we know about the biology of wild parrots to understand and eliminate the threats that endanger so many wild species. This knowledge helps us rehabilitate and release thousands of parrots caught up in the illegal trade, and it also guides every decision we make about the long-term care of birds in captivity. Beyond just knowing how to feed and house them, this knowledge helps us interpret their behavior, to know how and when to provide them with social partners, how to ensure that they get plenty of exercise and stimulation, and how to recognize serious problems with their health and welfare. Until the 1980s, shockingly little was known about wild parrots, which meant that much of this work on their conservation and welfare had to be based on a little bit of knowledge and a healthy dose of guesswork.

Happily, all this has changed; we now know vastly more about the biology of parrots than we did at that time. Over the past three decades, the number of published scientific papers on parrot biology has grown rapidly, exponentially in fact. We can no longer make statements like "almost nothing is known about wild parrots," because this is simply no longer the case. Whether you're a parrot lover who just wants to understand what makes your favorite bird tick (or squawk), a college student hoping to understand a large and diverse group of birds, or a professor of biology studying and publishing cutting-edge science on parrots, new discoveries about these complex creatures are coming at us too fast for all but the most dedicated to keep up.

As it turns out, Dr. Catherine A. Toft wore all these hats, and she spent much of the last decade pulling together, reading, and digesting just about everything currently known about the biology of parrots. Initially she was motivated to do so through her own wide-ranging scientific studies of parrots—including publications on their genetics, phylogeny, diet, and conservation biology—but she also noted a pressing need to assemble and synthesize this growing body of knowledge. So, starting over a decade ago, Cathy began writing a comprehensive book on the biology of parrots, a project that eventually developed into the volume now in your hands. Sadly, she was diagnosed with nonsmoker lung cancer in the summer of 2011, just as the first draft of this book was delivered to the editors at University of California Press. It proved to be an aggressive cancer, and on December 2, 2011, she passed away, at just 61 years of age. The bittersweet task of producing the final version of the book was taken up by Cathy's long-time collaborator, Tim Wright.

Naturally, such a parrot book could have evolved in any of several different directions—a purely scientific endeavor targeting an academic audience, a coffee-table book focusing on their stunning beauty and impressive diversity, or a popular treatment with an aim to entertain and enlighten some of the millions of parrot enthusiasts around the world. What Cathy attempted and what she and Tim together have achieved is all of these and more.

Whether the reader turns to a specific topic of interest, or starts from the beginning and systematically devours the book chapter by chapter, all will quickly learn just how complex and intriguing parrots are by nature. And by artfully walking us through such a wide range of subjects, Cathy and Tim's decades of teaching shine through to enlighten us about how interconnected so many of their traits are—how, for example, the fact that they "talk" relates to their breeding and social systems, which relate to their cognitive function, which relates to their longevity, and so forth. Even the most common species—birds like budgies and cockatiels which many of us overlook as no more than a small bird in a cage—are themselves remarkably complex, and in fact many of the most intriguing findings come from studies of such familiar species.

When Cathy and I first discussed the idea of this book, one of the goals I hoped it might achieve was to give readers a newfound appreciation for all parrots—that these creatures are so much more than just a bird in a cage which may or may not amuse us with a whistle, a song, or a word or two. There is a great deal more going on with parrots than first meets the eye, which surely explains why they are the world's most popular non-domesticated pets. It's hard to imagine anyone reading *Parrots of the Wild* and not seeing all aspects of these birds in a whole new light, and beyond that, having a new and deeper appreciation for their complexity and their needs, whether captive or wild. For me as a parrot researcher and conservationist, having all this information together for the first time is more than convenient. Reading this book has already inspired new questions and suggested promising new avenues of research.

And while I'm certain that this book will inform and inspire all manner of people to understand, appreciate, and explore the many remaining mysteries of the lives of parrots, it will also directly help save parrots themselves, both wild and captive. From its inception, Cathy chose to work closely with the World Parrot Trust in the planning and writing phase, and in gathering images from photographers around the world, and she encouraged us at the Trust to be collaborators. She also generously agreed to donate all royalties from the sale of this book to a special conservation and welfare fund, administered by the Trust. So this book will not just give you the reader years of pleasure; your purchase will also support our work to save and celebrate all the parrots of the world.

James D. Gilardi
Davis, California
December 2014

MOTIVATION

The raucous screams of macaws and amazon parrots were my alarm clock for the months that I (Cathy Toft) spent as a graduate student doing my field research in the lowland rainforests of Peru, at the westernmost reaches of the great Amazon Basin. The break of dawn in the tropics happens just before 6 A.M. and is heralded by choruses of parrots and primates, through which sleep is impossible. That is, parrots and primates join forces, in the few parks and reserves where the large vertebrate fauna has not been virtually eradicated (figure 1). At my other field site, in northern Amazonian Peru, the forest looked the same but was strangely quiet.

My habit was to put my pup tent out in the forest a small distance from the main camp and cookhouse of our remote field station. At six o'clock every morning, screams pierced my sleep, and branches, leaves, and discarded fruit rained down on my tent, which was nestled at the base of a large tree in the forest. The rustling of leaves and the barrage of debris alerted me to a group of macaws more than 50 meters above me. I could not see them for the layers of canopy between us. It sounded as if there would have to be five hundred of them to account for all the ruckus, but at most they were a family trio of parents and chick, or maybe two trios, totaling perhaps a half-dozen birds. There was no point in pretending to sleep, and the twelve-hour tropical day would be short enough, as I had much to do that I preferred to do in daylight. Once night fell, the poisonous snakes, the big cats, and the large caimans in the oxbow lake, all common in Manú National Park, made wandering around in the forest a delicate enterprise.

The lowland rainforest that has not been touched by "civilization" is difficult to describe adequately. As a teenager, I read every book I could on pre-European North America, including the accounts of pioneering naturalists such as John Charles Fremont, Lewis and Clark, and John James Audubon. When I first arrived in the remote Manú National Park, the memories of these accounts flooded back to me as I saw the same biological wealth unfold before me. Most impressive was the unimaginable abundance of large vertebrates, at every turn, completely unafraid of

FIGURE 1 Red-and-green Macaws, *Ara chloropterus,* congregate in large numbers at a clay lick in Peru's Manú National Park. Photo © Luiz Claudio Marigo.

humans, a stark contrast to other lowland rainforests I had visited in Panama, Gabon, and farther north in Peru. In 1973, the forest was packed with curassows, guans and trumpeters—the large fowl that are the first to go when guns arrive. Monkeys of thirteen different species lounged in the treetops, and Harpy Eagles patrolled above the dense canopy, hoping to catch one of them off guard. Parrots were everywhere, numbering at least sixteen species. Mixed flocks of up to three species of macaws (Scarlet Macaw *Ara macao,* Red-and-green Macaw *Ara chloropterus,* and Blue and Yellow Macaw *A. ararauna*) often gathered in the afternoon in the enormous, open canopy of a single ceiba tree. Throughout the day, the chatter of parakeets and parrotlets (in the genera *Aratinga, Psittacara, Pionites, Pyrrhura, Brotogeris, Nannopsittaca,* and *Forpus*) filled the air with cascades of sound like an ever-present waterfall. Their calls worked their way into my subconscious until they were inextricable from the smell of the decaying earth, the caress of damp, warm air, and the oppressive greenness—the sheer magnitude of the rainforest. This veritable Garden of Eden still waits, as yet unspoiled, for the enterprising ecotourist to experience close at hand the daily life of wild parrots.

Few places like Manú National Park remain on this earth, since the expansion of one species, *Homo sapiens,* to every corner of the planet. The Manú River drainage contains one of the last ecosystems to remain virtually unaltered as our species has learned to extract resources for its own needs everywhere on earth, and the density of our human population has soared to unprecedented levels. Parrots have taken the brunt of our exploitation of the biosphere, and today the parrot order Psittaciformes comprises among the most endangered and threatened species of the twenty-eight (or so) orders of birds. Ironically, parrots are also among the birds most beloved by their human aficionados, and yet this love has endangered parrots far more than the destruction of their habitats or hunting them for food or as pests.

One summer day in 2002, Jamie Gilardi and I met over lunch to flesh out an idea that Jamie had to produce a book on the natural history of wild parrots. Jamie and I were gravely concerned about the welfare of both wild and captive parrots. Perhaps a book telling a story of all that is known about parrots in the wild could improve the lives of parrots worldwide, whether existing wild and free or with human companions.

Our concern about wild parrots was fueled in particular by a study that we had co-authored, led by our colleague Tim Wright, of New Mexico State University, and in collaboration with our many fellow scientists studying parrots in the wild. Our paper presented data on the rates of poaching of parrot nestlings for the international trade in wild-caught birds. Without a doubt, the legal plus illegal harvest of parrots for the trade in wild-caught birds was larger than any source of natural mortality. Moreover, the cause of this unsustainable exploitation of wild parrots was the demand created by parrot lovers worldwide. Jamie and I wondered whether a book that communicated all that is amazing and wonderful about wild parrots—getting to know about them in their natural lives unfettered by humans—could make a strong case for leaving them alone to live out their lives in the wild, with the richness that only nature can provide them.

At the same time, Jamie's work at the World Parrot Trust and my experience with aviculture and my own captive parrots opened our eyes to the potential for parrots to suffer in living their lives with us, as pets, companions, and as breeding livestock. Research done by the World Parrot Trust estimates that as many parrots exist in captivity as in the wild worldwide, albeit a different mix of species. With these impressive numbers, surely parrots in captivity are as worthy of our concern as those in the wild.

Regardless of how much love, concern, and caring we humans may have for our companion parrots, parrots are not as yet domesticated, in the sense that they have not been bred over many generations for traits that benefit humans. Many captive parrots were born in the wild, and most are only a generation or two removed from their wild ancestors. Their physical and mental needs have evolved through natural selection in the wild, in the habitats they have occupied for thousands of years. Like all of the animals that humans exploit, parrots have little or no choice about how they live with us. We provide them with an environment relatively safe from predators and food scarcity, yet we deprive them of much, much more. Parrots live with us in a highly altered and potentially barren social existence compared to their lives in the wild. They get little exercise and have few activities to occupy their minds for long. Unlike their wild parrot companions, we humans as substitute flock-mates do not provide companionship for our pets 24/7. Furthermore, their diets in captivity have little resemblance to the foods that they are adapted to eat—to some extent for the better but in far too many ways for the worse.

The answer to improving the lives of captive parrots, however, is not necessarily to return them to the wild, even if they were wild-caught. Yes, successful reintroductions of confiscated parrots are becoming more common as scientists work out the formulae for that success. But most pet parrots now are born in captivity or taken from the wild as naive nestlings. As highly social, intelligent beings, they will forever lack sufficient knowledge of how to cope on their own, and they lack the physical conditioning for a strenuous wild existence. Nonetheless, these animals do have a role to play in conservation. As recognized by a growing number of conservation biologists, companion and captive wild animals keep humans in touch with nature as the civilized world that we have created for ourselves becomes increasingly more stressful and hostile to our own needs. Being charmed by a pet parrot often causes people to care even more about what happens to parrots in the wild.

In other words, keeping parrots as companions and as a hobby or business is here to stay and has many benefits for both people and parrots. The ticket then is to seek the highest quality of life for the parrots that live with us and to ensure that no more parrots are removed from the wild to fill the needs that parrots already existing in captivity can supply instead. This book is dedicated to the dual goals of improving the lives of both wild and captive parrots. I hope to accomplish these goals by relating to you the fascinating tale of parrots' lives in the wild, as we now know it, thanks to the work of many hard-working scientists who study them there.

To enhance our success in achieving these goals, all royalties due the author and the photographers and artists, who graciously donated the wonderful images gracing this book, will go to the World Parrot Trust to support research on and conservation of wild parrots.

ROADMAP

To write this book, I took on the enjoyable task of reading every published scientific study done on parrots in the wild, and those done in the laboratory with the aim of better understanding the biology of wild parrots. This book presents to readers a compilation of all that I learned, after digesting and organizing all of this information for you. If I am successful, you, as my reader, will fully appreciate the diversity of the world of wild parrots. Such an understanding is critical at a time when much of this diversity is in peril.

The scientific study of wild parrots is truly daunting. Parrots may be noisy, colorful and conspicuous, but—as we will learn—they fly far each day to find food, shelter, and flockmates and to care for their young. They live high in forest canopies, in remote locations, and often at low population densities. Few animals are as challenging to study in their wild haunts as are parrots. When Jamie Gilardi and I first started in 1992 to compile a comprehensive and exhaustive library of all peer-reviewed scientific studies on wild parrots or relevant to wild parrots, we found 650 or so published since the late 1800s. We continued to compile our list after 1996, when Jamie graduated from my lab and went on to do great things for wild parrots in the world of conservation. By 2011, our library of the scientific literature on wild parrots contained more than 2,400 publications (and counting). It was buttressed by a burst of studies of wild parrots done in the last dozen years mostly by intrepid, enterprising young scientists undaunted by the challenges presented by wild parrots (figure 2). These new studies span the globe, from the Neotropics through Africa, Asia, and

FIGURE 2 A feral Red-crowned Parrot, *Amazona viridigenalis*, part of the large introduced population in the United States. Photo © Mike Bowles and Loretta Erickson.

Australia (where there is a long tradition of thorough study of the wild native parrots), and now Europe and North America, where formerly wild and captive bred parrots are establishing feral populations.

Thanks to this flurry of recent science, we now know enough about wild parrots worldwide to paint a fairly detailed portrait of the order Psittaciformes. We learn in this book about the formerly secret lives of wild parrots—about who they are, what they do, and why they do it. Many mysteries have been solved, but as is so often the case in science, new ones have emerged to tantalize new generations of scientists.

Throughout this book, I have selected topics for which a substantial peer-reviewed literature exists in primarily the basic sciences. The vast literature of applied studies, mostly in clinical veterinary science and aviculture, is not included. Some topics that fit my criteria—such as parrot flight and olfaction—did not fit easily into the organization or length limits, and will have to be presented elsewhere. In my presentation, I strive to reach audiences of both my colleagues and curious laypersons, that is to say, any and all of you who are attracted to parrots and want to know more about them. My approach is not to recapitulate other excellent reviews of the natural history of parrots. Rather, each of the seven main chapters centers on a conceptual field to which the scientific study of parrots has contributed significantly. In other words, this book also hopes to showcase what parrots teach us about important scientific questions in roughly seven conceptual fields of science. Some chapters also include boxes that delve into detail about specific species of parrots or specific studies, topics that might interrupt the logical development of ideas within the chapter or topics that integrate over multiple sections of the book.

I have chosen not to cite in the traditional manner (e.g., Gilardi and Toft 2012) the numerous studies both of parrots and of supporting scientific concepts. Not only would this double the length of the text, but the many citations required (I imagine at least one for virtually every sentence) would make the book unreadable. Likewise, I have chosen not to cite with distracting superscripts. Instead, each chapter contains a final section with literature notes that cite background references for the information presented in that chapter. Within the chapters I highlight specific studies in the text by focusing on the scientists themselves, to provide the reader with a more personalized approach to how science is done. My approach risks the appearance of not giving sufficient credit to other scientists, and for that I sincerely apologize—all of these hard-working scientists made this book possible for me to write, and without their great contributions it could not have been done.

All scientific and common names used in this book are based on the Sibley and Monroe *World Checklist of Birds*, except for nomenclature changes published after 1993, which affect some genus and species names.

A CHANGE IN NARRATORS

Now we come to a difficult change in narrators in which I, Tim Wright, take over from Cathy Toft in telling this story. In spring of 2011, Cathy was diagnosed with nonsmoker lung cancer. The prognosis was poor; by the time of its discovery the cancer had spread beyond her lungs to other parts of her body. I (Tim) was already moving to the University of California, Davis, with my family that year for a sabbatical leave from my academic home of New Mexico State University. I had chosen Davis for my sabbatical in large part for the chance to work with Cathy again. Early in my career she and I had collaborated on a paper documenting the destructive effects on parrots of capture for the pet trade, an experience that had taught me much about how to conduct both science and conservation. I had reviewed initial versions of many chapters of this book for Cathy, and was excited for the opportunity to delve with her more deeply into the many interesting questions it raised about parrot biology. Little did I suspect just how deeply enmeshed I would become in this book.

After Cathy's diagnosis the nature of my involvement in this book changed dramatically. At that point a preliminary draft of the book had been submitted by Cathy to the University of California Press for external review. The reviews were positive, but as is so often the case with scientific writing, the reviewers and editor had many suggestions for areas of improvement. In particular, they requested a complete rewrite of the initial chapter, on the evolution of parrots. Cathy was excited about the positive reception and eager to revise her work. As her illness rapidly progressed, though, it became apparent to her that she lacked both the energy and the time to continue writing. In November of 2011 she asked me to join her as an author and see the book through its final stages. On December 2, 2011, she passed away, at home, with her family at her side.

It then became my bittersweet privilege to see Cathy's work through to completion. My contributions have included writing a new version of the first chapter, and judiciously editing and condensing the remaining chapters. While I am the narrator in Chapter 1, I have kept closely to Cathy's original narrative voice in the remainder of the book. I also reorganized the boxes, compiled the literature notes, and helped select and caption the photographs and figures. In this latter task I have worked closely with Steve Milpacher of the World Parrot Trust, whose encyclopedic knowledge and fine photographer's eye was crucial to gathering the wonderful figures that bring this book to life. Throughout this process I have been aided by Jamie Gilardi, a former student of Cathy's who was instrumental in the conception of this book and whose work with the World Parrot Trust was a continual source of pride and inspiration to her.

ACKNOWLEDGMENTS

Every book represents the contributions of many people. Sadly, Cathy died before she could complete her own acknowledgements and personally thank her wide network of friends and colleagues who supported her through the long process of writing this book. In the absence of her own words, I am left with only the certain knowledge that she was deeply appreciative of the many people whose interest in parrots made this work possible. She found great inspiration both in the work of scientists and conservationists with wild parrots, and in the efforts of pet owners to provide fulfilling lives for their captive companions. It was your fascination with these remarkable animals that moved her to write this book.

Any acknowledgement of the specific contributions of some people runs the unfortunate risk of omitting those of others out of pure ignorance. Nonetheless, there are many people who I know Cathy would want to personally thank here. Her faculty colleagues and administration at the University of California, Davis, especially those within the Section of Evolution and Ecology, provided essential support through the long process of writing. Parrot biologists Leo Joseph, Rowan Martin, and Juan Masello provided valuable comments on early versions of some chapters, while Joanna Burger and John Marzluff provided astute reviews of the entire draft. She was deeply appreciative of the judicious guidance of Blake Edgar, Merrik Bush-Pirkle, and the staff at University of California Press through the editorial process. Finally, her family, both human and feathered, was a great source of love and support throughout her life.

For my own part, I am also in debt to the parrot biologists whose work is profiled here; every new finding of theirs only serves to increase my appetite for more. My students at New Mexico State University have been a source of inspiration, with my interactions with my graduate students particularly rich and intellectually inspiring. I'd also like to thank my colleagues at NMSU and collaborators around the world whose expertise and enthusiasm have allowed me to take on many exciting scientific challenges. My early interest in the natural world was nurtured by my parents, and their support has fostered an abiding passion for its preservation, not just in me but also in my brothers and their families. Most importantly, I'd like to thank my children, Nicholas and Madeleine, for their insightful questions that keep me walking down the road of discovery, and my wife, Kathy, for walking with me.

PART ONE

The World of Parrots

Introducing the Psittaciformes

What are the Parrots and Where Did They Come From?

The Evolutionary History of the Parrots

PHYLOGENY

CONTENTS

The Marvelous Diversity of Parrots   3

Reconstructing Evolutionary History   5

   Fossils, Bones, and Genes   5

The Evolution of Parrots   8

   Parrots' Ancestors and Closest
Relatives   8

   The Most Primitive Parrot   13

   The Most Basal Clade of Parrots   15

Other Major Groups of Parrots   16

   Box 1. Ancient DNA Reveals the
Evolutionary Relationships of the
Carolina Parakeet   19

   How and When the Parrots Diversified   25

Some Parrot Enigmas   29

   What Is a Budgerigar?   29

   How Have Different Body Shapes Evolved in
the Parrots?   32

THE MARVELOUS DIVERSITY OF PARROTS

The parrots are one of the most marvelously diverse groups of birds in the world. They dazzle the beholder with every color in the rainbow (figure 3). They range in size from tiny pygmy parrots weighing just over 10 grams to giant macaws weighing over a kilogram. They consume a wide variety of foods, including fruit, seeds, nectar, insects, and in a few cases, flesh. They produce large repertoires of sounds, ranging from grating squawks to cheery whistles to, more rarely, long melodious songs. They inhabit a broad array of habitats, from lowland tropical rainforest to high-altitude tundra to desert scrubland to urban jungle. They range over every continent but Antarctica, and inhabit some of the most far-flung islands on the planet. They include some of the most endangered species on earth and some of the most rapidly expanding and aggressive invaders of human-altered landscapes. Increasingly, research into the lives of wild parrots is revealing that they exhibit a corresponding variety of mating systems, communication signals, social organizations, mental capacities, and life spans. In a great many respects the 360 or so species of parrots represent a textbook

FIGURE 3 Scarlet Macaws, *Ara macao,* playing in a guanacaste tree (*Enterolobium cyclocarpum*) in Costa Rica. Photo © Steve Milpacher.

illustration of how the process of evolution can, over much time, lead to the diversification of many species from a single ancestral population.

At the same time, parrots are one of the most physically homogeneous groups of birds. Anyone with a passing familiarity with birds can instantly recognize a parrot by its sharply curved upper beak topped by a fleshy cere, muscular prehensile tongue, relatively big head and stout body, and distinctive zygodactyl feet with two toes pointing forward and two pointing back (figure 4). This combination of anatomical features clearly sets parrots apart from other birds. There are other, less obvious, commonalities in physiology, behavior, and ecology that tend to distinguish parrots from most other birds. These shared features illustrate another principal feature of evolution: that it tinkers with the materials at hand rather than starting anew with each species. In other words, major innovations are rare. What more typically happens is that features already present in an ancestor are slowly modified through natural selection over many generations to produce a constrained range of variations on the basic template as different lineages adapt to changing and localized environments.

In the following chapters we will delve deep into what recent scientific investigations have revealed about the lives of wild parrots. We will discuss how parrots perceive the world around them, how individuals go about their daily lives and interact with others, and how populations are adapting to a world that is rapidly changing. Our focus will be both on what

FIGURE 4 Nestlings of the Blue-fronted Amazon, *Amazona aestiva,* illustrating some of the basic morphological features like curved bills, big heads and short, sturdy legs that are shared by all parrots. Photo © Igor Berkunsky.

these investigations tell us about parrots in general, and on what can be learned from the interesting exceptions to these generalities. But before we start this exploration, we want to set the stage by summarizing the current state of knowledge of the evolutionary history of parrots: Where did they come from, how did they diversify, who among them is most closely related to whom, and what does this evolutionary history reveal about the process of evolution itself? To understand these topics, we must first understand how scientists explore what happened in the long-distant past.

RECONSTRUCTING EVOLUTIONARY HISTORY

Fossils, Bones, and Genes

Reconstructing the past history of life is both a historical exercise and a scientific one. Scientists typically illustrate evolutionary patterns as trees, with the common ancestor of a group of species placed at the root, and existing species at the tips of the branches. The branching points between the root and the tips represent points where a single lineage split to produce two new lineages, while the length of each branch represents the amount of time or evolutionary change between branching points. As an aside, this representation of evolutionary history in tree form was an innovation of Charles Darwin himself, appearing first in his scientific notebook and then popularized in his seminal work, *On the Origin of Species*. These trees, or *phylogenies*, as they are termed by evolutionary biologists, are best viewed as hypotheses of how evolution occurred in a particular group of species. As such, they represent a well-informed supposition as to who is more closely related to whom, and when and how current species diversified from a common ancestor. As we will see below, such phylogenies also furnish predictions as to what traits or attributes might be shared among which species. Like all scientific hypotheses, they are subject to a rigorous process involving the collection and analysis of data and a careful evaluation of whether these results support or contradict the particular hypothesis in question. If the data are consistent with the hypothesis, then it remains standing as our best estimate of how evolution proceeded, for now. But, like all hypotheses, it is always subject to further testing and investigation with new data, and such investigations may well lead to modifications of the hypothesis and a new understanding of the past.

What sort of data do evolutionary biologists use to reconstruct evolutionary history? There are three primary sources: fossils of ancient taxa, physical traits measured from the anatomy of current specimens, and genetic data sampled from living or preserved animals. Fossils have the great virtue of concretely demonstrating how specific lineages appeared in the past, including lineages that have become extinct. Importantly, the geologic layer in which fossils are found provides context and can pinpoint when and where the lineage with this trait existed. Such data can be invaluable for calibrating the timing of branching points in a tree and grounding the hypotheses of how evolution proceeded in a group. The downside to fossils is that they can be hard to find and are typically fragmentary in nature, and thus provide only a partial view of the evolutionary past of an entire group of species. As we will see below, such is the case with the parrots.

In addition to fossils, scientists can use data from species still in existence and look for patterns of shared similarities and differences. These data can then be used to reconstruct a phylogeny that best explains the patterns of shared similarities. In the past these trees were often based on the straightforward principle of parsimony, which assumes that trees that require the fewest evolutionary changes are more likely than those that require more changes; now more mathematically sophisticated approaches often are employed.

Scientists prefer to build such trees using traits that are easily and reliably measured. The reason for this is simple: Even a few species can be arranged into enormous number of alternative trees with different branching patterns, each one representing a different hypothetical evolutionary history. Distinguishing between these alternate branching patterns is best done with measurements of lots and lots of traits (also called *characters*). More characters generally leads to better discrimination of the small set of trees that fit the data well from among the enormous forest of possible trees that could be constructed for a given set of species. Making these distinctions is a job best left to powerful computers applying carefully developed algorithms; with large numbers of species it can still take these computers weeks to sort through all the billions of possible alternative trees. It is still up to the scientists, however, to choose and measure their characters carefully so that the trees generated are most likely to represent sound hypotheses of evolutionary history.

Historically, the most abundant characters available to scientists were those provided by gross anatomy and morphology. Museum collections have thousands of specimens that are used for just this purpose, and they are carefully curated in impressive collections of study skins, skeletons, whole bodies in alcohol, and even nests and eggs. These specimens can then be used to painstakingly measure obscure details of the size and arrangement of bones and organs and compare these characters within and among different species. Such careful work exemplifies the classical approach to systematics, the branch of science that aims to reconstruct the evolutionary history of all organisms or, as it is colorfully known, the Tree of Life. Such knowledge was hard-won, however, as even the most creative and careful scientist eventually ran into limits as to how many morphological characters they could reliably measure across an entire set of specimens. This problem was especially acute when trying to compare across very distant branches of the Tree of Life separated by long periods of time from their common ancestor. (Imagine how few characters could be reliably measured across jellyfish, honeybees, and sharks, three distantly related members of the kingdom Animalia.) At the other end of the spectrum, early systematists also had difficulty with homogeneous groups in which many members shared similar values for most morphological traits, leaving few characters that actually helped distinguish among different groups. Such was the problem with the parrots, as their conserved morphology provided few external or even internal characters that varied enough to be useful in building well-resolved evolutionary trees. It took a landmark scientific discovery to break this impasse and eventually provide new insights into the evolutionary history of parrots and the entire Tree of Life.

This breakthrough was the discovery of DNA and the rapid rise of modern molecular genetics it permitted. In 1953, James Watson and Francis Crick, along with Rosalind Franklin and others, described the double-stranded helical structure of a molecule called deoxyribonucleic acid (DNA for short) and proposed that it encoded the genetic information necessary for life. This landmark discovery led to an explosion of studies into how these encoded instructions were used to build organisms, and how these instructions changed as they were passed from one generation to the next. This understanding of the basic molecular mechanisms of inheritance has benefited virtually every field of biology and opened vast new fields of study. The beneficiaries have included systematists, who were quick to realize the insights that direct study of genes themselves could contribute to reconstructing the evolutionary past.

Among the first pioneers of this new field of molecular systematics were Charles Sibley and Jon Ahlquist, who worked together through the late 1970s and 1980s to apply genetic approaches to understanding the evolutionary history of birds (class Aves). Their work culminated in 1990 with the publication of their monumental *Phylogeny and Classification of Birds*, the first large-scale study to apply DNA evidence to avian relationships. There was, however, considerable debate among ornithologists regarding their general approach, which relied on large-scale comparisons of overall DNA similarity across the entire genomes of pairs of species, and about many of their specific findings that resulted from this DNA–DNA hybridization technique.

Nonetheless, Sibley and Ahlquist's groundbreaking study did spur others to follow in their footsteps, and it revitalized interest in the relationships among major groups of birds. This interest was facilitated by rapid advances in biotechnology that started in the 1980s such as the invention of the polymerase chain reaction (PCR) and the mechanization of DNA sequencing. These technologies allowed researchers to isolate a single stretch of DNA from a sample, amplify many thousands of copies of it, and then read out the sequence of nucleotide base pairs. This DNA sequence could then be compared between species to look for patterns of similarities and differences. With the help of ever-improving computers, these patterns of sharing could then be transformed into trees of evolutionary relationships using many of the same approaches developed for morphological traits. The main benefit for molecular systematists was that they could now compile information from hundreds or thousands of DNA characters, whereas they used to struggle to find a few dozen characters from painstaking examination of morphology. These new biotechnological approaches have led systematists into a golden age of studies aimed at uncovering the evolutionary past of birds and other organisms. It is a golden age that continues today and will no doubt stretch on until such time as a comprehensive and well-supported hypothesis for the entire Tree of Life is produced. And, importantly for us, it has cast new light into the previously obscure history of the parrots.

THE EVOLUTION OF PARROTS

Parrots' Ancestors and Closest Relatives

The origin of parrots themselves is an evolutionary enigma. The unique set of morphological features shared by all parrots sets them well apart from other groups of birds and has made determining the identity of their closest relatives a challenge. In the absence of series of well-defined characteristics shared with another group, avian systematists resorted to proposing a long list of possible candidates as relatives, usually on the basis of a single feature that each shared with the parrots. Various proposed relatives included the pigeons, based on similarities of the humerus bone in the wings; the owls, based on the shared presence of a fleshy cere over a curved bill and features of the skull; the woodpeckers and their relatives, based on the shared presence of zygodactyl feet; the cuckoos and relatives for the same reason; the falcons or the owls, based on the hooked bill; and the toucans, based on the sharing of powder down. Others have noted morphological similarities with the mousebirds, an obscure group of small African birds composed of only six extant species that are able to switch their toes between the zygodactyl formation and the anisodactyl formation, in which three toes point forward and one backward. Most dismissed the shared presence of curved bills in the falcons and the parrots as a sign of a close relationship, instead explaining it as an example of convergent evolution, in which similar selection pressures lead to the evolution of similar features in distantly related groups. Others pointed out that the same argument could be applied to any of the similarities noted between parrots and other groups of birds. Clearly, morphology was providing little resolution to this thorny question.

The first attempts to answer this question using modern molecular genetics were only somewhat more successful. The comprehensive phylogeny produced by Sibley and Ahlquist using DNA–DNA hybridization suggested that parrots were most closely related to the cuckoos and to a group composed of the swifts and hummingbirds. The actual number of DNA comparisons on which this conclusion was based was limited, however, and the relationships were generally considered provisional until such time as better data were available.

The question of which group of birds is most closely related to parrots was wrapped up in a larger question of how and when the major groups of birds had diverged from their common dinosaur ancestor. This was a big question for scientists, and one that many groups tackled as more genetic tools became available. These new approaches did provide clear answers to some parts of this bigger question. Numerous molecular studies agreed in finding a deep division between the Paleoagnathae, a group composed of the flightless tinamous and ratites (ostriches, emus, and rheas), and all other living birds, the Neoagnathae. Within the Neognathae there was also a clear division between a group called the Galloanserae, consisting of waterfowl and the chicken-like birds, and the Neoaves, a large group containing all other living birds and some 95 percent of avian diversity. It was relationships within this latter group, the Neoaves, that proved the toughest nut to crack.

Distressingly, for over a decade the question of who was related to whom within the Neoaves became less rather than more clear. Study after study proposed different relationships among the major branches of Neoaves, and postulated different closest relatives for the parrots. Why were there such discrepancies among these studies? Part of the issue lay with the use in different studies of different types of genetic markers, which evolve at a distinct rates and may be subject to various evolutionary constraints. Part of it was due to different samples of species and groups from study to study; if a group is present in one study but absent in another it is difficult to reconcile the resulting trees. But part of the disagreement was certainly due to the nature of the problem itself. All these studies did agree on one conclusion: that the diversification of the Neoaves happened in a relatively short period of time, perhaps around the end of the Cretaceous period and the beginning of the Paleogene period, some 65 million years ago (the Paleogene was formerly known as the Lower Tertiary Period, and the boundary between the Cretaceous and Tertiary as the K/T boundary). This was a time of great upheaval in the Earth's biological history, when nearly 50 percent of the world's species became extinct, including the dinosaurs. With such losses came great opportunities for the survivors, as many ecological niches became available to those who could rapidly evolve abilities to exploit them. The result was a period of rapid evolution and diversification for the birds and other lineages, including the mammals. Such *explosive radiations*, as they are termed, pose a particular challenge for molecular systematists. This is because the rapid splitting of several lineages from a common ancestor leaves little time for the genetic changes used to measure differences between lineages to accumulate along the short branches that connect one separation into separate lineages and the next. As more evidence accumulated, some even suggested that the problem was insoluble—that the divisions between the groups of Neoaves occurred so rapidly that rather than an elegant branching tree, the history of Neoaves should be represented by a squat bush, or comb, with many branches arising simultaneously from the base.

Happily, not all scientists took such a nihilistic view of the effort to resolve relationships among the groups (or *clades*) within Neoaves. In particular, a group of scientists from a number of institutions including the Smithsonian Museum of Natural History, the Field Museum of Natural History, the University of Florida, and Louisiana State Museum coordinated their efforts to sample a large number of genes from the same samples representing all major groups within Neoaves. This large-scale effort culminated in the publication in 2008 of a paper by Shannon Hackett and colleagues in the prestigious journal *Science* that represented the most comprehensive molecular study of birds to date, with 169 species sequenced at 19 different genes. This landmark achievement not only yielded a much clearer family tree for the Neoaves, but it also provided a surprising answer to the question of who was most closely related to the parrots. The authors' various analyses gave strong, if not unanimous, support to a novel grouping of the parrots with the passerines, a group also known as the songbirds. Working backwards in the tree, the next group to have split off was the falcons. In other words, a common ancestor gave rise to the falcons as well as a lineage that later split into the parrots and the passerines, making these latter two groups each other's closest relative (figure 5).

To say that this relationship between the passerines and the parrots was surprising to many would be an understatement. The songbirds had never before appeared on the long list of possible relatives of the parrots in studies based either on morphology or genes. There is an adage in science, made popular by Carl Sagan, that "extraordinary claims require extraordinary evidence." Here was certainly an extraordinary claim, at least for those interested in avian evolution. But was the evidence also extraordinary? Most ornithologists outside those involved in the study viewed it as solid, but perhaps not extraordinary. Many reserved judgment until such time as more evidence was available. They didn't have long to wait. In 2011 Alexandar Suh and colleagues in Germany published a study examining Neoaves relationships using retroposons, an entirely different type of genetic marker whose presence or absence in different lineages was thought to be an especially reliable indicator of shared ancestry. This new study provided further confirmation of a sister relationship between parrots and passerines, with the falcons again appearing as the closest relatives to this group. Further support was provided in 2012 by a study by Ning Wang and colleagues at the University of Florida that utilized a new dataset of thirty genes to explicitly test various hypotheses for the identity of the closest relative of the songbirds. Once again they found

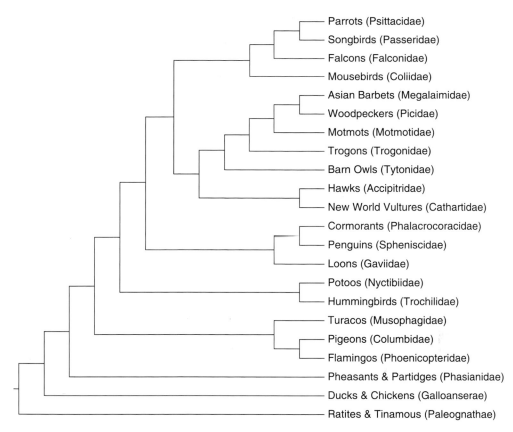

FIGURE 5 Phylogenetic tree illustrating the relationships among some families of Neoaves (the modern birds), including the close relationship between songbirds, parrots, and falcons. Figure by Tim Wright.

strong evidence in support of a grouping of the parrots and passerines. What was once extraordinary now was looking more like reality.

As I write this chapter,[1] the latest word in this debate is a work by John McCormack and others at a consortium of institutions led by Louisiana State University. They used sequence data from an astounding 1,541 independent genetic loci sampled from representatives of 32 clades within the Neoaves. This work, published in early 2013, used the largest dataset to date, and provided increased resolution for the explosive radiation of Neoaves at the end of the Cretaceous. It provided further support for the sister relationship between parrots and songbirds, with falcons as sister to this group. In just five short years the sister relationship between parrots and songbirds went from one viewed with wide skepticism to one that is emerging as rather solid. All the scientists involved are quick to point out that many uncertainties remain concerning the branching order of these avian groups that diversified so rapidly some 65 million years ago. They continue to collect more and more genetic data using

1. In this chapter the narrator is Tim Wright; in the rest of the chapters it is Cathy Toft.

FIGURE 6 Representatives of the parrots (left: Derbyan Parakeet, *Psittacula derbiana*) and the songbirds (right: Yellow-headed Blackbird, *Xanthocephalus xanthocephalus*). Based on DNA evidence, the parrots and the songbirds are now thought to be each other's closest relatives. Photo on left © James Gilardi. Photo on right © Steve Milpacher.

the new techniques from the toolbox of genomics to gain further resolution on this problem. At present, however, there is a growing consensus that the closest relatives to the parrots are indeed the songbirds.

One reason many scientists have had difficulty accepting a sister relationship between parrots and songbirds is that they do not look very much alike (figure 6). Both groups show a great deal of diversity in size, plumage, behavior, and ecological habitats, with the 5,000-plus members of the order Passeriformes showing an even greater diversity in most of these areas than the 360-odd members of the order Psittaciformes. (Recall that in the classical Linnaean hierarchy, species are grouped into genera, which are then grouped into families, which are then grouped into orders, classes, phyla, and kingdoms. The parrots and cockatoos constitute the order Psittaciformes within class Aves, phylum Chordata, and kingdom Animalia.) But in those areas where there is greatest consistency within each group, namely in the shape of the bill and the feet, the passerines and the parrots couldn't be more different. Virtually all the passerines have straight bills, slender legs, and anisodactyl feet, with three toes pointed forward and one pointing back. All the parrots have strongly curved bills, robust legs, and zygodactyl feet, with two toes pointing forward and two pointing back. The sharply different defining traits of these two groups had long obscured what now appears to be the true evolutionary relationship between them. This new finding has prompted the reexamination of fossil data by paleontologists in search of similarities between the ancestors of these lineages. As we will see in the next section, these paleontologists have found some evidence that ancient ancestors in each group differed from their modern descendants in some of these key features. While the picture is still unclear, it does suggest that the distinct differences between passerines and parrots in bills and feet may not always have been so clear-cut as they are today.

There is, however, one trait of special significance that is shared by parrots and at least some members of the passerines: vocal learning. Both parrots and songbirds are well known for their capacity to acquire vocalizations through vocal learning (see chapter 4). Evolutionary biologists had long inferred that this advanced behavior and the specialized neural pathways underlying it had evolved independently in the songbirds, the parrots, and a third group with vocal learning, the hummingbirds. The newly discovered relationship between songbirds and parrots is forcing a reevaluation of this assumption, and casting new light on studies that examine the neural basis of vocal learning in these three groups.

The studies of higher-order relationships among birds have enmeshed parrots in another debate: Exactly when and where did the parrots diversify? This is a topic we will take up below. But first we want to address another fundamental question: What did the ancestral parrot look like when it first branched off from its closest relative?

The Most Primitive Parrot

The ongoing debate about which group of birds is most closely related to the parrots has cast a new light on a related question, namely the identity and appearance of that most mysterious and ancient of all parrots, the ancestor that gave rise to all subsequent species. The question of the most primitive parrot is one that can be answered in a couple of ways. One method by which scientists sometimes infer the appearance of the "common ancestor" of a group of species is to focus on the first branching point in the evolutionary history of a group and examine any species that descend directly from that ancestor without further diversification. Such species located on long branches from the common ancestor are sometimes considered the most *primitive* species in the larger group, although this term is somewhat of a misnomer. It is based on the assumption that such an early-splitting clade group would more closely resemble the ancestor of that group than any of the other species. A brief inspection of any phylogenetic tree, however, would quickly reveal that all of the existing species (those found on the tips of the tree) would have had roughly the same amount of time to evolve differences from their common ancestor at the root of the tree. In some rare cases, such as the coelacanth fishes that are most basal group of tetrapods, such groups do seem to represent "living fossils" that retain many of the features found in their long-distant ancestors. Generally speaking, though, any one individual modern species does not necessarily provide any more information about the appearance of the common ancestor than any other within the clade.

A second approach that is on firmer logical ground is one called *ancestral state reconstruction*. This approach makes use of clever statistical algorithms to "reconstruct" the appearance of the common ancestor by taking into account the traits of all extant species and minimizing the amount of change in traits from the common ancestor to the present-day descendants. This approach, although often informative, can be led astray by extinctions of whole lineages, which can radically alter the perception of the characteristics of the larger group by their absence. As we will see below, such appears to be the case with the parrots.

A third, and more robust, way in which scientists try to establish the identity and appearance of the common ancestor of an entire group is through the fossil record. Paleontology, the study of the fossil record, is not without its challenges. Primary among these is the fact that the conditions for the process of fossilization and clear preservation of past life are somewhat rare, particularly for birds. This leads to gaps in the fossil record into which our knowledge must be interpolated. But as paleontologists continue their efforts to find and interpret fossils and systematically target gaps in the record, our knowledge of past life via the fossil record has become more and more comprehensive. Sometimes it provides surprising new insights into evolutionary history.

Given the general homogeneity in morphology of the parrots we see today, it is a reasonable supposition that the ancestral parrot also had the curved beak and other conserved features that characterize all modern parrots. Work by Gerald Mayr of the Frankfurt Natural History Museum, however, suggests that the ancestral parrot looked somewhat different from those of today. He and others have identified a series of fossils from European deposits laid down in the Eocene (35–55 million years ago) that form several distinct clades that have been given such names as Quercypsittacidae and Pseudasturidae. These now-extinct "stem" groups of parrots can be grouped with a modern "crown" group of parrots based on a number of shared similarities in the shape and proportions of their bones, particularly those that make up the zygodactyl foot shared by all these groups. Surprisingly, though, these extinct stem parrots all lack one of the most characteristic features of modern parrots, namely the long and deep upper bill, or maxilla, that curves strongly over a shorter lower bill, or mandible. Mayr has suggested that the curved bill so ubiquitous among modern parrots evolved as an adaptation for eating the larger fruits and nuts that gradually evolved during the early Cenozoic period, which started about 65 million years ago. That is, the diversification of these new food resources provided a driver for natural selection to promote the evolution and subsequent diversification of this new modern model of parrots with curved rather than straight bills. When and where this diversification took place is matter of continued debate that we will turn to below. Intriguingly, these stem parrot fossils are found exclusively in Europe, a region that now hosts parrots only as occasional, human-assisted invaders. It is worth noting that this is only one of several possible evolutionary scenarios; it is also possible that the curved bill found in both parrots and falcons was found in their common ancestor and then secondarily lost in the branch of the parrot family tree that settled in Europe in the Eocene. The discovery of further parrot fossils would surely help distinguish among these competing scenarios. But both the geography and the appearance of these fossils suggest that the evolutionary history of parrots is more dynamic than might be suggested from an examination of the modern parrots alone. As Mayr has aptly put it in a 2014 paper in the journal *Palaeontology*, "the benefits of a complimentary consideration of fossil taxa and molecular phylogenies are mutual," each providing context and insight for new discoveries of the other.

The Most Basal Clade of Parrots

Now that we have a clearer idea of how parrots are thought to appear at their origins, we can turn our attention to understanding relationships among existing groups of parrots. We will start our survey of these relationships at the base of the modern parrot family tree. The term favored by systematists for the clade that split off first from the common ancestor of an entire extant group is *basal clade*, meaning it split off at the base of the phylogenetic tree. So what is the most basal clade within the parrots?

For over two centuries, classical systematists have debated this fundamental question as they wrestled with the difficult task of sorting out relationships among the physically homogeneous parrots. This debate started with the naturalist Comte de Buffon in 1779, continued with work by Count Tommaso Salvadori (so many counts!) in the late 1800s, and stretched through the 1900s with many important contributions, perhaps the most notable being the first appearance of Joseph Forshaw's *Parrots of the World* in 1973. Throughout this long-running debate, there has been much arranging and rearranging of the parrot family tree. This work culminated in an exhaustive 1975 compendium by George Smith of characters measured from bones, muscles, organs, plumage, ecology, and behavior. As a whole, this work led to a fairly consistent view of relationships at the tips of the parrot tree (how species were grouped into genera and genera into tribes), but little consensus on the higher-order relationships that grouped tribes and families within the order Psittaciformes that encompasses all parrots. There was a general view that the cockatoos were the most basal clade in the parrot family tree given their unique combination of features, including an erectile crest, powder down, and an absence of the Dyck texture in the feather barbs that produces the stunning colors seen in other parrots (see chapter 3). But even Smith, who had compiled the most extensive dataset of his time, was circumspect in his arrangement of relationships and included several other tribes with the cockatoos in a group he placed at the base of the tree. Further resolution of this thorny issue had to wait for the advent of phylogenies based on molecular genetic characters.

Starting in the 1990s and on into the mid-2000s a raft of molecular phylogenies appeared that have vastly improved our understanding of the evolutionary history of parrots. These include studies by Leslie Christidis, Richard Schodde and colleagues, Cristina Miyaki and her students at the Universidade de São Paulo in Brazil, Rolf and Siwo de Kloet at the biotechnology firm Animal Genetics in Florida, Masayoshi Tokita and colleagues at Kyoto University in Japan, Nicole White and colleagues at Murdoch University in Australia, Miguel Schweizer and colleagues at the Naturhistorisches Museum Bern in Switzerland, and an international collaboration led by my own lab at New Mexico State University. These phylogenies all differ slightly in scientific and geographic focus, or in the species sampled and in the genes used to reconstruct a phylogeny. But when viewed together now they provide a much clearer and relatively consistent view of parrot evolution than that historically afforded by morphology (figure 7). In some cases they confirmed some long-held hypotheses of relationships based on morphological evidence, but in other cases the answers they provide have been novel and unexpected. Such is the case with the fundamental question of the identity of the basal clade of parrots.

These new molecular phylogenies were unanimous in pointing toward an unexpected clade as the sister group to the rest of the parrots. Instead of the cockatoos long favored by classical systematists, these new phylogenies identified a clade composed of some of the oddest parrots in the whole family tree: the New Zealand endemics the Kakapo (*Strigops habroptilus*), the Kea (*Nestor notabilis*), and the Kakas (*Nestor meridionalis* and the extinct *N. productus*). The Kakapo will be familiar to many readers as the largest, and for many years the most endangered, of all parrots. It is also one of the strangest, as it is completely flightless, and nocturnal, and has an unusual polygynous lek mating system, in which males advertise loudly to attract females to mating and females are responsible for all subsequent care of the chicks. As we will see in chapter 5, this mating system is definitely the exception to the general pattern of monogamy seen across the parrots. The Kea and Kakas have their own peculiarities (chapter 6). The Kea is one of relatively few parrots to be found at high altitudes, where it has adapted to the limited resources of the alpine zone by evolving a highly omnivorous diet and an unusual degree of curiosity and manipulative intelligence that sometimes sets it at odds with the human inhabitants of these regions (figure 8). The Kaka, which inhabits lower-elevation forests, is also distinctly omnivorous, feeding on fruits, nuts, berries, flowers, nectar, and small invertebrates. Kea and Kakas share extra-long and slim upper bills and tongues tipped with brushy papillae, both of which may be adaptations for extracting sap from trees (chapter 2). Neither of these species is in any way primitive; rather, they exhibit a series of advanced specializations evolved during the long isolation of New Zealand that help them exploit the unusual ecological niches presented by the historical absence of mammals. As a result, they are not thought to closely resemble the ancestral parrot that gave rise to all species that exist today. They are, however, the most basal clade of parrots, and as such are most distantly related to all other parrots alive today.

Other Major Groups of Parrots

Figure 7 illustrates a consensus phylogeny of the parrots built by combining the various molecular studies discussed above. To simplify the patterns of relatedness among major groups, I have combined closely related parrot genera into clades, represented by large triangles, which generally correspond to different families or subfamilies. So each triangle represents a group of species, all of which share a common ancestor with each other more recently than with any other such group. While these triangles simplify the visualization of the large groups they do obscure the relationships among genera within these groups. For details on these relationships I refer interested readers to the resources listed in the notes to this chapter.

To determine the pattern of relatedness among these groups we need to look far back in time at the base of the tree, on the left of figure 7, and then work our way through the branching pattern toward the modern parrots represented by the genera on the right side of

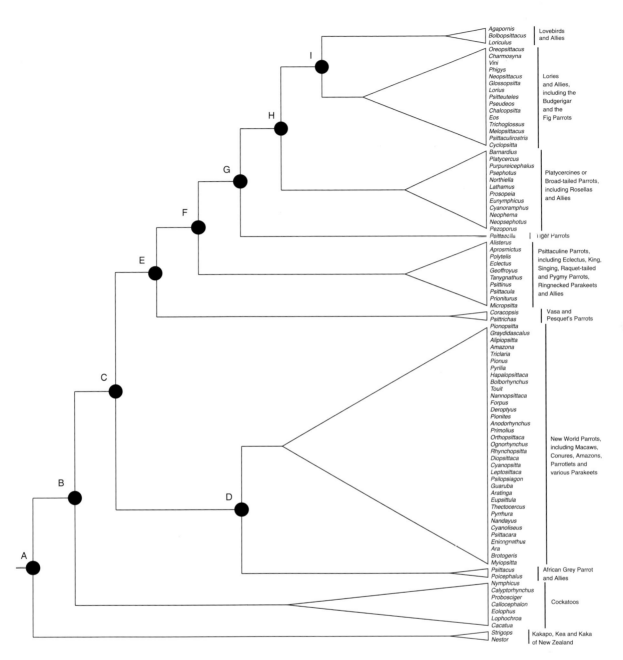

FIGURE 7 Phylogenetic tree illustrating the current scientific consensus on evolutionary relationships among major clades of the parrots. Figure by Tim Wright.

FIGURE 8 A Kea, *Nestor notabilis*, rests on a rocky outcrop high in the mountains of New Zealand. Photo © Brent Barrett.

the figure. The point at which the peculiar New Zealand parrots diverged from the common ancestor of all the extant parrots is marked as *A* on the far left of figure 7. When we follow the upper branch that leads to the rest of the parrots, we find that the next group to split from their common ancestor (*B*) is the cockatoos. So although cockatoos turn out not to be the most basal clade of parrots, they did diverge quite far back in the parrot family tree and are thus only distantly related to the remaining parrots on the tree.

As we continue to trace our path toward the present, the next branching point (*C*) indicates a split between two major groups of parrots. The lower branch leads to two groups: one an enormous assemblage of all New World parrots (i.e. those that live in North or South America), and the other a smaller group composed of two African genera (*Psittacus* and *Poicephalus*). The New World clade includes some 150 species of amazons, macaws, conures, parrots, parrotlets, and parakeets. It also includes the extinct Carolina Parakeet, a species whose newly discovered evolutionary relationships are described in box 1. The phylogeny clearly indicates that the enormous diversity of parrots found in the New World all evolved from a single common ancestor that first colonized the Americas, without any further colonization by other lineages of parrots. It also indicates that this ancestor diverged from a common ancestor (*D* in figure 7) that eventually also gave rise to the African Grey Parrot and its smaller *Poicephalus* relatives found only in Africa. Notably, this African lineage did not diversify to the same extent as the one that colonized the Americas, nor, as we will see, was it the only one to invade the continent of Africa.

One of the biggest thrills of uncovering the true phylogeny of the parrots has been learning where the Carolina Parakeet *Conuropsis carolinensis* fits into the parrot tree of life. This species became extinct over ninety years ago, well before any systematic study of its ecology, life history or behavior could be undertaken. Even the causes of its extinction remain mysterious; as we will see in chapter 7, many parrot species have become endangered, but most of these occur on islands. In contrast, the Carolina Parakeet was widespread and relatively common over most of the eastern United States before it underwent rapid range contraction and eventual extinction. Even the evolutionary relationships of this lone North American parrot were uncertain. Various candidate groupings had been proposed based on biogeography, plumage coloration, and behavioral adaptations to cold climates. There was a general sense that the closest relatives were probably conures from the genus *Aratinga,* but exactly which species within this large genus was uncertain, to say the least. Recently Jeremy Kirchman, Erin Schirtzinger, and I set out to find the closest relatives of the Carolina Parakeet as a first step toward a better understanding of this enigmatic species.

To accomplish this we first had to overcome a technical challenge: finding usable samples for DNA-based comparisons with existing species. To meet this challenge, Jeremy extracted ancient DNA from the toepads of scientific specimens of the Carolina Parakeet preserved in museums as study skins. He did this work in a lab dedicated to this task that is kept clean of any potential contaminating DNA from more modern specimens. We then used the polymerase chain reaction to make many copies of short fragments of the mitochondrial genome, a small DNA genome exclusive to the mitochondria. Since most cells have many copies of this organelle, mitochondrial DNA is found in relative abundance, even in older specimens where most of the nuclear DNA has been degraded. We then compared the sequences of the Carolina Parakeet to those we had amplified for a broad sample of other Neotropical parrots. With these data we were able to place this extinct species in its rightful spot in the parrot evolutionary tree (box figure 1.1). The results provided new insight not just into the relationships of the Carolina Parakeet but also into general relationships in Neotropical parrots, especially the large assemblage of species once placed in the genus *Aratinga.*

Unsurprisingly, the Carolina Parakeet was placed with confidence in the broader group of Neotropical parrots, reaffirming that all Neotropical species descended from a single common ancestor. Within this group, we found that the Carolina Parakeet was part of a clade that included the Nanday Conure (*Nandayus nenday*) and two

(continued on next page)

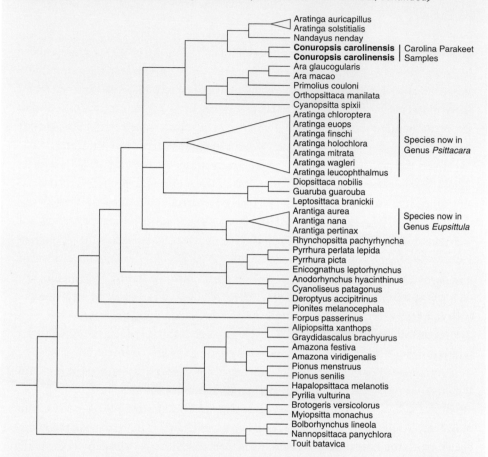

BOX FIGURE 1.1. A phylogeny illustrating the relationship of the extinct Carolina Parakeet, *Conuropsis carolinensis,* to other Neotropical parrots. This evolutionary tree also shows that several species classified at that time in the genus *Aratinga* are not each other's closest relatives; some of these species have now been reclassified into other genera. Figure by Tim Wright.

species of the genus *Aratinga,* the Sun Conure *A. solstitialis* and the Golden-capped Conure *A. auricapillus.* The latter two species, along with another likely member of this clade, the Jandaya Conure *A. jandaya,* all share to varying degrees the distinctive trait of bright yellow and orange plumage on the head, neck, and chest. The other member of this clade, the Nanday Conure, lacks this yellow coloration and instead has a black plumage on its head that may be produced with a pigment that bears some biochemical relationship to the yellow seen in its relatives. All of these species share a distinctive blue edging to their primary and secondary feathers. Thus, in retrospect, plumage provides a strong indicator of the close relationships between the Carolina

Parakeet and other members of this group. In contrast, biogeographical relationships were a poor guide, as other members of this new group are spread from northern Argentina through the Amazon Basin, but none have ranges anywhere near the former range of the Carolina Parakeet. How this species came to occupy its unique range in eastern North America remains a mystery.

Another unexpected finding in our results was that the genus *Aratinga* was not a natural assemblage of closely related species. Instead, our tree indicated that it was composed of at least three different groups, each of which was more closely related to other genera of parrots than they were to other members of their same genus. In addition to the species that grouped with the Carolina Parakeet, there was one clade that included the Brown-throated Conure *A. pertinax,* the Peach-fronted Conure *A. aurea,* and the Olive-throated Conure *A. nana,* and was most closely related to the Thick-billed Parrot *Rhynchopsitta pachyrhyncha* of northern Mexico. Another clade of *Aratinga* was composed of a number of species, including the Mitred Conure *A. mitrata,* Finsch's Conure *A. finschi,* and the Green Conure *A. holochlora,* and was most closely related to a collection of macaws and conures placed in other genera. Again, plumage patterns appeared to be a reasonable guide to distinguishing among these groups, with the Brown-throated Conure clade all sharing olive or brown plumage on their throats or breasts, while most members of the Mitred Conure clade share predominantly green plumage with occasional small patches of red. In contrast, biogeographical patterns were a poor guide, with both clades having members ranging from Central America and the Caribbean down through the Amazon Basin.

In a separate effort led by Van Remsen of the Louisiana State University Museum of Natural Sciences, my colleagues and I set out to realign the old taxonomy of these species with the new reality of parrot relationships. With taxonomic rules dictating that the Sun Conure and relatives remain in the genus *Aratinga,* we moved the Brown-throated Conure and relatives into the new genus *Eupsittula,* and the Mitred Conure and relatives into the new genus *Psittacara.* Other evidence suggests that yet another species, the Blue–crowned Conure *A. acuticaudata,* belongs in its own new genus, *Thectocercus.* While such taxonomic revisions certainly can be a hassle for people accustomed to the previous names, the pain of change is offset by the pleasure of having scientific names that accurately reflect the current knowledge of evolutionary relationships. Bad taxonomy can obscure interesting evolutionary patterns, while good taxonomy can highlight previously unappreciated ones, as with the shared plumage of the Carolina Parakeet and its newly identified relatives.

If we now trace the upper branch leading from *C* we come to the point *E* at which another group of peculiar parrots branches off. These are Pesquet's Parrot (*Psittrichas fulgidus*), found in New Guinea, and the Vasa and Black Parrots of the genus *Coracopsis*, found on Madagascar and neighboring islands off mainland Africa. These two genera are found on opposite sides of the Indian Ocean and bear few resemblances to each other, or indeed to most other parrots. Their relatively close relationship was never suspected before molecular studies. Even now the molecular studies to date have disagreed on how closely they are related and where they fit on the parrot tree, so their placement here in this consensus phylogeny should be considered provisional and, given their geographic remoteness from each other, somewhat of a mystery.

The upper branch leading from *E* leads to several other major groups of birds that comprise the remainder of the parrots. The first to diverge from the common ancestor, indicated by *F* in figure 7, is a diverse group collectively called the psittaculine parrots that includes such well-known parrots as the Eclectus Parrot (*Eclectus roratus*), the King Parrots of the genus *Alisterus*, and the parrots of the genus *Tanygnathus*, which have disproportionately large bills (figure 9). Collectively, this group has one of the broadest geographic ranges of any clade, with representatives spreading from southern Australia through the island archipelagos of Southeast Asia and across India. It also includes one of the most widespread of all parrot species, the Rose-ringed Parrot (*Psittacula krameri*), which has natural populations across the Sahel of Africa and the Indian subcontinent and has established invasive populations in over thirty-five countries outside its native range.

If we take the upper branch from ancestor *F* we come to a branching (*G*) that leads to a small group of four species known as the tiger parrots (genus *Psittacella*), found in northern Australia and New Guinea. The next group to split off (*H*) is a speciose group known as the platycercine or broad-tailed parrots, best characterized by the familiar rosellas (genus *Platycercus*), which are distributed in a rainbow of contrastingly colored species across Australia. It also includes the highly cryptic ground-dwelling parrots of the genus *Pezoporus*, one of which, the Night Parrot (*Pezoporus occidentalis*), is so rare that many thought it extinct through most of the twentieth century until a road-killed carcass was found in 1990 in Queensland, Australia. Most of the other platycercines also are found in Australia, but the species in the genera *Cyanoramphus* and *Eunymphicus* have dispersed as far as the islands of New Zealand, New Caledonia, and Tahiti.

Our final branching point (*I*) is the common ancestor between two other groups that both contain relatives that were not recognized as such until united by molecular studies. One is the large group that is dominated by a peculiar group of birds known as the lories or lorikeets but also includes the well-known Budgerigar (*Melopsittacus undulatus*) and the fig parrots in the genera *Psittaculirostris* and *Cyclopsitta*. We discuss the evolutionary implications of this odd grouping below. In addition to having spectacularly varied plumage and unusual brush-tipped tongues, the lories rival the psittaculines in their ability to colonize distant lands. From their likely origins in New Guinea they have colonized south into Australia, west into Sulawesi and Bali, north into the Philippines, and

FIGURE 9 A Great-billed Parrot, *Tanygnathus megalorrhynchus*, preens itself in a tree in Indonesia. Photo © Mehd Halaouate.

far eastward into the Pacific Ocean to settle the remote islands of Fiji, Tahiti, and the Marquesas.

The final group of parrots branching off from ancestor *I* is another group that was only recognized as such through molecular data. It is a smaller group that contains the lovebirds of the genus *Agapornis,* the hanging parrots of the genus *Loriculus,* and the Guaiabero *Bolbopsittacus lunulatus*. It is a geographically disparate group, lovebirds being found across Africa, the Guaiabero restricted to the Philippines, and the hanging parrots stretching between them from India to the Philippines and south into the islands of Indonesia and even New Guinea. Despite this geographic disconnect, the group does share a physical homogeneity, with small, stocky bodies and unusually short tails.

Though the new molecular phylogenies of the past decade have greatly clarified relationships among the parrots, some areas of uncertainty remain. One example of this is the cockatoos, and in particular the relationships of two of the most familiar species, the Cockatiel *Nymphicus hollandicus* (figure 10) and the Palm Cockatoo *Probosciger aterrimus*. The first to address relationships among cockatoos using genetic approaches was a pioneering study published in 1999 by Cathy Toft, the primary author of this book, and her student David Brown. They isolated a short stretch of a single gene of mitochondrial DNA from the small standalone genome of the mitochondria, an organelle found in the cells of all animals (and plants and fungi, too). Mitochondrial DNA has been a popular target for molecular studies, in part because many tissues have multiple mitochondria in each cell, making it easy to isolate high

FIGURE 10 A Cockatiel, *Nymphicus hollandicus*, the smallest member of the cockatoo clade. Photo © Georgina Steytler.

concentrations of its DNA, and in part because the markers used to target specific genes in one species are often found in other species. Using differences in the basic sequence of this one gene, Brown and Toft reconstructed relationships among the cockatoos. They argued that the Palm Cockatoo was the most basal lineage in the cockatoo family tree, and that the next lineage to split off was a clade containing the Cockatiel, the Gang-Gang Cockatoo *Callocephalon fimbriatum*, and the "black cockatoos" (species in the genus *Calyptorhynchus*). The remaining clades were composed of the "white cockatoos" of the genus *Cacatua* and the Galah *Eolophus roseicapilla*. So, in this early molecular tree the grey Cockatiel was most closely related to a clade composed of the black cockatoos and one of the grey species, while the black Palm Cockatoo, in its basal position, was most distantly related to all other cockatoo species.

A decade later my colleagues and I revisited the issue using additional evidence as part of our larger study of the relationships among parrot genera. We used a somewhat larger dataset, with two different mitochondrial genes and three stretches of DNA from the nuclear genome that comprises most of our DNA complement. We found somewhat different relationships among the cockatoos, in which the Cockatiel was in the most basal position and with the black cockatoos of *Calyptorhynchus,* the Palm Cockatoo, and then the white cockatoos splitting off in succession. But different analyses of our dataset using different tree-building algorithms revealed some uncertainty in these relationships, suggesting there was more work to be done.

The latest word on this subject is provided by a 2011 study by Nicole White and colleagues from Australia, the land of cockatoos. They undertook to sequence the entire genome of the mitochondria for six cockatoo species plus a subset of genes for most of the rest of the cockatoos. With this massive dataset in hand they had another go at reconstructing cockatoo evolutionary history (figure 11). Their results firmly supported our previous tentative placement of the Cockatiel in the most basal position in the cockatoo family tree, with the black cockatoos next to branch off, and the Palm Cockatoo now sister to a clade composed of the Gang-Gang Cockatoo, the Galah, and the white cockatoos of *Cacatua.* This well-supported arrangement has some interesting implications for some of the physical features of cockatoos. It suggests that the relatively immobile crest of the Cockatiel is ancestral in the cockatoos, and that the fully erectile crest found in the rest of the cockatoos evolved after this lineage split from the common ancestor with the cockatiels. It also suggests that traits such as the black plumage and slender bill shared between the Palm Cockatoo and some other cockatoos have evolved multiple times within different lineages, an evolutionary pattern that we will return to below.

How and When the Parrots Diversified

Throughout this discussion of the past evolutionary history of parrots we have ignored some of the most basic contextual questions, namely when, where, and how the parrots diversified from a single common ancestral species to the over 360 species that exist today. Here again we encounter a larger debate in avian systematics, this one regarding which evolutionary processes led to the diversification of Neoaves and exactly when this diversification took place. It is a vigorous debate, in which the challenge is to marry the evidence from the fossil record with results from modern molecular phylogenies and additional insights from patterns of biogeographic distributions. And once again the parrots are right in the thick of it.

The heart of the debate lies with the question of whether the major clades of modern birds diversified before or after the extinction of their dinosaur relatives 65 million years ago at the boundary between the Cretaceous and the Paleogene periods. In an influential series of books and papers published from the mid-1990s to the mid-2000s, Alan Feduccia of Yale

distribution because the bulk of parrot diversity is found in South America, Africa, and especially Australasia (meaning the great island archipelagos of Southeast Asia, New Guinea, and Australia). Thus in the parrots, as with birds in general, there existed alternative, and strongly divergent, hypotheses for when the major diversifications took place.

This was the state of the science in the mid-2000s, as my collaborators and I were preparing our first paper on parrot molecular systematics. We were interested in using our new phylogeny, which was the most comprehensive to date, to address the question of when and where parrots diversified. Since both sides of the debate seemed to have valid evidence, and we didn't have strong *a priori* reasons to prefer one side over the other, we decided to test both alternative hypotheses, the one of an ancient Cretaceous divergence and the other of a more recent Paleogene divergence. We tested these hypotheses by using two alternative calibration points for the time at which the most basal lineage split from the rest of the parrots (*A* in figure 7) and then evaluating how well the resulting divergence patterns matched what was known about the geologic history of the continents over the same time frame. In essence, we applied the biogeographic approach of Cracraft to evaluate two alternative hypotheses. We tested the Cretaceous scenario by calibrating the split between the basal New Zealand clade and the rest of the parrots at 82 million years ago, which corresponds to when the islands of New Zealand are thought to have first commenced their protracted split from Gondwana. We tested the Paleogene scenario by calibrating the same split at 50 million years ago, corresponding to the dating of the oldest known parrot fossils, found in Eocene deposits in Europe. We then used the molecular dating approach to estimate divergence times across the rest of our parrot phylogeny.

The resulting "chronograms" suggested very different scenarios for how parrots diversified. The older calibration point gave us a pattern of diversification, the timing of which matched fairly well with geologic events such as the initial separation of Australia from New Guinea and East Antarctica, the separation of South America from West Antarctica, and the relatively recent close approaches of Australia to New Guinea and both of these to Southeast Asia. This scenario suggests that after an origin in Australia, the major splits among modern parrot lineages were due to vicariance, that is, physical separation of populations as the daughter continents of Gondwana broke apart. Subsequent diversification at the level of species or genera occurred within these different continents, perhaps driven by habitat changes caused by global alterations in temperature and sea level.

The more recent calibration told a different story. It also suggests that modern parrots diversified largely in Australia and New Guinea, but only 40 million years ago, perhaps after colonization from the northern continent of Laurasia, where most of the older parrot fossils are found. Colonization of different regions like Africa and South America must then have occurred through dispersal across the oceans separating the continents, with subsequent diversification driven by habitat changes. Overall, we considered this scenario less likely than that suggested by the more ancient divergence, although both were plausible.

Additional insight into the evolutionary diversification of parrots was subsequently provided by Manuel Schweizer in a series of papers in which he too reconstructed phylogenies of parrots using a different set of genes. He calibrated the divergences within the parrot evolutionary tree by coupling divergence dates derived from fossils of non-parrot bird groups that have richer fossil histories than parrots with phylogenies of the relationships of parrots with these groups. This approach allowed him to estimate the timing of divergences with much greater confidence than our single parrot fossil approach. The resulting time-calibrated trees (chronograms) were much more closely aligned with our second scenario, a later divergence in the Eocene. They provided strong support for the idea that many of the modern lineages of parrots became established in their current locations through an active process of over-sea dispersal, rather than a passive riding of different continental blocks as they split apart. The debate could be reignited by the discovery of a single parrot fossil from the Cretaceous, perhaps in a little-explored area like Antarctica, which was once much warmer and more welcoming to parrots. At present, though, the weight of the evidence suggests that the initial divergence of the modern parrots occurred in the Australo-Papuan region about 58 million years ago in the Eocene, and was followed by the colonization of South America, Madagascar, and Africa by dispersal over oceans.

With the stage now set for where and when the parrots diversified, I will focus for the remainder of the chapter on a few specific questions, about specific branches of the parrot family tree, that illustrate some interesting general principles concerning the winding paths taken by evolution.

SOME PARROT ENIGMAS

What Is a Budgerigar?

In some senses the question "What is a Budgerigar?" is immediately answerable: it is the most popular captive parrot species in the world. This small parakeet is native to the interior of Australia. There it roams widely, often in large flocks, breeding opportunistically when the intermittent rains of that dry land fall sufficiently long and hard to produce enough of its favored grass seeds to sustain a clutch of chicks. The Budgerigar was first described as a species in 1805 by George Shaw, and the first live bird was brought to Europe in 1840 by noted ornithologist John Gould. Shortly thereafter aviculturists began breeding Budgerigars in captivity. The birds proved readily adaptable to captive life and quickly grew to enjoy the wide popularity as pets that they still enjoy today. This popularity can be attributed to their playful personalities, intelligence, relatively mellifluous voice (for a parrot), and dazzling assortment of color varieties. These last have been bred by aviculturists, who carefully selected any mutations that diverged from the wild-type coloring of yellow face, brilliant green breast and belly, and mottled black-and-yellow head that verges into a scalloped yellow-and-black pattern on the back. It is this scalloped patterning that gives rise to another name for the Budgerigar, the shell parakeet. The origins of the name "Budgerigar" are unknown but are thought to derive from corruptions of a name used by Aboriginal inhabitants of Australia.

The small size and adaptable nature of the Budgerigar have also made it popular as a subject for scientific studies. As will be discussed in later chapters, a substantial amount of what we know about such diverse topics as how parrots feed their young, how the brilliant colors in parrots feathers are formed, and how parrots produce, learn, and perceive their varied communication calls have come from studies of captive Budgerigars. Paradoxically, the life of wild Budgerigars remains relatively unstudied, apart from the pioneering studies by Ed Wyndham and a few other intrepid Australian ornithologists (see chapter 6). This paradox arises from the nomadic nature of Budgerigars, which makes sustained study of the same populations well-nigh impossible. The difficulties are compounded by the enormous flock sizes that are sometimes seen—just imagine the difficulty in following the movements of one marked bird within the kaleidoscopic whirl of a flock of thousands as they fly across the scrublands of Australia (figure 12).

Until recently our ignorance of the wild Budgerigar extended even to the fundamental question posed at the beginning of this section: What is a Budgerigar? Or, to state it in evolutionary terms, where does this species fit in the parrot Tree of Life, and who are its closest relatives? For many years this question was approached in the same way all of parrot systematics was approached: by comparing anatomical, morphological, and behavioral characters, adding in considerations of geographical distributions, and making considered judgments as to which species most closely resembled each other and were thus most closely related. The conclusion that was drawn by early workers in parrot systematics was that the Budgerigar was part of a collection of parrots with long wings and long, broad tails collectively known as the platycercines that, as we have seen, is distributed across Australia and neighboring islands. As described above, other parrots historically thought to be in the platycercines include the brightly colored rosellas of the genus *Platycercus,* the migratory swift parrot *Lathamus discolor,* the widely distributed parakeets in the closely related genera *Eunymphicus* and *Cyanoramphus* that are found on New Zealand and the far-flung islands of the southern Pacific Ocean, and the ground-dwelling parrots of the Genus *Pezoporus* (figure 13). In particular, the latter were judged to be the closest relatives to the Budgerigar given their similar mottled green-and-yellow plumage, their shared habit of feeding on the ground, and their overlapping distributions in the arid interior of Australia.

This picture changed, however, when scientists started using DNA evidence to reconstruct the evolutionary history of parrots. In a study published in 1991, Les Christidis and colleagues included the Budgerigar in a phylogenetic study based on variation in allozymes, or protein variants. Their work suggested that Budgerigars are most closely related to lorikeets, a relationship so novel that the authors were inclined to dismiss it as an experimental artifact. The first DNA-based study to include Budgerigars was the pioneering study by de Kloet and de Kloet discussed above. Their phylogenetic trees, built with sequences from the sex-linked spindlin genes, confirmed the close evolutionary relationship between the Budgerigar and the lorikeets, extensive differences between lorikeets and the Budgerigar in diet, habitat, and appearance notwithstanding (figure 14). This new grouping was later supported by our study that included wider sampling of different parrot groups and genes, by subsequent studies by Manuel Schweizer and colleagues, and by Gerald Mayr's anatomical studies. Support is now robust for the hypothesis that the Budgerigar is the closest relative of the lorikeets and that

FIGURE 12 A superflock of Budgerigars, *Melopsittacus undulatus,* in their native grasslands near the Karratha area, Australia. Photo © Jim Bendon.

FIGURE 13 A Western Ground Parrot, *Pezoporus wallicus flaviventris,* blends in to its favored habitat in the heathlands of southwestern Australia. Photo © Brent Barrett.

FIGURE 14 An endangered Red-and-Blue Lorikeet, *Eos histrio,* from the Talaud Islands, Indonesia. Photo © Mehd Halaouate.

both belong in a broader clade that also includes the fig parrots of the genera *Cyclopsitta* and *Psittaculirostris.* This firmly established relationship puts a novel twist on thinking about the evolution of the lorikeets themselves, and in particular how the unusual lifestyle of nectar-ivory (feeding on nectar), and its anatomical peculiarities like brush tongues, evolved from the seed-eating habits of the Budgerigar and fig parrots. In chapter 2 we will examine in more detail this lifestyle and the consequences it has for lorikeet anatomy and physiology. As a final addendum, a later study I conducted with Leo Joseph and colleagues showed that the ground parrots of genus *Pezoporus* that were formerly linked with Budgerigars were most closely related to the grass parrots in the genera *Neophema* and *Neopsephotus,* all of which live in Australia and some of which are brightly colored with patches of blue, orange, pink or yellow. It appears that the mottled green-and-yellow plumage seen in both the ground parrots and the Budgerigar is not a trait derived from a shared common ancestor, as formerly thought, but instead a product of convergent evolution driven by similar selective pressures imposed by their shared lifestyle of feeding on the ground and the need for cryptic plumage that helps them blend into the grasses and hide from aerial predators.

How Have Different Body Shapes Evolved in the Parrots?

Anyone who has paged through Joseph Forshaw and William Cooper's masterpiece *Parrots of the World* or another book illustrating parrot diversity will immediately notice the immense variety among the species. Each species differs from all others, whether it is the arrangement of its brilliant plumage patches, the color or shape of its bill, the size and shape of its wings, the overall size of its body, or some unique combination of all these attributes. Continue paging through the book, however, and you will begin to notice that some features are shared across multiple species. Cockatoos have erectile crests. Some of them have white plumage, while others have black or grey. Lorikeets have brushy tongues with many raised papillae. Macaws of the genera *Anodorhynchus* and *Ara* have particularly massive bills, even given their large body size. Several species in the Neotropical genus *Aratinga* have brilliant golden plumage on their head or bodies, while all of the species in the genus *Prioniturus* found in the islands of Southeast Asia have distinctive racquet tails, in which the two elongated central tail fathers have extended shafts that are bare but for tufts of feather barbs at the tips. The list of such shared characteristics becomes longer and longer as one continues to examine the book.

The question of why two different species might share a given trait is a fundamental one in evolutionary biology. At its most basic, there are three alternatives for such shared similarities. One is that the species in question are closely related, and share the trait because at some point in the evolutionary past their common ancestor had the same trait. A trait that is shared between species due to their shared ancestry is called a *homologous* trait. A likely example of such a trait is the erectile crest of the cockatoos, since all members of the group except the most basal species, the Cockatiel, share the trait and the cockatoos are well supported as a cohesive group in a variety of studies, as described above. In other cases, two species may look similar because similar selective pressures have promoted the evolution of similar traits *de novo* in each species. Traits that arise from shared selective pressures but are not present in the common ancestor are called *analogous* or *convergent,* and the process that produces them is called *convergent evolution.* A likely example of convergent evolution is the cryptic plumage of the Budgerigar and the *Pezoporus* parrots described above, which is not due to shared ancestry (they are not closely related to each other) but instead may result from natural selection working on both species to reduce their visibility while feeding on the ground. Similar selection may have been at work on the ancestors of the ground-dwelling Kakapo of New Zealand to produce their mottled brown-and-green plumage; although mammalian predators are absent from New Zealand there is abundant fossil evidence of avian predators that became extinct following the arrival of the Maori. A third possibility is that the trait was present in an ancestor further back in the tree. Hummingbirds and parrots, for example, both have feathers not because they share a more recent common ancestor with each other than either does with any other bird group, but because all birds have feathers.

Importantly, a trait's being homologous and shared by close relatives does not mean it is without adaptive value for the species that have it; traits present in an ancestor may be maintained through natural selection in descendent species because they confer a benefit to these species. Conversely, as our ability to uncover the genes involved in producing specific traits improves, we are beginning to discover that traits that evolved convergently sometimes are produced by the same genetic variants or collections of genes, which have been

FIGURE 15 Comparison of a Red-cheeked Parrot, *Geoffroyus geoffroyi* (left), and a Red-lored Amazon, *Amazona autumnalis* (right), two distantly related parrot species with similar morphotypes. Photo on left © Bent Pedersen. Photo on right © Mike Bowles and Loretta Erickson.

independently recruited during natural selection to produce similar traits in distantly related species. Puzzling out the genetic pathways underlying phenotypic traits and determining the extent to which traits are homologous or convergent is one of the most exciting and challenging frontiers in evolutionary biology today.

The body shapes of different parrots present just such an intriguing challenge to evolutionary biologists. Some parrots, like members of the genera *Amazona, Psittacus, Geoffroyus* and *Cacatua*, have stocky bodies, short, square tails, and broad, rounded wings (figure 15). Others, like species in the genera *Forpus* and *Cyclopsitta*, have small bodies with broad pointed wings and short wedge-shaped tails. Still others, like members of *Polytelis* and *Psittacula*, have narrow pointed wings and long narrow pointed tails. As with all birds, such differences in wing, tail, and body shape, sometimes termed *morphotypes*, strongly affect the flying abilities of the birds, with some morphotypes having greater speed and others having greater maneuverability. As such, they can strongly affect the fitness of individuals in a given habitat and are likely the product of natural selection. But to what extent are these morphotypes homologous versus convergent?

One of the first to tackle this question in parrots using modern genetic methods was Cristina Miyaki and members of her research group at the Universidade de São Paulo in Brazil. They investigated the evolutionary relationships among the parrots of the Neotropics using mitochondrial DNA and found that, broadly speaking, the nine species they sampled fell into two clearly separate clades. One was composed of the Blue-fronted Amazon *Amazona aestiva* and the Blue-headed Parrot *Pionus menstruus*, both species with short tails and broad, rounded wings. The other was composed of the Hyacinth Macaw *Anodorhynchus hyacinthinus*, the Painted Conure *Pyrrhura picta*, the Blue-and-yellow Macaw *Ara ararauna*, the Golden Conure *Guaruba guarouba*, Spix's Macaw *Cyanopsitta spixii*, and the Hawk-headed Parrot *Deroptyus accipitrinus*. For the most part these species have long, graduated tails and broad, pointed wings. In recognition of the strong concordance between morphotypes of the species sampled and their independently derived evolutionary relationships, Miyaki and colleagues labeled these the short-tailed and long-tailed clades. Subsequently, other studies by Miyaki's group, de Kloet and de Kloet, Schweizer and colleagues, and my own research group have extended this work by sampling more species and more genetic markers, and in the process have created a family tree for the Neotropical parrots that contains nearly all the species. The picture that has emerged has confirmed and amplified on Miyaki's early work. It now appears that these parrots fall into five distinct clades, most of which correspond to a distinct morphotype. In addition to the long-tailed and short-tailed clades described above, there is now thought to be a clade composed of the genus *Brotogeris* and the Monk parakeet *Myiopsitta monachus*, both of which have broad pointed wings and graduated tails of moderate length, and a clade composed solely of the genus *Forpus*, which are small parrots with short, pointed wings and extremely short wedge-shaped tails. The exception to the strong correspondence between morphotypes and phylogenetic clades is a clade composed of the genera *Touit*, *Nannopsittaca*, *Bolborhynchus*, and *Psilopsiagon*, most of which have broad, pointed wings but which sport a variety of tail types, from the short squarish tail of the members of *Touit* and *Nannopsittaca* to the long graduated tails seen on the members of *Psilopsiagon*. Such differences may be driven by selection from the different habitats inhabited by these different genera.

Overall, though, the picture that emerges from our improving knowledge of Neotropical parrot evolutionary history is that body shape is fairly conserved within the different clades, such that most species tend to be shaped like closely related species. This pattern provides good evidence that within the Neotropical parrots these morphotypes are homologous traits, derived from ancestors that had similar shapes. A recent paper by Manuel Schweizer and colleagues suggests that the diversification into these different ancestral morphotypes occurred shortly after the common ancestor of all Neotropical parrots colonized the Americas, probably about 30–35 million years ago. This initial colonization was followed by a rapid diversification into the five morphotypes as the ancestral parrot populations adapted to the local conditions they found. These ancestral morphotypes then persisted and further diversified into the more than 150 species we see today as they moved through South and Central America, into the islands of the Antilles, and, in a few cases, into North America.

The picture becomes more complicated, though, when we broaden the scope to look across all parrots. Each of the morphotypes discussed so far has representatives not just in the Neotropics but also in species seen in other parts of the world. For example, the same stocky bodies, broad rounded wings, and short square tails seen in *Amazona* and *Pionus* in the Neotropics are also seen in the African Grey Parrot *Psittacus erithacus* of Central Africa, the Kea *Nestor notabilis* of New Zealand, and the Galah *Eolophus roseicapillus* of Australia. Yet the clear picture that has emerged in the last decade from phylogenetic analyses of parrots is that these species are all in different evolutionary lineages. This pattern tells us that on a broader scale, similar morphotypes have evolved convergently in different lineages of parrots. Whether similar mutations in the same or different genes that produce these body plans have been promoted by natural selection in different lineages remains an exciting question for future investigation.

Now that the stage is set and the players have been introduced, it is time to move into the main act. In the following chapters we delve deeper into the lives of wild parrots. We start with aspects of physiology and behavior: how different species have adapted to different diets, how parrots perceive the world around them and communicate with others, and how they use their impressive brains to process this information. We will then move on to examine questions of parrot life history: how they find mates, raise their offspring, and live their long lives. We will end with the population biology of parrots, focusing on how populations are affected by the many changes we humans are making in the natural world. Throughout, we will draw on the rich and rapidly growing body of scientific literature that is shedding an ever-greater light on the previously obscure and mysterious lives of parrots in the wild.

PART TWO

The Functional Parrot

Physiology, Morphology, and Behavior

The Thriving Parrot

The Foods and Beaks of Parrots

FORAGING ECOLOGY

Dis-moi ce que tu manges, je te dirai ce que tu es.

—ANTHELME BRILLAT-SAVARIN (1826)[1]

CONTENTS

Introduction to the Diets of Parrots   39

The Seed Eaters   41

    Granivory: Life as a Seed Predator   41

    What Generalist Granivorous Parrots Eat   42

    Box 2. The Right Tools for the Job   49

    Specialized Granivores: Living on Ancient, Cone-Bearing Trees   53

    Geophagy: How Eating Soil Helps Parrots Cope with Toxic Plant Defenses   57

Other Diets   62

    The Diets of Ground-Feeding Australian Parrots   62

Nectarivory: Parrots That Consume Nectar   64

The Challenges of Consuming Nectar and Pollen   66

The Only Frugivorous Parrot: Pesquet's Parrot   69

    Box 3. Herbivores and Nitrogen   71

How the Only Folivorous Parrot Survives on Leaves   74

Truly Omnivorous Parrots   75

Parrots as Mesoscale Foragers: The Ecosystem Effects of Seed Predation   79

INTRODUCTION TO THE DIETS OF PARROTS

For the parrots, Brillat-Savarin's catchy saying rings true. Organisms within a clade often commit to a common method of getting their food, because diet evolves to increase survival and reproduction (as do all traits). If we know a parrot's diet, in other words, we may be able to identify its clade correctly, along with its relatives who eat the same things. Likewise, "you are what you eat" applies to what we now understand about the evolution of parrots from their ancestor shared with the songbirds and falcons (chapter 1). Parrots, like falcons, are rapacious predators, but they feast primarily on the reproductive parts of plants, that is

---

1. The literal translation from the French: "Tell me what you eat, I will tell you what you are."

FIGURE 16 A White-fronted Amazon, *Amazona albifrons*, foraging on fruit.
Photo © Mike Bowles and Loretta Erickson.

flowers, fruits, and seeds (figure 16). Thus, if we know what it eats, we can tell you whether it is a parrot or a falcon.

All parrots are primarily herbivorous, meaning that they eat plants. Within the Psittaciformes, different groups of parrots have diverged in the focus of their herbivory. From what we consider to be the more generalized diet of parrot ancestors, four specialized diet categories have evolved: granivory (eating seeds), nectarivory (eating nectar and pollen), frugivory (eating fruit pulp), and folivory (eating leaves). For some parrots, we consider a fifth and generalist category, omnivory, which is one that combines one or more of the four specialized plant-food categories with significant animal protein.

A major component of any plant is cellulose fiber, which cannot be digested by most vertebrate animals. Herbivores can be ranked by the amount of this indigestible cellulose they consume. Diets with the smallest percentage of indigestible cellulose fibers are ranked no. 1 on Langer's herbivory rating, and those with the highest percentage as cellulose are rank 6. The lowest level of herbivory comprises the omnivores, which eat significant amounts of plant and animal tissue. Perhaps the diets of a few species of parrots in New Zealand and surrounding islands qualify as true omnivory (under "Other Diets," below), in that these species routinely eat small vertebrates. The highest level of herbivory comprises the specialized obligate grazers, such as the ruminant mammals. Parrot diets largely fall into Langer's rank 2, which includes mostly plant parts such as fruit, tubers, seeds, buds, flowers, leaves, and sap, supplemented with some animal material. This category is referred to as a *concentrate selector* in Hofmann's scheme of feeding types, a term that captures how these animals seek the most concentrated nutritional parts of the plants, with the least indigestible cellulose. Only the Kakapo *Strigops habroptilus* creeps up to level 3 or 4 in tackling large amounts of cellulose as its sole diet, without supplements from animal protein.

In this chapter, we will consider each of these four diets of parrots in turn, exploring the benefits that each provides and the challenges that must be overcome. The first section of the chapter is devoted to granivory, its variations, and the implications of such a diet. The second section covers the other diets found in the Psittaciformes.

THE SEED EATERS

Granivory: Life as a Seed Predator

Parrots may be difficult to see in the shrouded canopies of rainforests, but an experienced field biologists knows how to locate them—you follow the noise. If their collective screams are difficult to pinpoint, the rain of debris beneath foraging flocks makes the job of navigation easy. These consumers of plant reproductive parts are obscenely wasteful, taking what seems like one bite out of each fruit before tossing it to the ground. A suite of ground-scavenging species such as White-winged Trumpeters *Psophia leucoptera* and agoutis of the genus *Dasyprocta* in the Neotropics take advantage of the arboreal orgies of parrots to glean the abundant leftovers. One could understandably ask of the parrots: "What *are* these animals eating?"

From a plant's perspective, it hardly matters what part of the plant went down the hatch, if the plant's seeds are destroyed in the process of foraging. Plants invest in costly strategies to entice animals to eat entire fruits but leave the seeds intact. They pack the structures holding the seeds with many dietary goodies, primarily energy but also essential micronutrients intended expressly for animals to make part of their diets. On the other hand, they lace the seed coats and unripened fruits with bitter or toxic chemicals, and cover them with hard cases, to discourage animals from eating the fruit before the seeds are ready for dispersal. They also place these toxins in the flowers and leaves to deter herbivory. The plants do so in the hope that animals will consume the entire ripe fruit but later defecate the unharmed seeds, ideally somewhere far from the parent. Such animals are doing plants a favor by dispersing their seeds, something that is generally difficult for immobile plants to accomplish by themselves.

Granivorous parrots therefore face a formidable array of plant weaponry that they must overcome to get nutrition from the plant. The better part of valor might be for the parrots politely to eat the parts of the plant that the plant has prepared for them, namely the nectar and the fruit pulp. The problem with that way around the plant–animal warfare is that fruit and nectar have little more than sugar in them, with maybe some lipids in the fruit, depending on the species of plant. Seeds on the other hand are packed with both protein and lipids, especially when the seed is not yet ripe, which the plant embryo is depending on for its own growth when the seed germinates. Any sensible animal would want this nutrition for itself and its own offspring. A parrot might wait until the seeds are ripe, at least, before eating them, but while the toxins are usually less abundant in the ripe seed, often so is the nutrition, especially the protein. This trade-off prompts parrots to dine on the nutritionally richer unripe seeds, taking various measures to ameliorate the effects of the toxins.

First, let us find out what granivorous parrots actually ingest, exploring how they manage their nutrition, and then we will discover how they avoid plant defenses.

What Generalist Granivorous Parrots Eat

We begin with the parrots that live in forests and feed largely from the canopies of trees and shrubs. Most of these species live in tropical and subtropical forests. In these areas, mild climate and an abundance of rainfall keep plants of a wide variety of species fruiting throughout the year. A host of studies reveal that seeds form the foundation of the diets of most opportunistically feeding parrots. These granivorous parrots eat seeds in all stages of ripeness, along with a variety of other plant and animal foods (figure 17). Just how much the parrots depend on seeds versus other foods depends on the species and the environment.

As they would for any herbivorous diet, biologists have wondered whether seeds—even the more nutritious unripe seeds—provide parrots with enough protein (in particular) to meet their needs. One answer to this question is to observe what foods parrots select relative to those available. Jamie Gilardi and I[2] asked this question of a guild of parrots in the Peruvian lowland rainforest in Manú National Park and the nearby Tambopata National Reserve, which included at least eighteen species of generalist granivorous parrots in eleven genera: *Ara, Aratinga, Psittacara, Nannopsittaca, Forpus, Amazona, Pionus, Pionites, Pionopsitta, Brotogeris,* and *Pyrrhura.* The diets of species in this guild varied significantly but stayed within the theme of generalized granivory. Over 80 percent of the plant parts selected by these species were seeds and fruit in various stages of ripeness, with fruit pulp itself not being an obvious target.

---

2. In this and all subsequent chapters the narrator is Cathy Toft.

FIGURE 17 An Australian Ringneck, *Barnardius zonarius,* feeding on grass seeds in Australia. Photo © Georgina Steytler.

Nevertheless, these parrots in southern Peruvian Amazon did not simply eat whatever plant reproductive parts were available at the time. First, we found that, as expected, seeds were higher in protein and fat and lower in fiber than any other plant part that the parrots ate. We then focused on diet selection in the three species of *Ara* macaws (Scarlet Macaw *A. macao,* Red-and-green Macaw *A. chloropterus,* and Blue-and-yellow Macaw *A. ararauna*). We compared the items (whole fruit containing seeds) available but not eaten to those that the macaws actually ate. The macaws indeed included in their diet items that were higher on average in protein and fat and lower in indigestible fiber than items that they did not consume but that were available on fruiting trees at the time. While these data do not tell us that macaws received enough protein to meet their needs and those of their growing young, they do reveal that macaws were seeking higher sources of protein and not foraging either randomly or for energy alone.

Hosts of studies in the New and Old Worlds confirm what Jamie Gilardi and I observed with our collaborator Charlie Munn in the parrots of Manú. Throughout the tropics, rainforests and seasonal forests are inhabited by a mixed-species guild of mostly granivorous parrots. The genera that we studied in Manú National Park occur throughout Central and South America. In addition to those species included in our study, biologists in the Neotropics have encountered other generalist granivorous parrots, including the Hawk-headed Parrot *Deroptyus accipitrinus,* the *Forpus* parrotlets, the Blue-bellied Parrot *Triclaria malachitacea,* the Burrowing Parakeet *Cyanoliseus patagonus,* the Yellow-eared Parrot *Ognorhynchus icterotis,* Spix's

FIGURE 18 A Nanday Conure, *Nandayus nenday,* eating a mango fruit. Photo © Luiz Claudio Marigo.

Macaw *Cyanopsitta spixii* (now extinct in the wild), the Nanday Parakeet *Nandayus nenday,* and the Short-tailed Parrot *Graydidascalus brachyurus.* Similarly, in mainland tropical and sub-tropical Africa, the guild of generalist granivores includes the Vasa Parrots *Coracopsis nigra* and *C. vasa,* the Grey Parrot *Psittacus erithacus,* the various species of *Poicephalus,* and the lovebirds *Agapornis,* of the species studied thus far in their native environments. The Australasian realm also is home to many species in the guild of generalist granivorous parrots, but because of the unique history and environments in those areas, we will cover this region separately.

The scientists who have studied the diets of these species of granivorous parrots all tackle the questions that we posed earlier. Just what are these parrots actually eating (figure 18)? In the rain of fruit and plant parts and other debris, how can we discern what the parrots have ingested? Why are the parrots wasting so much apparent food? Do they have so much food available that it is not worth being careful to eat it all? Or, do the parrots have a deliberate strategy to maximize their nutritional benefits, or perhaps minimize the amount of toxins ingested? Exactly how does this food provide them with the nutrition they need?

We are interested in how these granivores obtain their nutrition, of course, but tropical biologists also seek to understand the relationship that parrots have with the plants they use for food and the ecosystems in which both partners are embedded. The scientists strive to determine whether parrots are entirely predators that digest seeds and other nutritious parts of the fruit and flowers, destroying them before they can complete their mission in the reproductive cycle of the plant. Alternatively, if parrots should function in a more congenial role with the plants, dispersing their seeds and pollen, then they would play a very different role in forest ecosystems. Whether parrots are predators or helpful partners matters hugely in rainforest ecosystems and determines many ecological patterns, including why there are so many species of plants and animals in tropical rainforests.

The jury now seems to be in, and by far the preponderance of parrot dining is destructive to the plant and its offspring. How rapacious parrots can be has been impressively documented in a number of studies. Mercival Francisco, Vitor Lunardi, Paulo Guimarães, and Mauro Galetti investigated Canary-winged Parakeets *Brotogeris versicolurus* preying on the seeds of trees in the Brazilian Cerrado, focusing on the semi-deciduous trees *Erica gracilipes* and *Pseudobombax grandiflorum*. In their efforts to extract the seeds from the fruit, the parakeets destroyed at least two-thirds of the fruit on average, and up to 100 percent of the fruit on some trees, in a matter of a few days. In the carnage that ensued when flocks of parakeets invaded the trees, no whole seeds could be found falling to the ground. The parakeets also attacked the fruits when they were still unripe. The scientists concluded that virtually no seeds were left to germinate, and called the predation event "massive," implying that the reproductive efforts of the trees were entirely foiled by the parakeets.

Elsewhere in Brazil, José Ragusa-Netto, Paulo Antonio da Silva, Sandra Paranhos, Carlos Barros de Araújo, and Luiz Octavios Marcondes-Machado in three different studies documented the closely related Yellow-chevroned Parakeets *B. chiriri* feeding on a wide variety of tree species, consuming fruits, fruit pulp, arils,[3] seeds, nectar, and flowers. These parakeets ingested quite a bit of pulp in their diet, but nevertheless consumed the seeds destructively as well. In those studies, up to 12 percent of the seeds were destroyed, as observed by da Silva for the tree *Chorisia speciosa*.

In far-away Costa Rica, the generalist fruit-and-seed-eating guild of parrots is also found, represented by the same genera, if not also the same species. There, the indefatigable curiosity of tropical biologist and natural historian Daniel Janzen compelled him to shoot an Orange-chinned Parakeet *Brotogeris jugularis* while it was engaged in eating the fruit of the fig tree *Ficus ovalis*. Janzen wanted to know whether the parakeets are friend or foe of the fig, either eating the pulp while discarding the seeds or vice versa. Sure enough, he found only fig seeds inside the downed parakeet and an abundance of fig fruit pulp, stripped of seeds, on the ground beneath the foraging flock. Furthermore, virtually all of the fig seeds found in the bird's crop had been cracked by the bird as it picked them out of the flesh of the fig, rendering them more easily digested by the parakeet but useless for the plant. In a less invasive study, Pedro Jordano kept vigil for five days at the foot of a single fig tree *F. continifolia* to record in detail how many animals arrived at the fruiting fig, how many figs each individual ate, and what they seemed to do with the figs. A suite of ravenous frugivores exhausted the estimated 100,000 figs on the tree in those five days, with 95 percent being removed by

3. Arils are sometimes called "false fruit." You could also think of an aril as a subfruit; for example the individually pulp-wrapped seeds of pomegranates are arils.

the end of the third day. Of 44,000 figs removed each day of the heaviest feeding, birds took 65 percent of the fruits, and parrots of three species, Orange-fronted Parakeets *Eupsittula canicularis* (formerly in the genus *Aratinga*), Orange-chinned Parakeets, and White-fronted Parrots *Amazona albifrons* consumed the bulk of those (78 percent of birds' share). Jordano estimated that parrots destroyed 36 percent of the seed crop. Why not more? Most seeds, as it turns out, had already been consumed by the seed-eating larvae of small parasitic wasps. The wasp larvae probably provided more nutrition for the parrots than did the seeds one step down the food chain. The parrots, mostly the smaller parakeets, could just as easily strip the delectable grubs out of the fruit pulp as they could the seeds. From the tree's perspective, the little wasp was the villain, and if anything, the parrots were assisting the tree by damping the next generation of wasps.

Back in Brazil, Paulo Martuscelli found Maroon-bellied Parakeets *Pyrrhura frontalis* taking this tree aid one step further. These parakeets fed heavily on insects inhabiting galls of the massaranduba tree *Persea pyrifolia*. Larvae of many insects hijack the tissue growth of trees to trick the tree into growing them a shelter and safe place to feed on the carbohydrates produced by the plant's photosynthesis. These pests drain plant resources for their own use, and high infestations of them could greatly reduce the fitness of the host plant. Martuscelli found that parakeets treated the galls just as if they were fruits, digging out the insect larvae as if they were seeds and discarding the bitter gall just as they would the palatable fruit pulp. In this case, foraging parakeets are helping their food plants instead of harming them.

Across the ocean, in Africa, a similar picture emerges of generalist parrot granivores that prey on and destroy seeds before they are dispersed from the parent plant. Remarkably, the diet of the common and well-known Grey Parrot has not been formally studied in the wild (at least nothing has been published, to my knowledge), surely reflecting the difficulty of studying free-ranging parrots in the dense forests of Central and West Africa where this species resides. Studies of their habitat selection and movements, however, suggest that the diet of this species in the wild is that of a generalist granivore and may be well approximated by that of the better-studied species of *Poicephalus*. The prolific studies by scientists in the African Parrot Research Centre, including Mike Perrin, Steve Boyes, Margaret Hunter, Richard Selman, Craig Symes, Stuart Taylor, and J. O. Wirminghaus, reveal rich information on the biology of the *Poicephalus* parrots. The more common *Poicephalus* species fit the profile of granivorous generalists, including the Brown-headed Parrot *P. cryptoxanthus,* the Brown-necked Parrot *P. fuscicollis fuscicollis,* the Grey-Headed Parrot *P. f. suahelicus,*[4] Meyer's Parrot *P. meyeri* (figure 19), and Rüppell's Parrot *P. rueppellii*. Parrots of these species range far and wide, following the seasonal pulses of fruiting trees and gleaning nutrition not only from the seeds, but also flowers, leaves, fruits, and the insect herbivores (mostly the larvae of

4. A recent taxonomic revision has elevated some former subspecies of *Poicephalus robustus* to species status, and in so doing has required the Cape Parrot *P. robustus* to retain the original specific name. The other species are now assigned to *P. fuscicollis* and then recognized as subspecies under this name.

FIGURE 19 A Meyer's Parrot, *Poicephalus meyeri,* uses its hooked beak to extract seeds from the fruit of the sausage tree (*Kigelia africana*) in the Okavango Delta of Botswana. Photo © Ian and Kate Bruce.

beetles, moths, and flies) that share the botanical repast with the parrots. Symes and Perrin found that seeds included in the diet of the Grey-headed Parrot could be quite high in protein, and that feeding and movement patterns varied seasonally, as did fruiting of trees and breeding of the parrots. The parrots appeared to be feeding on higher-protein foods when they were provisioning their young. Like many of the medium-to-large Neotropical parrots, Meyer's Parrots are able to crack unripe seeds to reach the higher-quality nutrition within the kernel.

Does Size Matter?

A team of biologists in Costa Rica studied a guild of six species of granivorous parrots ranging from the small parakeets in the genera *Eupsittula* (Orange-winged Parakeets *E.*

canicularis) and *Brotogeris* (Orange-chinned Parakeets *B. jugularis*), through the medium-sized *Amazona* (Yellow-crowned Parrot *A. ochrocephala*, Red-lored Parrot *A. autumnalis*, White-fronted Parrot *A. albifrons*), to the Scarlet Macaw. Among other findings, scientists Greg Matuzak, Bernadette Bezy, and Donald Brightsmith noticed that the larger the parrot, the more individuals included seeds to the exclusion of other plant parts in their diet. Conversely, the smaller the parrot, the more individuals ingested of the fruit pulp, nectar, and flowers, consistent with the studies we just reviewed. In their study, more than half of the observations were of parrots eating seeds destructively, that is, cracking and digesting them, versus dispersing whole seed. They noted, however, that this percentage is biased by the large proportion of macaws in the sample. Looking at species individually in their data, the large macaws ate seeds in 70 percent of the observations and fruit pulp in 10 percent. The two species of parakeets ate seeds alone in at most 25 percent of their observations and at least some fruit pulp in nearly 50 percent of observations, a pattern consistent with our results in Peru. Matuzak, Bezy, and Brightsmith hypothesized that this difference in diet reflects in part higher protein requirements in the larger-bodied macaws and amazons (more tissue to build and maintain) versus the higher energy requirement of the smaller parakeets (with an expected higher metabolic rate per gram).

Although this plausible hypothesis needs to be tested, another consideration is that the larger size of these parrots may be required for them to eat the seeds in the first place. Recall that another plant defense is mechanical rather than chemical; a common strategy is to make the seed containers so hard that herbivores might be persuaded to eat something else. To explore this idea a bit more, consider that the largest granivorous parrots are the *Anodorhynchus* macaws, the Hyacinth Macaw *A. hyacinthinus* (figure 20), the Lear's Macaw *A. leari*, and the extinct Glaucous Macaw *A. glaucus*; all are specialists on arguably the hardest of seeds, palm nuts. What size brings to bear on eating the hardest of seeds is the mechanical force required to crack them. Everyone who has had a close encounter with a macaw surely has noticed its massive beak and head. Elsewhere, we consider how this machinery for crushing strength actually works (box 2), but no matter how clever the design of the lever system, sheer size brings force, as we know from human athletes such as sumo wrestlers and throwers of weapons such as discus, javelin, and caber.

The *Anodorhynchus* macaws in particular possess the most massive chisel structure of any parrot beak, rivaled only by the *Ara* macaws and the Palm Cockatoos *Probosciger aterrimus*, aptly named because of their diet. The gnathotheca, the large horny surface of the upper beak, serves as an anvil against which the lower beak can wedge and split open the rock-hard seed coat. Only the largest macaws can split the hardest of the hard, the seed coats characteristic of palms. Although passage through the gut of a mammalian herbivore makes this job easier, the macaws do a fine job of cracking the fresh nuts on their own. In addition to growing as it were "the right tool for the job," Hyacinth Macaws both in captivity and in the wild have shown how handy they are with crafted tools, after the fact. Andressa Borsari and Eduardo Ottoni describe how both adult and juvenile Hyacinth Macaws living in a family group regularly used bits of wood that they chiseled off branches to hold nuts in wedge-

BOX 2   THE RIGHT TOOLS FOR THE JOB

Anyone who has had a close encounter of the parrot kind can attest to the crushing strength of parrot jaws. Even the miniscule Budgerigar *Melopsittacus undulatus* is formidable. I always have to take a deep breath and gird myself when I need to capture and handle a budgie. Cockatiels *Nymphicus hollandicus*, at only 100 grams, are completely unassailable with my bare hands, and as the parrot gets larger, ever more gruesome injuries can result to unprotected human flesh. In contrast, I can grasp a songbird with impunity, as long as I can dodge having something vulnerable such as my eye pecked at with a sharp point. I need not fear any damage from the grasp of a songbird's beak, which can deliver little more than a pinch.

Parrots, as it turns out, have anatomy that is unique for birds and that seems to have evolved for crushing very hard objects. Like all birds that are accomplished fliers, parrots require a lightweight but strong skeleton to support their bodily needs. Low weight and high power are contradictory, but clever application of leverage goes a long way to overcome the limitation on the cross-sectional density of bird bones. Evidence points to the divergence of parrot ancestors from those of falcons and songbirds along a path of preying on seeds, which requires overcoming the defenses of plants. Plants defend themselves from the depredations of granivores in part by making their seeds hard, stowing the vulnerable embryo inside an impenetrable case. Parrot ancestors may already have been equipped with empowered jaws and muscles to quell animal prey, but as evolution proceeded, this ability was enhanced by the recrafting of the skull and its musculature.

Natural selection crafted two morphological innovations in the parrot lineage: the suborbital arch of the skull and the muscle associated with it, known as the musculus pseudomasseter, or M. pseudomasseter for short. The suborbital arch is a bony reinforcement in the skull, attaching to other skull bones on either side of the ventral aspect of the orbit, containing the eye. This bony structure arises from ossifying a major subocular ligament. The M. pseudomasseter, when present, is a massive muscle that overlies the suborbital arch or other suborbital processes of the skull and attaches to these structures dorsally. These reinforcements for both leverage and power enable both crushing strength and mobility in the jaws. Parrots with these structures have more ability to raise the upper jaw, as opposed only to depressing the lower jaw, when the individual widens its gape. They also have enhanced ability to move the jaws (and the attached upper and lower beaks) from side to side, allowing finer control of slicing, cutting, and prying with the beak.

Although these parrot novelties were discovered long ago by morphologists using traditional methods to study anatomy, our understanding of their function has

(continued on next page)

improved greatly with modern molecular and comparative methods. In a fascinating study overlaying phylogeny with morphological traits, Masayoshi Tokita, Takuya Kiyoshi, and Kyle Armstrong discovered that the evolution of the suborbital arch and the M. pseudomasseter was surprisingly independent in the parrot lineage. These novelties both arose multiple times in the parrot phylogeny (and were sometimes later discarded) by natural selection.

The suborbital arch may have been possessed by the ancestral parrot, but in the basal-most clade of parrots, only the Kakapo has it today. The Kea and Kaka go without. Likely the need to grind massive amounts of plant fiber is made easier by buttressing the skull. Nevertheless, the M. pseudomasseter does not develop in the Kakapo, perhaps implying that the extra power it provides is not necessary for grinding. All cockatoos, also representing a basal clade of parrots, possess both the suborbital arch and the M. pseudomasseter, and the two structures appear to have evolved in tandem in this family. Whatever their mode of sustenance, the cockatoos have considerable strength as well as dexterity in the functioning of their jaws and beaks.

Along the rest of the parrot lineage, the suborbital arch and M. pseudomasseter make spotty appearances. Both arch and muscle show up in the *Cyanoramphus* parakeets and in the clade containing the lories, the Budgerigar, and the fig parrots *Psittaculirostris*. Perhaps not surprising is the apparent loss of the suborbital arch in the one taxon of lorikeets represented in their study, *Lorius*. Lorikeets need dexterity but not power to acquire their food. Therefore, alternatively, perhaps the suborbital arch never developed in the lorikeets, but arose secondary to the appearance of the M. pseudomasseter in that clade.

The suborbital arch also appears in some of the Neotropical parrots, in the amazons and close relatives and in the macaw-conure branches. Only the *Amazona* and *Pionus* parrots possess a well-developed M. pseudomasseter. Interestingly, the larger *Anodorhynchus* macaws are inferred to have evolved from an ancestor possessing the suborbital arch, but these largest of parrots, preying on the hardest of nuts, have divested themselves of this buttressing. Evidently other roads lead to Rome, and these macaws solved the need for massive crushing strength with other morphological structures.

Morphologist Dominique Homberger has maintained that the function of the hefty M. pseudomasseter is to permit lateral movements of the beak and not simply to provide more power to crush. In fact, because of the need for opposition in applying force, there may be a trade-off between lateral dexterity and crushing force. Her hypothesis is consistent with the findings of Tokita and his colleagues. She derives her hypothesis in part through detailed study of the beaks of parrots and how these are used in foraging. Homberger recognizes two types of beak morphology in the parrots, which she labels *calyptorhynchid* (after the black cockatoos in the genus

Calyptorhynchus) and *psittacid* (referring to the major branch of the parrot phylogeny following the cockatoos).

Homberger designates the calyptorhynchid as the ancestral type, yet not even all species in the genus *Calyptorhynchus* possess this beak type. The cone-specialist *Calyptorhynchus* cockatoos and the Gang-gang Cockatoo *Callocephalon fimbriatum* have the calyptorhynchid type of beak, but not the ground-feeding Red-tailed Blaco Cockatoo *C. banksii*. Still, her hypothesis is intriguing and plausible. She suggests that ancestors of parrots made their livings by stripping insect larvae out from under bark, with dexterous movements of a laterally flexible beak. Today, two basal-most members of the parrot phylogeny in the genus *Nestor* employ this mode of foraging. This anatomy then lent itself well to the predatory harvesting of seeds, and the rest is parrot history. Many terrestrial birds are gleaners of insects who subsidize their diets with plant offerings such as fruit and sap, as do many of the songbirds. The parrots then followed another evolutionary path, by opening up the ability to exploit seeds that were difficult to eat and to digest, thus creating for themselves a niche with rich nutritional resources for which there was relatively little competition.

Homberger's extensive study of parrot beaks also reveals another insight. To keep their beaks sharp and able to cut, parrots need to keep their edges honed, particularly those of their lower beaks. The palate of the upper beak has a serrated roof, and parrots use this surface to sharpen the edges of their lower beaks. Anyone with pet parrots, but particularly cockatiels, has heard the sounds of little grinding noises after nightfall, when the birds are at roost. A flock of sleepy cockatiels can make a considerable racket of beaks grinding away. We owners of cockatiels anthropomorphically consider this sound one of contentment. While that may be the case, clearly the cockatiels are using their "down time" to catch up on an important maintenance activity, keeping their beaks sharp.

like fashion, using one foot and their beak, so that the nut was less likely to move as the macaw applied opposing forces with its upper and lower beak.

Like Hyacinth Macaws, Palm Cockatoos are the largest-bodied species in their family and also consume the larger, very hard, and highly nutritious seeds produced by tropical trees. This largest cockatoo forages on the massive seeds of the Ti tree *Terminalia impediens,* as well as other species of *Terminalia, Cerbera floribunda,* and various species of canopy trees. Christopher Filardi and Joshua Tewksbury describe how the Ti tree protects itself from depredations of Palm Cockatoos, and how in turn the cockatoos circumvent the tree's defenses. To disperse its seeds, the tree embellishes each of the 15-gram goliaths with a deep covering of fruit flesh. To deter the cockatoos, the unripe fruit covering is hard and bitter. Because neither hardness nor bitterness protects the seed well from parrots, however, the

FIGURE 20 Hyacinth Macaws, *Anodorhynchus hyacinthinus,* using their enormous bills to crack open palm nuts in Brazil. Photo © James Gilardi.

tree also drops the fruit to the forest floor while it is still unripe, expecting to escape parrots through flight rather fight. Once on the forest floor, the fruit flesh ripens, when it is then eaten by Cassowaries, large flightless birds of the genus *Casuarius*. Cassowaries provide the same dispersal role as the large herbivorous mammals in South America and Africa. The big birds consume the fruit, swallowing the viable seeds whole. They then disperse the seeds from the parent and their siblings and deposit them with a modified seed coat, so that the seedling is prepared to germinate in a heap of fertilizer. Palm Cockatoos frequently forage on the ground, digging through piles of Cassowary excrement to find the cleaned and softened seeds, which they then crack open to scoop out the protein- and lipid-rich pulp. The cockatoos' beaks are well designed for this task. Not only are they massive and able to apply considerable crushing force to the seed coats, they also possess a long, tapered, sharply pointed tip, which they use to scrape out the seed contents (figure 21). Palm Cockatoos also use tools to aid them in this task. Famous evolutionary biologist Alfred Russell Wallace described in 1869 how a Palm Cockatoo would roll a *Canarium* nut up in a leaf to prevent the nut from slipping as the bird cracked it open.

As we will encounter again in subsequent chapters, parrots and primates have much in common. New World monkeys and Old World chimpanzees also use tools, in this case made of stone, to crack open very hard palm and other tropical nuts for the sumptuous nutrition inside.

FIGURE 21 A Palm Cockatoo, *Probosciger aterrimus*, in Cape York, Australia using its long, tapered bill to crush nuts and then scrape out the nutritious contents. Photo © Ian Montgomery.

Why go to all of this trouble to eat palm nuts? Evidently the benefits in balanced and abundant nutrition outweigh both the evolutionary investment in big bills and the immediate time-and-effort costs of opening these very hard nuts. These trees load up their large seeds with high-value nutrition to send their own offspring out in the world. They invest mightily in the rock-hard seed coat surrounding the embryo, and indeed few animals are able to overcome this defense. As a result, from the parrots' perspective, there are few competitors for this wealth.

In contrast, Blue-throated Macaws feed on palm fruits but do not eat the nuts. They scratch deep scars into the seed coat as they peel off the oily, nutritious mesocarp (pulp) of palm fruits. These scratches also facilitate the germination of the embryo. Here we see another of the relatively rare examples where parrot feeding habits may be beneficial to the plants on which they feed.

Specialized Granivores: Living on Ancient, Cone-Bearing Trees

A diet of only palm or other very hard nuts is certainly specialized, but frequently these macaws and Palm Cockatoos eat from a variety of species of plants over their home ranges. Some other species of parrots take specialization to the ultimate degree. They consume the seeds and perhaps some other items from only one species of plant, for extended periods of time, or maybe dine from only that species and a few close relatives over their lifetimes.

The most extreme specialists known so far include certain cockatoos in the Australian genus *Calyptorhynchus* (Glossy Black Cockatoo *C. lathami* and one population of Red-tailed Black Cockatoo *C. banksii*), which eat seeds from the cones of the sheoak and bulloak genera *Casuarina* and *Allocasuarina,* and the South African Cape Parrots *Poicephalus robustus*, which feed on seeds of trees in the coniferous genus *Podocarpus* for 70–80 percent of their diet. Another group of specialists is the two species in the primarily Mexican genus *Rhynchopsitta* (Thick-billed Parrot *R. pachyrhyncha* and the Maroon-bellied Parrot *R. terrisi*), which feed heavily on the seeds of conifer species, particularly those in the genus *Pinus*. We can also include here some populations of Austral Parakeets *Enicognathus ferrugineus* studied by Soledad Díaz and Thomas Kitzberger. Austral Parakeets consume just about anything and everything found on one or another species of tree in the genus *Nothofagus,* the so-called southern beeches such as *N. pumilio,* in extreme southern South America. Exactly what they eat depends on the season and ranges from the trees' pollen (sometimes heavily, like lorikeets), seeds, honeydew from the tree's herbivorous insects, and lichens and parasitic mistletoe growing on the trees; so in one sense their diets are generalized but in another sense highly specialized on one or just a few species of trees in one genus. One can argue that Austral Parakeets are habitat specialists rather than food specialists, I suppose, but nevertheless, this focus on one species of tree is dietary.

These far-flung examples of highly specialized diets have some common themes. First, the particular tree taxa on which these specialists depend are either basal in the plant clades, such as the gymnosperm genera *Pinus* and *Podocarpus,* or closely tied to the same ancient Gondwanaland origins (chapter 1) as parrots such as *Notofagus, Allocasuarina* and *Casuarina.* Although concurrent with the long-ago origins of parrots, we cannot jump to the conclusion that these specialist diets represent that of parrot ancestors. Rather, these specialists have more in common with island-inhabiting species, in which isolation or other restriction to particular habitats has resulted in the evolution of a narrow and locally adapted diet, dependent on the plants with which they share this isolation. Second, like many island-dwelling species of parrots, all these specialist populations are threatened to some degree.[5] Fine-tuned specialization in diet or habitat could be expected to result in populations that lack resilience in the face of major environmental change, such as that propagated by our own species.

The Glossy Black Cockatoo's utter reliance on the seeds of one or two closely related species of tree, the sheoaks *Allocasuarina verticillata, A. diminuta, A. littoralis,* or *A. gymnanthera,* depending on the location, seems both foolhardy and improbable. Being among the largest of parrots, Glossy Black Cockatoos must have a reasonably high protein requirement, and as it happens, the seeds of sheoaks are impressively rich with nutrition. They pack a whopping 27–40 percent protein by dry weight, topped off with up to 38 percent lipids, the plant equivalent of power food for parrots. If nitrogen- and lipid-containing foods are universally limiting for herbivores, then the lucky parrots need look no further once they have located a patch of these trees. But that's the catch. To support themselves entirely on sheaok seed, the parrots are relegated to searching for a non-stop, year-round supply. Unfortunately for the parrots, the plants usually do not oblige, because they are constrained by their own seasonal requirements for resources, such as sufficient moisture and soil nutrients, and ideal growing tem-

5. These populations may not be listed as threatened or endangered by CITES or BirdLife International if they are taxonomically considered subspecies of species with more abundant populations.

FIGURE 22 A Glossy Black Cockatoo, *Calyptorhynchus lathami,* feeding on the seed cone of a black sheoak (*Allocasuarina littoralis*). This cockatoo species is threatened in part because of its dependence on seeds from this genus of tree. Photo © Mehd Halaouate.

peratures. These constraints usually mean that plants set fruit seasonally, when conditions are just right for the growth of the seed and fruit, for pollination, and for dispersal.

Matt Cameron and others, including Mick Clout, Tamra Chapman, David Paton, Gabriel Crowley, Stephen Garnett, Leo Joseph, John Pepper, T. D. Male and G. E. Roberts, documented in a variety of studies from 1982 to 2007 that breeding and non-breeding Glossy Black Cockatoos fed on sheoak seeds nearly exclusively year-round (figure 22). Clout observed that Glossy Black Cockatoos fed their growing young a diet that is nearly 90 percent seeds of the sheoak *A. littoralis.* Clout noted that the sheoaks' own reproductive strategy fortuitously provides the parrots with a reliable source of seeds virtually without seasonal interruption. The sheoaks are adapted to frequent burning, as is common in seasonally dry Mediterranean climates found in regions of Australia. Sheoak cones are tough and fire-resistant, but fire also causes the cones to open. The trees then shed their seeds immediately after a sufficiently intense fire, which just happens to be a perfect time for the seedlings to emerge from the seeds. Competitors for light and water were just incinerated by the fire, releasing the nutrients that were stored in their bodies for unfettered use by the growing seedlings. For this reason, sheoak cones full of ripe seeds hang out at the ready for extensive periods, year-round, waiting patiently for the occasional fire to sweep through and release the seeds. This plant strategy of waiting for the inevitable next fire to release seeds is known as *serotiny,* and the cones are described as *serotinous.*

But this plan for regenerating sheoaks after a fire is thwarted by the foraging parrots. With their massive beaks, the parrots can easily crack open the cones and access the seeds. The cockatoos use a fine-tuned lateral motion permitted by the morphology of their beaks to extract the seeds in a pincer-like action and scoop out the most nutritious part of the kernel. This foraging method is as exacting as it sounds, and the cockatoos spend prodigious amounts of time foraging, four or five hours a day minimum for non-breeding birds and six to seven hours a day for cockatoos provisioning their young. These cockatoos occupy up to 93 percent of their waking time with finding and processing food. With such a costly livelihood, the cockatoos seek to maximize their efficiency. Glossy Black Cockatoos forage in small flocks of at most two to three birds, probably pairs foraging with an older chick from a previous breeding season. They choose to search for food only in the trees most laden with fruit, and they favor groups of such trees over singletons. Once in a tree, the cockatoos concentrate on the branches with the densest burdens of cones. They also direct their attention to the younger, russet-colored cones with the freshest seed. In applying these tactics, they overlook branches and trees with only scattered cones, as edible as these might be. This foraging strategy presumably allows them to maximize their profit, gaining the most nutrition for the least effort traveling between trees, searching for cones, and meticulously removing individual seeds from the cones. The number of seeds that one cockatoo can harvest in a day is enormous. Crowley and Garnett recovered 2,000 seeds from the crop of one Glossy Black Cockatoo chick, along with a good measure of insect larvae.

Thick-billed and Maroon-fronted Parrots specialize on seeds held in the cones of pine trees that, although not closely related to the sheoaks favored by Glossy Black Cockatoos, share close physical resemblance in foliage and reproductive strategies. Noel Snyder, Susan Koenig, Ernesto Enkerlin-Hoeflich, Javier Cruz-Nieto, and Tiberio Monterrubio-Rico and their colleagues have studied the Thick-billed Parrot in Arizona and Mexico extensively. These biologists document the heavy dependence that the Thick-billed Parrots have on pine forests, comprising only a few species of pines at any given location (more on this in chapter 6). Thick-billed Parrots migrate over long distances and encounter many habitats and differing vegetation with various available foods along their routes. Nevertheless, the parrots depend heavily on pine seeds as food, supplemented by acorns, juniper berries, terminal buds, and agave flowers, and other parts of the trees that make up the pine-juniper-oak woodlands of the mountainous areas of the Sierra Madre Occidental and southern Rocky Mountains. Thick-billed Parrots tend to nest around the time that the large seeds of Arizona Pine *P. arizonica* and Durango Pine *P. durangensis* become mature. The parrots feed these seeds in abundance to their chicks, making up 75 percent of the chicks' crop contents. The rest of the crop contents include other seeds, insect larvae, and apparent debris such as pine needles and bark.[6]

6. When animals consume "debris" of no apparent value as food, one always wonders whether such items are consumed intentionally or only incidentally. We should remain open to these items' being consumed intentionally. For example, bark is not infrequently found in the crops of parrots and may have some nutritional or medicinal value.

The coniferous forests in which Thick-billed Parrots live are frequently burned, but it is not clear whether the cones of *P. durangensis, P. ayacahuite,* and other local pines are formally serotinous. The pine-seed diet of Thick-Billed Parrots depends on pine species that produce cones annually and shed their seeds as soon as they are ripe, as well as on pine species that have long-maturing cones, up to two and three years. The parrots eat both unripe and ripe seeds, as the majority of granivorous parrots do. The latter species of pines therefore provide a reasonably reliable supply of this one foodstuff year-round. The number of pine seeds consumed by each parrot daily rivals that of the sheaok-seed-eating Glossy Black Cockatoos. One parrot may eat seeds from over 100 cones per day—2,000 seeds or more. This labor-intensive mode of foraging requires that the parrots be as efficient as possible, and like the Glossy Black Cockatoos, Thick-billed Parrots seem to select the trees with the highest yields, especially in the morning, when they are most hungry. They may sequentially go from one species of pine to another during a single day, as they appear to hunt systematically for the most profitable foods.

Thick-billed Parrots feed destructively, as implied by the label of granivore (figure 23). They typically remove the cone from the branch before they begin to eat it, and anything that interrupts the meal will cause them to drop the unfinished cone to the ground. For pine seeds that are wind-dispersed, falling to the ground prevents them from dispersing from each other and from their parent tree, even if the dropped cone is ripe. Some species of pine respond to this insult by engaging in *masting*. This is a reproductive strategy in which the pines produce very large cone crops irregularly. This makes their cone crops unpredictable to predators, so that seed predators will be challenged to find the seeds, and if they find them, unlikely to be able to eat more than a small fraction of the crop. The parrots respond by roving far and wide, searching for abundant cone crops. This combination of the trees' habits and the parrots' habits makes the conservation of endangered species such as the Thick-billed Parrot exceedingly difficult, complicating both local protection in Mexico and efforts to reintroduce them into Arizona.

Cape Parrots also travel in search of food, as we would expect from a destructive food specialist, as opposed to the usual generalist granivores. The parrots need to range far and wide, and they stay on the move to find unexploited crops of *Podocarpus* cones. Trees of the genus *Podocarpus* are conifers, as are pines, but with a strictly Gondwanaland distribution. Similar to the Glossy Black Cockatoos and Thick-billed Parrots, Cape Parrots depend on species that have extended fruiting periods, *P. falcatus* and *P. latifolius,* and like the palm and sheaok seeds, podocarp seeds are rich in protein and lipids. Cape Parrots also eat the cones when they are unripe and hard, probably beating their competitors to the punch. And, like the Thick-billed Parrots, Cape Parrots are now listed by the IUCN as endangered. Their dependence on one species of tree, which happens to be heavily logged, and their need to range widely in search of food, is an unlucky combination when the parrots are confronted with massive land-use conversion by a rapidly growing population of humans in Africa.

Geophagy: How Eating Soil Helps Parrots Cope with Toxic Plant Defenses

In our review of granivorous parrots so far, we have learned that most feed liberally on unripe and chemically defended fruits, either for the fruit flesh itself or more likely for the

FIGURE 23 A pair of Thick-billed Parrots, *Rhynchopsitta pachyrhyncha,* near their nest in Madera, Mexico. This species is another dietary specialist that is endangered, in part, because of loss of the pine trees on which it depends. Photo © Steve Milpacher.

precious seeds inside. Why are parrots able to extract nutrition from toxic parts of plants, when other herbivores avoid the chemical defenses and wait until the fruit is ripe, when it is ready for dispersers to consume it?

Biologists have repeatedly observed parrots digging into the dirt of riverbanks and cliffs and appearing to ingest the stuff. The most famous of these clay licks (or *colpas*) are the much-photographed sites along the meandering Manú River, in Manú National Park in Peru and near Tambopata, where many biologists have worked, myself included. Jamie Gilardi, Charles Munn, Donald Brightsmith, Romina Aramburú Muñoz-Najar, Luke Powell, Thomas Powell, George Powell and others have studied this mysterious behavior there and what it might mean for the parrots. Since this work on geophagy in Amazonian parrots, biologists have reported other parrots regularly and deliberately consuming clay-based soils, including the Maroon-fronted Parrot in the Sierra Madre Occidental of Mexico and the Palm Cockatoo in Papua New Guinea.

Jamie Gilardi and his colleagues Sean Duffy, Charles Munn, and Lisa Tell outlined five non-mutually exclusive hypotheses to explain why parrots might make such an effort to include substantial amounts of dirt in their diet. One is that the dirt helps them break down the hard seeds they consume. We are familiar with birds that consume grit, which lodges in the gizzard (the ventriculus) and helps process food mechanically. Parrots, however, do not eat their seeds whole but masticate them using their powerful beaks and muscular tongues, which precludes the need for mechanical help in the gizzard.[7] Another hypothesis is that the dirt provides essential nutrients; we are all familiar with the universal need for minerals in the diets of all vertebrates, especially herbivores. In addition to these two more obvious hypotheses, Gilardi et al. considered the following three. First, the parrots may benefit from buffering gastric pH, which could occur because of mineral ions in the dirt—that is, they act as antacids (calcium is a familiar example found in many of our own medications). Second, parrots may gain gastrointestinal cytoprotection. In this scenario, the fine clay particles coat the lining of the gut, physically shielding the cells from either physical or chemical damage or having some other protective effect. This protection is an important task necessary for proper digestion, and familiar hormones such as the prostaglandins play a vital role by inducing the secretion of protective mucus in vertebrate guts. Third and last, the clay may function to absorb toxins preemptively. In other words, perhaps the clay has particular physical properties that cause it to bind to certain molecules and thus prevent their being absorbed through the gut, allowing their safe and innocuous passage to the outside of the organism.

To test these hypotheses, Gilardi et al. used a simple but powerful study design. A key element is that they collected dirt at the lick sites from sedimentary layers that the parrots fed on and from those that they did not, allowing a comparison of what might be in the "preferred" sediments, controlling for many other factors related to location (figure 24). They combined tests of these two classes of soils with *in vitro* and *in vivo* studies of organisms, including the Orange-winged Parrots *Amazona amazonica*, at the University of California, Davis, Psittacine Research Project facility.

The grit and pH-buffering hypotheses were discarded as unlikely to explain the parrots' dirt-eating behavior. At least, the soils at the colpas in Peru that the parrots chose were particularly devoid of sand and other potentially abrasive particles. Instead, the parrots fed differentially on the finest (smallest) particles, the sizes in the range that soil scientists call clay. We expected this result because parrots do not swallow unhusked or whole seed. The gastric-buffering hypothesis was rendered irrelevant when Gilardi et al. discovered that the Peruvian clays selected by the parrots had no pH-buffering capacity whatsoever and had no

7. Some wild parrots may retain a need for grit. Burrowing Parrots (including nestlings) have been observed to have grit in their crops (Juan F. Massello, personal communication).

FIGURE 24 A flock of Blue-headed Parrots, *Pionus menstruus,* feeding on clay at the *colpa* on the Madre de Dios River in Peru. Photo © Manfred Kusch.

effect on the pH of any tested substrate. This finding removed the necessity of showing that parrots need antacids in the first place.

Evaluating the mineral-source hypothesis was less straightforward. Although the clay soils chosen did have significant mineral content, including minerals known to be necessary for metabolism, there was a great deal of variation in the quantities of these minerals biologically available to the animals eating them. One clue was that the parrots' natural diet contained all of these minerals, and at levels comparable to those in commercially prepared animal feeds. Therefore, the parrots did not appear to have a great need to supplement their diets with minerals from clay. Complicating matters is the fact that mineral release by clays goes hand in hand with adsorption of other charged particles like toxins, so clays with capacity to perform one feat are necessarily capable of the other feat as well.

The toxin-absorption and cytoprotection hypotheses, in contrast, were strongly supported by several complementary lines of evidence. Jamie Gilardi and I discovered in a separate study that the plant parts on which the parrots fed were chock-full of toxic phenolic compounds. When these compounds were leached from the seven most toxic seeds eaten by macaws in Manú, they killed brine shrimp in the biological assay portion of the study. Similarly, the shrimp died in the presence of quinidine, a commercially available alkaloid derived from quinine, in turn originally extracted from the tropical tree genus *Cinchona*. When both the extracts of seeds eaten by macaws and the quinidine were first treated with clay, however, the mortality of the brine shrimp was greatly reduced.

Next, the toxin-absorbing properties of the Manú clay were tested in live Orange-winged Parrots in the laboratory. These experimental parrots were tube-fed quinidine, which was chosen for its low potential to harm the parrots compared to quinine, with and without addition of the Peruvian clay. Those parrots given the clay chaser showed much-reduced levels of the alkaloid in their blood. These tests were followed up with more chemical assays that demonstrated the powerful adsorptive properties of the clay, with high affinity (high cation exchange capacity) for the alkaloid and phenolic compounds that are produced by the plants to protect their leaves and seeds.

Gilardi et al. could not test the cytoprotective hypothesis with parrots, but these studies have been amply done in other species, including humans. For reasons as yet not well understood, the clay particles (like prostaglandins) increase the secretion of mucus by the cells lining the gut. The toxins in the seeds that the parrots eat can act as strong corrosives when in contact with the sensitive and absorptive gut lining. Therefore, the two mechanisms probably work in tandem. Initially, the clays bind the toxins with their high cation-exchange capacities, allowing them to be physically expelled from the gut. At the same time, the clay induces mucus, which shields the sensitive cells lining the gut from damage by exposure to the toxins. This double protection may ensure that parrots can eat the most toxic of plant parts with impunity.

The picture seems to come full circle, then. Without question, parrots eat unripe and chemically defended seeds and fruit. We assume that they do so to obtain their ample nutrition, rich in protein and lipids. Parrots access a nutritionally more balanced diet than many other herbivores can by eating these foods. Moreover, their ability to eat unripe, toxic seeds eliminates competitors for this food, which parrots would encounter in droves if they were to wait until the fruit was ripe. Foraging on chemically absorptive clays may be a small price to pay for this considerable benefit.

There is however another price to pay. The suitable clay layers are scattered about in the landscape and not all that common, much like the seeds and fruits themselves. Indeed, some parrot populations do not seem to have the proper soils available to them, or at least have never been seen eating dirt. How these parrots counteract the toxins they ingest remains a mystery. Those parrots that do have clay licks accessible have to travel there regularly, and this is just one more task that they have to include in their regular schedules. The parrots gather in large, noisy groups at the licks. Although parrots do not avoid conspicuous aggregations (see chapter 6), the birds are particularly exposed in the open on the riverbanks and not shielded from sight by vegetation. This exposure leaves them vulnerable to predators, more so than when they forage in the tree canopies. Nevertheless, despite the price, these ingredients allow parrots to fill a niche occupied by few organisms in the forests, undoubtedly explaining much about how and why parrots evolved into such voracious seed predators.

OTHER DIETS

The Diets of Ground-Feeding Australian Parrots

Australia is home to a wide diversity of parrot species representing different parts of the phylogenetic tree (chapter 1). We have already considered the diets of a few Australian species that are specialist granivores; in the next section will introduce ourselves to the many species that are nectarivores. Here we consider the generalist herbivorous parrots in Australia in a section of their own, because their diets reflect the unique and marvelous environments on this most isolated of continents.

Although the Island Continent has earned its name by being both small and isolated relative to the other continents, Australia does host a remarkable range of vegetation types. In part because of its coastal Mediterranean climate and dry interior, Oz is covered with widespread savanna, a vegetation type characterized by open-canopy woodlands carpeted with a grass understory. Parts of Australia support closed-canopy forests that require wetter conditions, and true rainforest occurs as a thin band on its northern reaches. This continent does not sport the vast expanses of tropical rainforests typical of the other continents, filled with granivorous parrots such as those we just covered. As a result, Australia is populated with parrots that are more generalist herbivores. Most make their livings in the Eucalyptus woodlands and savanna or other open-canopy shrublands, heathlands, and grasslands. Their diets therefore are significantly different from those of rainforest parrots.

These generalist parrots include those familiar species ubiquitous in the Australian landscape, such as cockatoos in the genera *Cacatua, Eolophus,* and *Nymphicus,* the *Platycercus* parrots (rosellas and ringnecks), and the Budgerigar *Melopsittacus undulatus.* These Aussie generalists also include an impressive diversity of other species, most of which have eluded formal study of their diets; the Red-capped Parrot *Purpureicephalus spurius,* the Hooded Parrot *Psephotus dissimilis,* and the elusive Ground Parrot *Pezoporus wallicus* are notable exceptions.

What all of these well-studied species have in common is that they routinely search for food on the ground. Notably, this terrestrial habit seems to be common only in open habitats that allow good visibility for detecting predators and multiple escape routes. Only one species of parrot lives and forages on the ground in dense forests. This is the Kakapo *Strigops habroptilus,* whose lifestyle was made safer by the fact that New Zealand was historically mammal-free (box 7).

Ground living thus opens up somewhat different fare for parrots. Most still glean much of their diet from the seeds of plants, but instead of tall trees most of these seeds come from

FIGURE 25 A Blue-winged Parrot, *Neophema chrysostoma,* one of the smallest Australian parrots to feed on grass seeds. Photo © Alan Milbank.

low herbaceous plants like shrubs and especially grasses. Even the smallest of seed-eating parrots, the Budgerigar and other small parrots in the genus *Neophema* (the aptly named Grass Parakeets, figure 25), are often too heavy to perch on the grain-bearing culms of grasses. Grass seeds are usually gleaned either from the ground or by standing on the ground while dining on low-hanging swards, making them available to parrots of all sizes. The smallest cockatoo, the Cockatiel *Nymphicus hollandicus,* specializes on the seeds of grass and low herbs, often preferring the soft immature seeds over the hard ripe ones. Daryl Jones found that Cockatiels in New South Wales were eating the seeds of at least seventeen species of native grasses and herbaceous ground-dwelling plants. The Cockatiels have also readily accommodated to the land-use changes made by the recent European settlers by feasting on the abundant cereal grain crops they planted.

The larger Western and Long-billed Corellas *Cacatua pastinator* and *C. tenuirostris,* and Galahs *Eolophus roseicapillus,* feed extensively on the ground, as well as on the convenient agricultural crops, just as do Cockatiels and Budgerigars. Their diet includes many grass seeds, both introduced and native to Australia. They also eat the underground storage organs of plants, for example those of the common introduced onion grass *Romulea rosea* (in the iris family), and historically those of native orchids. As do the tree-dwelling parrots, these cockatoos forage on insects with the same behaviors and methods they use for plants. Cockatoos root up grubs, which are the ground-dwelling larvae of beetles and other insects. They seek out grubs and consume them in large numbers, thus precluding any notion that they find them incidentally while rooting for corms and other underground plant parts. Apparently they use smell to locate the corms and the grubs underground, so that they can focus on profitable places to dig. Remarkably, no one has published on the diets of wild Sulphur-crested Cockatoos, to my knowledge, other than reports of their use of agricultural plants. Matt Cameron reports in his book *Cockatoos* that the diet of this species in native habitats is similar to that of the corellas and Galah.

The Ground Parrot specializes in ground living, as its common name implies; its haunts are the heathlands and shrublands with grass understories. These parrots live where trees do not occur, as do their even more elusive relatives, Night Parrots *P. occidentalis,*[8] so living and foraging in trees is not an option. Their ground-feeding habit, then, reflects necessity, and their night-living habit reflects prudence, as a defense against predation in such exposed surroundings (chapter 7). David McFarland's extensive study of Ground Parrots (figure 26) reveals that they eat the seeds of whatever plant is available and reproducing in the heathlands, whether it be monocotyledons (grasses, sedges, and their relatives) or dicotyledons (herbs and shrubs). They also eat other parts of the plants, including seedpods, small fruits, flowers, buds and stalks, and frequently various kinds of invertebrates, such as beetle larvae, true bugs, and egg sacs, all likely part of the herbivorous fauna living on the heathland plants. McFarland notes the lack of mammalian granivores, such as small rodents, that are so common in arid lands on other continents. Their absence probably allowed parrots to occupy this common niche. One potential group of competitors for these seeds is the granivorous ants. The ants take dropped seeds on the ground surface, however, whereas Ground Parrots take the seeds directly from the plants before they drop to the ground, reducing the potential for competition between the ants and the parrots.

Nectarivory: Parrots That Consume Nectar

Australia is also the land of lorikeets, those absurdly colorful parrots that grace Oz and the surrounding islands of the southern Pacific with their dazzling beauty. Perhaps some natural law dictates that animals that feed on nectar and pollen should be as glorious as the flowers that supply their food, in another twist on "You are what you eat."

8. A molecular phylogeny recovered by Leeton et al. supports assigning *occidentalis* (formerly *Geopsittacus*) to the genus *Pezoporus,* as do later studies by Murphy et al. (2011) and Joseph et al. (2011).

FIGURE 26 A Western Ground Parrot, *Pezoporus wallicus flaviventris*, vocalizing in its native heathlands of Australia. Photo © Brent Barrett.

At last we encounter parrots that cooperate with plants and partake of what the plant prepared especially for them—at least at first glance. Just as with seed dispersal, plants face a challenge when trying to move their pollen (the reproductive equivalent of animal sperm) to meet individuals' seed (which includes the animal equivalent of the egg, plus additional nutrients and protection). To entice animals into performing this task, the plants prepare delectable fare for them, at great caloric investment. Flowers meant to attract animal pollinators are filled with a solution of simple sugars, the carbohydrate products that plants make directly from their interception of sunlight. The animals come to feed on the nectar and are tricked into getting pollen somewhere on their bodies, which they then transport to other plants of that species, to fertilize the seed ovules and propagate another botanical generation.

From an animal's perspective, nectar offered for the taking might seem irresistible. As a liquid solution of simple carbohydrates, nectar provides readily processed energy. With small molecular weight and few carbon bonds in each molecule, simple sugars are easily absorbed by digestive tracts and only a few enzymes are required to release the energy for the animal to use for its own needs. In addition, water is also a necessary and often scarce resource, and that comes free in nectar. Nectar may have some traces of other nutrients, but it completely lacks the indigestible fibers made of cellulose that plant tissue usually contains, making it easily digestible. Thus, nectar would seem to be a high-profit source of energy for animals with few downsides.

Were nectar-eating that simple, true nectarivores—animals that depend on nectar for a substantial portion of their sustenance—might be more common than they are. For

Which pollinator the plant is targeting will determine, among other things, the concentration and content of nectar sugars.

Most plant nectars are a cocktail of different sugars, including the disaccharide sucrose, familiar to us as common table sugar, and its two simpler monosaccharide building blocks, glucose and fructose (both *hexose* sugars, with six carbon atoms). "Average" nectar is about 50 percent sucrose and 50 percent its hexose subproducts, that is, 25 percent glucose and 25 percent fructose. Nectar of any given species of plant, with a given primary pollinator, and in any given environment, however, can vary significantly around this average, and plants can put other hexose sugars into their nectar. Plants also regulate how much water to provide in the nectar for a given investment of sugar. The more concentrated the nectar, the less water is used and the less evaporates, saving the plant precious resources when water is limited. On the other hand, the more concentrated the nectar, the less often the animals visit for a given number of calories that the plant invests. Therefore, plants will dilute the nectar to a concentration that maximizes pollination at the minimum cost, hoping to attract sufficient pollinators with these stingy rewards.

From the parrot's perspective, on the other hand, the more concentrated sucrose nectars are preferable to dilute hexose nectars. Sucrose contains more energy than the simpler hexose sugars in the unbroken chemical bond linking the glucose and fructose. In an experimental study on nectar preferences, Patricia A. Fleming and her colleagues found that Rainbow Lorikeets and other nectarivores could not meet their energy requirements on hexose sugars, even when such nectar was as concentrated as the preferred sucrose nectar.

Fleming and her colleagues discovered that another, subtler factor is critical to lorikeets: the concentration of sugars in nectar puts demands on water balance for the birds. The lorikeets they studied preferred higher concentrations of hexose than other avian nectarivores in the passerines (Passiriformes) and hummingbirds (Apodiformes). This pattern suggests that the parrots they studied had a greater tolerance of high sugar concentrations than other avian nectarivores.

For even high-concentration sucrose nectars, birds need to process large amounts of nectar as quickly as they can to get enough energy from it. The digestive tracts of lorikeets and Swift Parrots reflects this necessity of their nectar diets. In these nectarivores, the crop is larger than that of granivorous parrots, serving as a holding tank for the large amount of nectar that they must consume. After nectarivores fill up their crops, they perch in safety while emptying it and digesting the nectar. The gizzard of nectarivorous parrots is less muscular, reflecting the low fiber in such a diet and the lower need for mechanical processing of cellulose in the seeds, buds and other plant parts.

The Problem with Pollen

Another layer to exploiting nectar for energy then comes in with the birds' need for protein, which nectar does not contain. Conveniently, nectarivores get both nectar and protein-rich pollen out of the same trough. The pollen of the common foods of the *Trichoglossus* lorikeets studied by Fleming and her colleagues contains 25–33 percent crude protein and eighteen amino acids, including ten of the essential ones.

Unfortunately, this nutrition is not simply there for the taking. Pollen is encased in an armored shell, designed to protect the contents until the grain arrives at its destination. Some animals trick the grains into incubating and expelling their contents as happens on the receiving flower. The vulnerable pollen tubes are easy to digest. Parrots, however, seem to break down the coat in the acid environment of the proventriculus, which corresponds to the stomach of mammals. The proventriculus of lorikeets and Swift Parrots is long and lined with rows of glands that acidify the digesta as it passes through. This emptying of the acidified food from the proventriculus is apparently the rate-limiting step for nectarivores. The crop upstream cannot pass on its contents until the prepared digesta is sent downstream to the small intestine. In lorikeets, the contents of pollen grains appear to be extruded when they get to the duodenum, the upper part of the small intestine, where most of the digestion and absorption of nutrients takes place in vertebrates.

Although the lorikeets and the Swift Parrot rely most heavily of all the parrots on a diet of nectar and pollen for their nutrition, this combination may still not supply sufficient energy or protein year-round because of the seasonality of flowering. Variation in diet exists within the lorikeets, with some species, even congeners, taking a wider range of food than others. All of the nectarivorous parrots eat insects, and not just incidentally.[9] Like pollen, insects are rich in protein and lipids but they are also covered with an indigestible shell made of chitin—the arthropod equivalent of cellulose. Vertebrates cannot digest chitin, so they resort to crushing instead, which allows the nutritious contents to leak out. Birds that eat insects therefore maintain some degree of muscle in the ventriculus (gizzard) for this task. Swift Parrots have a more muscular ventriculus than that of Musk Lorikeets, but similar to that of Green Rosellas, suggesting that Swift Parrots retain more diversity in their diets, eating more insects and seeds than do Musk Lorikeets.

Far from all species of lorikeets have been well studied, but we see variation even within genera. Rainbow Lorikeets eat a more generalized diet than do Scaly-breasted Lorikeets *Trichoglossus chlorolepidotus*, which stay with a more nectarivorous regimen. Rainbow Lorikeets seem to eat more seeds, fruits, and buds and exploit a wider variety of species of plants than do Scaly-breasted Lorikeets. As anyone who has spent any time in Australia knows, Rainbow Lorikeets thrive in urban and agricultural environments and so are spreading and increasing in numbers as more natural habitat is converted to human uses. Their dietary variety is surely the root cause for their wider distribution in the face of human-induced changes in the environment.

Next we explore two more feeding strategies that present parrots with particular challenges in getting enough nitrogen to meet their dietary protein needs.

The Only Frugivorous Parrot: Pesquet's Parrot

Parrots ranging in size from the tiny Pacific-island fig parrots of the genus *Loriculus* and the parakeets of the genus *Brotogeris* to the large parrots of the genera *Amazona* and *Ara* of the

9. They also eat lerp, which is the sugary secretion of certain types of herbivorous insects.

FIGURE 28 A Pesquet's Parrot, *Psittrichas fulgidus,* in Indonesia. This odd-looking parrot species is unusual in the degree to which it consumes the pulp of fruit rather than the seeds. Photo © James Gilardi.

Neotropical rainforest include fruit pulp in their diets. Unlike nectar, fruit can contain a substantial amount of lipids in the form of essential fatty acids. On the downside, fruit also contains cellulose as a component of both pulp and seeds. These indigestible components make the nutrition from fruit pulp more costly to extract. Worse yet, this fiber causes protein to be lost when it abrades the cellular lining of animal guts (box 3). These downsides of fruit consumption may be a reason why only one species of parrot has evolved into a strictly obligate fruit-fresh eater—a true frugivore.

Pesquet's Parrot *Psittrichas fulgidus* is an odd parrot that occurs in the montane rainforests of New Guinea and Irian Jaya, Indonesia. This parrot is odd for a number of reasons, including a mostly bald head and an elongated beak (figure 28). These features have earned

BOX 3   HERBIVORES AND NITROGEN

Animals that eat plants and the plants that they eat share a common dietary challenge, namely procuring enough nitrogen, an essential component of proteins. Plants have a corner on getting enough energy from an inexhaustible supply of sunlight, leaving herbivores to face the dual challenge of getting both energy and nitrogen from their environment. Among herbivores, nectarivores and frugivores live the most nitrogen-deprived lives, followed by folivores and lastly the least-deprived granivores.

Such animals may choose among several strategies to cope with limited nitrogen availability. First, and simplest, they can eat more to get sufficient nitrogen. This strategy is not commonly employed, however, because most nectarivores and frugivores are already processing food as fast as they can. Also, getting enough of a rare ingredient means getting too much of the others. Second, they can search out food with more nitrogen, and all parrots use this strategy, regardless of diet. Parrots seek higher-protein foods, such as pollen (if they can harvest it and digest it, which lorikeets can), insects, and other sources of animal protein, and plant parts with more protein, like seeds or, to a lesser extent, leaf buds. Parrots of all dietary persuasions seek out protein-rich food when their own protein demands are higher, as they are when raising young or molting feathers.

Several other strategies focus on using protein sources more efficiently. In these strategies, animals evolve mechanisms for digesting a higher proportion of nitrogen-containing molecules from the crude sources that they ingest and for retaining more of the nitrogen-containing molecules that are generated from the breakdown of these molecules, a process called catabolism. Commonly, animals employ the use of symbiotic microbes able to obtain nitrogen from the atmosphere and both nitrogen and carbon from metabolic pathways that their hosts do not possess. Two recent studies suggest that parrots have a microbial arsenal in their crops to help them get more out of their food. Fowl in the order Galliformes and other birds are well known to use pouches off the lower intestine, known as cacae, to host bacteria that aid in the digestion of their food, but parrots lack these pouches. Although no nectarivorous parrots have yet been studied, Adreína Pacheco, Alexandra García-Amado, Carlos Bosque, and María Domínguez-Bello discovered that Green-rumped Parrotlets *Forpus passerinus* host a rich microbial flora in their crops. Part of this flora consists of bacterial species able to produce the starch-hydrolizing enzyme amalyse. These bacteria share the easily digested, decomposed product, glucose, with their hosts in exchange for abundant food and a safe environment.

More intriguing are the signs of nitrogen-fixing bacteria in the low-oxygen environment of parrot crops, created by the metabolic activity of the starch-feeding bacteria. These anaerobic bacteria can use energy supplied by the digestion of the parrots' food to convert the abundant atmospheric nitrogen gas (N_2) into another nitrogen-

(continued on next page)

containing molecule that is biologically usable, such as ammonia (NH_3) and its derivatives, which eventually end up as amino acids and proteins. Pulchérie Gueneau and her colleagues hypothesize that these nitrogen-fixing bacteria found in the crops of parrots are not incidental, opportunistic inhabitants, but rather may play a significant role in balancing the nitrogen budgets of herbivorous parrots.

With or without microbial help, animals on especially low-nitrogen diets have a few other tricks up their physiological sleeves. For example, they can make better use of the nitrogen they already have circulating around in various biochemical pathways. One tried-and true strategy is to recapture and recycle the nitrogen-containing wastes, primarily ammonia, at the cellular level. When protein is catabolized for various uses by animals, its waste products are typically excreted in the urine. The immediate product of this deamination (when amino acids are broken apart) is ammonia. This molecule is so reactive with other molecules that it is toxic to normal cell functions. Unless ammonia is immediately recycled into needed molecules, it is converted as rapidly as possible into less toxic waste molecules and sent to the kidneys for safe passage outside the animal's body in urine. Bird kidneys work mostly with uric acid, and mammal kidneys with urea; each of these nitrogen-containing waste products is less toxic than ammonia.

Parrots and other birds that subsist on low-nitrogen diets seem to have a greater capacity to recycle ammonia before it is expelled to the environment. This capacity is detected as low endogenous nitrogen loss, which can be measured by analyzing the urine and feces of individuals on a standardized diet with known nitrogen content. So far, exactly how nectarivores and frugivores save so much of the nitrogen that their relatives release is still a mystery. The most specialized of parrots with these diets, such as the lorikeets, lose an order of magnitude less nitrogen in their urine than do granivorous parrots.

Physiologists have various ways to estimate minimum protein requirements, based on estimates of endogenous nitrogen loss and of nitrogen equilibrium. In one study, Gregory Pryor estimated the minimum intake of crude protein that allowed the parrots not to catabolize their muscle to replace lost protein. He found that the minimum protein requirement was 1 percent of dietary intake for nectarivorous Red Lories *Eos bornea*, 3 percent for frugivorous Pesquet's Parrots *Psittrichus fulgidus*, and 8 percent for granivorous Budgerigars *Melopsittacus undulatus*. These figures are low compared to that of developed commercial diets and low for most probable levels in the diets of wild parrots. Free-ranging parrots would need more protein than the minimum requirements to maintain their body mass if they are highly active or are growing, molting, or feeding young.

We have hints about some of the mechanisms that allow nectarivorous parrots to subsist on such low levels of protein, but those for our frugivorous Pesquet's Parrot are less certain. Conveniently, nectar contains a great deal of water and not much fiber. Fibrous diets consumed by frugivores and granivores abrade the gut lining, causing cells to be sloughed off and excreted, taking the nitrogen contained in their cell membranes and contents with them. Nectarivores can avoid this loss of nitrogen because neither nectar nor pollen contains such abrasive ingredients. In addition, the massive amounts of water that must be excreted from consuming sufficient nectar allow nectarivorous birds to employ a metabolic shortcut. If the solution to toxicity is dilution, then birds can avoid the expense of producing uric acid and can dump ammonia directly into the urine. Not only does this save investment in proteinaceous enzymes and energy, but also some birds may be able to reabsorb the smaller ammonia molecules and shunt them back into amino-acid-producing metabolic pathways before the urine leaves the body. Some hummingbirds can excrete urine that is up to 25 percent ammonia, a feat called ammonotely; but we still do not know to what extent nectarivorous parrots may exploit ammonotely to recapture excreted nitrogen. Birds that can obtain nitrogen this way probably do so with the aid, again, of microbes, rather than being able to reel in ammonia molecules from their urine themselves. Because parrots and most other nectarivorous birds do not possess cacae, the most likely site for such activity is the lower distal end of the large intestine. These mechanisms for nectarivorous parrots, however, remain intriguing hypotheses for now.

Thriving on a nectar-and-pollen diet seems to work only for small parrots. The hypothesis that smaller parrots require more energy per gram is consistent with our observation that all nectarivorous parrots weigh less than 300 grams, and two-thirds under 100 grams. In other words, a high-energy, sugar-rich diet brings more benefit to smaller parrots than to larger ones, all else equal.

The hypothesis that larger parrots require too much protein to be supported on a low-protein diet of nectar or fruit is not consistent, however, with Pesquet's Parrot. This truly frugivorous parrot is among the largest of parrots at an average weight of 800 grams, ranking in the top 5 percent of parrots by weight. Moreover, studies of protein requirements for maintenance in a wider variety of parrots suggest that some of the Amazon parrots may subsist in the wild on a lower-protein and thus more fruit-pulp-rich diet than other large, rain-forest-dwelling parrots that are primarily granivorous. Claudia Westfahl and her colleagues discovered that endogenous protein loss in captive parrots of the genus *Amazona* is midway between that of macaws and Grey Parrots on the one hand and lorikeets on the other. These findings surely reflect evolutionary adaptations of the parrots to their diets. The amazon parrots are medium-to-large parrots, ranging in weight up to that of Pesquet's Parrot. For now, the protein-requirement hypothesis receives no support from the patterns we see in parrots with different diets. Instead, we confront an unsolved mystery of how fruit-eating parrots can be so large.

it another name, the Vulturine Parrot, but rather than eating carrion, this parrot has a diet composed almost exclusively of fruit pulp.

Andrew Mack and Debra Wright discovered that Pesquet's Parrot feeds exclusively on the fruit pulp of three species of figs at their study site in Papua New Guinea. Observations of captive Pesquet's Parrots reveal that they eat their fruit fare very differently from other parrots. They do not use their beaks to grind the food, as do the granivorous parrots that eat fruit pulp. Rather, Pesquet's Parrots slice out large chunks of fruit with their more pincer-like beaks. Then, while holding the fruit pulp, they push their brush-tipped tongues into the pulp, moving the tongue back and forth to bring the pulp into their mouths to be swallowed. Dominique Homberger surmised, from a study of the morphology of beak and head muscles of parrots, that this method of feeding is an adaptation to the specialized diet of Pesquet's Parrot. Homberger documented in her study how the anatomy of Pesquet's Parrot seems to be derived from that of granivorous parrot ancestors, evolved to function more efficiently for a frugivorous diet. Her hypothesis, put forward in 1981, has been fully supported by recent studies of molecular phylogeny (chapter 1). Descended from generalist granivorous ancestors, Pesquet's Parrot has recrafted the morphology and physiology inherited from them for its peculiar diet. Pesquet's Parrot has also evolved a low requirement for protein in its diet, through a variety of mechanisms (box 3). These evolutionary adaptations reflect the immediate livelihood of Pesquet's Parrots and not that of their ancestors.

How the Only Folivorous Parrot Survives on Leaves

The ultimate herbivory is the eating of leaves. A few parrots venture into folivory, namely the Kakapo *Strigops habroptilus* and to a lesser extent some island-dwelling *Cyanoramphus* parakeets.

This plant fare has little to recommend it for easy nutrition (box 3). As a whole, birds only dabble in the eating of leaves; after all, it is hard to imagine flying with all that digestive machinery. Of the few bird species that have adopted this lifestyle, geese are the best-studied. The consensus is that geese are not so good at getting much out of cellulose, compared to animals with proper fermenting guts, but they are better suited to folivory than are parrots. Whether parrots can be counted as truly folivorous depends on how you define "eating of leaves." Parrots show some evidence of having bacterial partners in their crops to help them decompose starches with amalyze. Nevertheless, parrots lack the complex digestion of ruminant mammals, the ultimate folivores.

So, how good are parrot folivores? The Kakapo shares the basic requisites of folivory. It is the largest parrot, bar none, weighing between one and a half and three kilos. It cannot really fly. The Kakapo has a low basal metabolic rate and the lowest daily energy expenditure known for any bird. In chapters 5 and 7 we find out more about this most unusual of parrots and how its diet intertwines with the rest of its behavior, ecology, and evolution. Without a doubt, subsisting on a diet heavy with leaves is an essential part of what a Kakapo is.

Kakapo make do with what nutrition they can get from leaves without being proper fermenters of leaves. Because they are critically endangered (and always have been since

Western science became interested in them), no one has killed a few just to see what is going on inside their guts. Rather, Kakapo diet and digestion are studied by biologists watching what goes into them and what comes out. Richard Gray and a host of Kakapo biologists after him have discovered Kakapo eating leaves and the pithy parts of stems and twigs of many plants that grow low in the understories of lush New Zealand forests, including those of ferns, sedges, club mosses, podocarps, myrtles, orchids, and grasses. These ingredients of their diet are chewed extensively. Kakapo possess not only a massive beak but also a large muscular tongue enhanced with a rough, keratinized lobe. A Kakapo presses the plant material against its upper mandible, which is constructed with a serrated plate in its roof (as is that of most parrots), using both the tongue and the edges of the lower mandible. Kakapo therefore masticate their food thoroughly, extracting the digestible bits and spitting out the spent fiber—the cellulose, hemicellulose, and lignan—in the form of chews. In other words, Kakapo do not attempt to digest the indigestible, and they save themselves the trouble of moving it through the gut. A universal cost for any herbivore is processing roughage from plants, all the worse when they cannot digest it at all or only slightly. This refractory material slows down and otherwise interferes with herbivores' ability to extract nutrition they can use from plants. The Kakapo are clever to plumb what they can on the front end and expel from the get-go what their bodies cannot use. Nevertheless, Kakapo possess a muscular ventriculus that may aid them in more mechanical processing of their relatively fibrous foods.

Adult Kakapo get by on this low-quality and time-consuming diet by being inactive, energetically thrifty, and in physiological slow motion (chapters 5 and 6). Their young, unfortunately, cannot be equally thrifty given the physiological demands of growth. Being a chick in a nest is risky for a ground-dwelling bird, even in prehistoric New Zealand (chapter 7). Therefore, it behooves them to grow quickly, a task requiring more protein. The adults solve this problem by timing their reproduction to intercept the mass fruiting of the podocarp trees that dominate in their forest homes. These podocarp fruits are much higher in protein than the Kakapo's usual cuisine. Nevertheless, Kakapo chicks grow normally on a diet of only 6 or 7 percent protein by dry weight, a feat impossible for other species of parrots.

The Kakapo is the best folivore that the order Psittaciformes has to offer. Although it belongs to the most basal clade of parrots, folivory is most likely a dietary experiment arising from long isolation in a strange land, and the diet of the Kakapo does not remotely resemble that of its parrot ancestors.

Truly Omnivorous Parrots

A thread throughout this book is that islands are odd. Organisms that live on islands often offer insight into those everywhere; other times, they are just strange. No matter what, island life fascinates and frequently astounds, and so it is with island-living parrots.

Subarctic Parakeets

One of the harshest places a parrot could end up would be the remote islands of the far-southern Pacific Ocean. Parakeets of the genus *Cyanoramphus* live on these improbable and mostly inhospitable tips of drowned volcanoes. These parrots must have arrived by air from the neighboring and relatively much larger islands of New Zealand, when lower sea levels made such events more likely (chapter 1). However unlikely their reaching these remote and hostile outposts lost in a vast, cold ocean, once the parakeets arrived they were there to stay.

Necessity is the mother of invention, and *Cyanoramphus* parakeets eat whatever they can, literally. Their menu therefore includes some unusual dishes for parrots. Antipodes Island is home to not one but two species of *Cyanoramphus*: the endemic Antipodes Island Parakeet *C. unicolor* and the Red-fronted Parakeet *C. novaezelandiae*. The two species coexist by dining on somewhat different foods selected from the limited fare that the island has to offer. Much of Antipodes Island is covered with the tussock-type grasses and low herbs that characterize tundra, a vegetation type characteristic of high-latitude environments. Trees are not a dominant feature of tundra, but some small shrubby versions of plants that are taller elsewhere grow in protected areas. Thus, the parakeets forage heavily on grasses and sedges, eating leaves as well as seed, and on berries and leaves of herbs and low shrubs.

As we have seen, leaves are hard to digest and almost bereft of nutrients, especially protein (box 3). To augment this diet, both species eat some distinctly novel foods, at least for a parrot. Red-fronted Parakeets have been seen grazing on the protein-rich larvae of the marine flies that lay eggs in shoreline flotsam. Parakeets of both species, but especially Antipodes Parakeets, scavenge on the bodies of dead seabirds, which occur in some abundance on these outpost islands. Although the land is barren, the adjacent oceans are rich in food resources. Therefore, seabirds of many species nest in dense colonies on the islands. This fortunate fact has more implications for the diet of the Antipodes Parakeet, which evolved on the island and is apparently more specialized in making a living in this challenging environment.

Surely one of the most intrepid of parrot biologists, Terry Greene discovered a remarkable instance of parrots deliberately and regularly preying on other vertebrates, as opposed to hapless invertebrates that act like seeds. Greene could not directly observe the acts of killing, but this predicament faces all biologists—predation is a common event that is rarely observed, probably because it is over so quickly. Greene's first clue was when he discovered an Antipodes Parakeet feeding on the corpse of a Grey-backed Storm Petrel *Garrodia nereis*. This discovery might not be especially notable, except that Greene found the petrel's body to be still warm and bleeding. He then stumbled on an Antipodes Parakeet leaving one of the numerous burrows in which the petrels nest on the island. Investigating the burrow, Greene found a carnage of crushed eggs and the dead petrel that had been incubating them. Thus alerted to a crime scene, Greene investigated more of the petrel corpses that he came across as he traversed the island while studying parakeets. He discovered that all had the same peculiar injuries. To make a long story short, apparently Antipodes Parakeets seek out petrels that are incubating eggs in their burrows and kill them by ripping out their throats. Once the petrel is dead, the parakeet then skins the petrel's back and breast so that it can feast on the large flight muscles there. Like most vertebrate predators, Antipodes Parakeets do not let the viscera of their prey go to waste. Sometimes the parakeet dines in the safety of the burrow, but just as often it drags out the remains to feed on them out in the open, where biologists had earlier observed them and assumed that the petrel had died of other causes and was simply being scavenged. After all, how could a gentle herbivore have wreaked such butchery on another bird, one that, like it, nests in holes?

Kaka and Kea

On the main islands of New Zealand reside two more species of enigmatic parrots, the Kea *Nestor notabilis* and the Kaka *N. meridionalis*. Although the main islands of New Zealand are practically continents compared to the tiny rocks that are home to the *Cyanoramphus* parakeets, both Keas and Kakas must improvise to make a living. Both species are observed today to be quite omnivorous, eating animal and plant foods in equal abundance and exploiting whatever is edible that they can catch or digest. Some biologists speculate that their diets in New Zealand's pre-European past were not quite the same as today's, but we can still deduce their original dietary proclivities from a variety of observations.

One notable clue to their dietary origins would be their specialized beaks. Compared to the average parrot, the beaks of both Keas and Kakas look like they are badly in need of a trim. Both species have longer, narrower, and more strongly curved beaks than their close relative the Kakapo and other more generic species of parrots. The beak of the Kaka is more recurved than that of the Kea, slightly but noticeably. Likewise, the tongues of both species are proportionately long and narrow compared to that of Kakapo and most other parrots (save the lorikeets, the Swift Parrot, and Pesquet's Parrot). The tongues of both species sport brushy tips, being covered with small appendages known as fimbriae. The tongues of Kakas are somewhat narrower than those of Keas and have a more spoon-like shape. These slight differences between the two closely related species reflect differences in their diets, foraging behavior, and overall ecology that characterize their divergence into somewhat different ecological niches on the isles of New Zealand.

Both species feed on a wide variety of plant fruits and seeds. The morphology of their beaks and tongues allows them to slice apart fruits to get the seeds or to consume pulp as Pesquet's Parrots do, probing and retracting their brushy tongues. Thus far, their diets are ordinary parrot fare, which they consume as a considerable proportion of their food, and for which they do not seem to need such specialized tongues and beaks.

My suspicion is that their beak and tongue morphology evolved to enable their habit of sap sucking. Unlike the woodpeckers of North America that fill this niche, Kakas and Keas use their finely pointed, recurved beaks to dig through the bark of trees and then peel it deftly off in layers, rather like using a can opener (figure 29). Once they have peeled back the bark layers, they can expose the phloem, the layer just inside the bark that carries the vessels that transport substances around the plant, mainly sucrose that the plant has created through photosynthesis. The stuff in these vessels is commonly known as sap, and it is a ready and abundant source of energy if animals can exploit it on a large enough scale. Not only can Kakas and Keas peel back the bark neatly and efficiently, their brushy-tipped

FIGURE 29 The omnivorous Kaka, *Nestor meridionalis,* of New Zealand uses its long bill to forage under bark and through flowers for sap, nectar, and grubs. Photo © Bent Pedersen.

tongues can lap sap up fast enough to make feeding on this liquid worthwhile for such large birds.

Once so equipped for efficient sap sucking, Kakas and Keas could put their oral anatomy to use feeding on other novel items for parrots. After all, these parrots inhabit islands, albeit large ones, and they have to make do with the more limited choices available to them on an island. In addition, we have already seen how sap (like nectar) is deficient in protein. The feeding apparatus of Kakas and Keas is also handy for digging for grubs burrowing through the wood of trees and in the ground. With beaks so built, the parrots are adept at piercing eggshells, through which their brush-tipped tongues can quickly lap up the eggs' contents. Moreover, a pointy-tipped, razor-sharp beak operated with force by strong jaw muscles and good leverage from special skull bones (box 2) is a pretty formidable weapon. If it can strip bark off a tree, the skin and bones of vertebrates are no match. Keas have also been observed feeding on mice and on the chicks of ground-nesting sea birds, which they kill in the same manner as Antipodes Parakeets do, by ripping out their throats.

The Kea is also the subject of mythology because of its foraging behavior, which is more flexible than that of the Kaka. The most infamous of the deeds attributed to the Kea is that of eating sheep alive, with the same utensils it uses to imbibe the sap of trees. On the one hand, it is only a small step to extend this method from trees to vertebrates. Although the tree obliges by standing still, some vertebrates you can imagine (such as sheep) would be relatively helpless to prevent the Keas from peeling back their skin to lap up fat, blood, and muscle. Despite some sensational reports, such heinous acts have not been observed on any large scale in formal scientific studies of the Kea and there is no support for the idea that mutton on the hoof is a major dietary item for them. Perhaps the extinct moas could have suffered the parasite-like depredations of Keas before domestic livestock made the New Zealand scene. We do know that the diet of the Kea is surely more omnivorous than that of any other species of parrot save the Kaka and the Antipodes Parakeet. Omnivory notwithstanding, these parrots are not carnivores and do not depend to any great extent on flesh of large vertebrates.

Parrots as Mesoscale Foragers: The Ecosystem Effects of Seed Predation

Whatever their dietary proclivities, all parrots are faced with finding food that is flung far and wide. Dependence on the reproductive activity of plants requires that parrots search out and find individual plants that are flowering, growing their seeds in ripening fruit, or dispersing those seeds when the time is right. Plants are constrained by the seasonal availability of their own resources and also employ a strategic shell game of hiding their get from persistent herbivores. In wet lowland rainforests, other forces cause individual plants of each species to be highly dispersed, creating for animals serious challenges of finding sufficient food even in the face of lush, tropical abundance. In less productive environments such as the high-latitude islands, food is hard to find simply because there is not much of it, and a parrot has to work hard and travel far to get enough.

Thus, over a wide variety of diets and habitats, parrots are faced with finding their food, often over vast areas. Parrots meet this challenge by being strong fliers with a relatively low metabolic rate, so that they cover much ground quickly and efficiently. They also meet the challenge of finding dispersed and seasonally variable food by being social, as do primates, with which parrots often share this food. As we will explore soon (chapters 4 and 6), one major function of sociality in animals is to share information on the location of food, and this may be the primary reason for the communal roosts common in parrots.

Ecologists have coined the term *mesoscale* to refer to ecological processes occurring on intermediate spatial scales, particularly as scaled with the movement of individuals within a population. These movements depend on both the size of the organism and the mode of locomotion, as well as the need for such travel. Parrots are mesoscale foragers in many environments, moving around to meet daily and annual needs on a scale of up to 1,000 hectares, according to a scaling system devised by tropical biologist John Terborgh. This area is roughly several kilometers on each side and presumably represents an average area over which a pair of parrots can provide for themselves and their offspring. This scaling seems appropriate for parrots in tropical rainforests, as well as the dry savannas of Australia and seasonal tropical woodlands occurring in parts of South and Central America and Africa. Home ranges of parrots living on large islands, such as New Zealand, New Guinea and surrounds, or Puerto Rico (chapter 6) may necessarily be scaled downward somewhat compared to those living on continents.

In chapters 5 and 6 we explore more how parrots arrange their days around the quest for food and manage their need to reproduce, including finding the right foods to accomplish reproduction.

THREE

The Sensible Parrot

How Parrots Perceive and Use Information

SENSORY BIOLOGY AND COMMUNICATION

What is called "having a sensory perception" is . . . simply thinking.

—RENÉ DESCARTES, 1641

CONTENTS

Introduction to Sensory Biology   81

Visual Communication   82

   How Parrot Eyes and Brains Perceive
Color   82

   How Pigments and Molecular Structures
Create the Color of Parrot Feathers   84

      Box 4. The Physics and Biology of How
Animals Make Color   87

   Why Parrots Are Colorful   89

Vocal Communication   94

   A Natural History of the Sounds Wild
Parrots Make   94

Sound Perception in Parrots   97

Communicating Acoustically: The
Contact Call as an Important Social
Signal   103

Introduction to Vocal Learning in Parrots   104

Some Functions of Vocal Learning in Wild
Parrots   107

Vocal Learning for Social Cohesion   111

Individually Specific Labels and How Parrots
Use Them   113

INTRODUCTION TO SENSORY BIOLOGY

In this chapter, we explore how parrots can experience properties of the physical world that we humans can only imagine, because our species lacks the sensory capabilities that parrots possess (figure 30). Philosopher-scientist René Descartes famously struggled with the quandary that provides a common thread through this chapter: To what extent does the reality we perceive with our senses correspond to the reality of others? We will find that what we, or parrots for that matter, perceive as real is not purely the physical universe. Rather, our perception of reality is an interaction between external physical entities, our senses, and our brain.

Those who persist in the illusion that what we humans perceive is the only reality will miss knowing many wonderful parts of the universe that we share with many beings distinctly different from ourselves. Fortunately, with an open mind and with the instruments

FIGURE 30 The aptly-named Rainbow Lorikeet, *Trichoglossus haematodus,* displaying its brilliant plumage. As discussed in this chapter, our perception of these colors does not entirely match what the parrots themselves see. Photo © Steve Milpacher.

of modern science, scientists can reveal to us an entirely different world, one that is experienced by parrots and not necessarily fully experienced by humans. We humans, however, can measure what parrots perceive with our instruments, and so we can detect that they literally see, hear, and smell the world differently from the way we do.

This chapter examines the ways in which parrots interact with their world through their primary senses of vision and hearing. In the respective sections, we first discover the particulars of the sensory organs of parrots, how those organs communicate information to their brain, and what their brain interprets from that information. Then, after building a foundation of what parrots know and how they know it, we can explore how they put this information to use in their daily lives.

VISUAL COMMUNICATION

How Parrot Eyes and Brains Perceive Color

Descartes was perhaps the first to write so eloquently on what can be described as the conundrum of all scientists, and of anyone who struggles to understand the world around us. To do our work as scientists, we assume that an external reality exists that we humans can perceive objectively, when in fact we cannot. This fallacy is nowhere better revealed than in the realm of color and color vision. Only recently have sensory biologists been able to meet the scientist's conundrum head-on, by admitting its existence. Andrew Bennett and Marc Théry, two leaders in the field of avian vision, remarked that a breakthrough in their research was possible only once scientists had rid themselves of "the misunderstanding that color as perceived by humans represents an objective reality." In the apt words of scientist Susana Santos and colleagues, in their study of the plumage of Blue-fronted Parrots *Amazona aestiva*, "Colour is not an inherent property of the object. It is a property that a certain visual system awards to the object. As the spectral composition is a process that engages physical, physiological, and psychological processes, colour is a very complex subject to investigate." To that I would add that color is a difficult phenomenon for humans to understand as colors arise by this interaction between the physical world, sense organs, and the brain. The subject becomes particularly challenging when we consider this phenomenon in species with sensory capacities different from ours.

Parrots, for example, experience four dimensions of sensitivity to light wavelengths rather than the three dimensions our eyes possess. That is, parrots are *tetrachromatic*: their eyes have four types of cone cells that detect light wavelengths, while humans are *trichromatic* with only three types of cones. Tetrachromatic birds such as parrots have three differentially sensitive cones in the long-, medium-, and short-wavelength portions of the human-visible light spectrum, roughly corresponding to the sensitivities of the three types cones most humans have (called LWS, MWS, and SWS). In addition, they possess a fourth type, sensitive to even shorter light wavelengths. Depending on the type of bird, this fourth cone is either violet-sensitive (VS, around 400–420 nm) or ultraviolet-sensitive (UVS, less than 400 nm, typically around 370 nm). Only a few types of birds have so far been shown to have UVS cones: species in the orders Passeriformes (songbirds), Struthionioformes (the ratites, specifically *Rhea americana*), the Cicioniiformes (shorebirds, specifically some gulls in the genus *Larus*), and last but not least, the Psittaciformes.

The terms *tetrachromatic* and *trichromatic* might make it sound as if parrots can see one more color than we humans can, but the difference is much greater than that. Birds can see colors arrayed over one more dimension than we can: four dimensions instead of three. This fourth dimension adds another order of magnitude of different colors that parrots can perceive compared to the paltry millions in humans' three-dimensional color universe.

We base our knowledge of color vision in parrots mostly on studies of Budgerigars *Melopsittacus undulatus,* which have been shown to possess four types of cones in their eyes. Their LWS cones are most sensitive at around 565 nm; their MWS cones, at around 508 nm, and their SWS cones at 445 nm. Then Budgies have UVS cones, which peak at 370 nanometers and thus detect light wavelengths to which humans are completely blind. Our short-wavelength vision peters out around 400 nanometers; anything smaller than that is invisible to us. Budgie vision, however, reaches almost to 320 nm at the short end of the wavelength continuum.

Showing that Budgerigars (and other parrots by implication) have the visual machinery to *detect* fine variations of light wavelengths from 700 to 320 nm is not the same as proving that they can *discriminate* colors over these wavelengths. Color (light wavelength) discrimination occurs mostly in the brain, as opposed to in the eyes. Timothy Goldsmith and Bryon Butler asked just that question: whether the visual acuity of the color vision of budgies matches that theoretically possible, given the anatomy of their eyes. In a series of experiments, the scientists trained budgies using operant conditioning to choose illuminated panels to receive a food reward. They would offer budgies two panels of smoked glass, beneath each of which were a perch and a food hopper. Food would dispense from whichever panel was back-illuminated with a specific wavelength of light; the reward-light wavelength depended on the exact experiment.

The trained budgies (named Dalton, Palmer, Lashley, Porter, Spot, and Morgan) quickly learned which wavelengths were productive and which not. Some of the experiments were designed to test whether budgies could indeed perceive ultraviolet light invisible to humans. Others were designed to test their visual acuity over the entire range of light that the budgies should theoretically be able to see. That is, how close can two reflected wavelengths be, with budgies still able to perceive the difference between them? The upshot of these experiments was that, yes, the anatomical structure and physiology of their eyes was a perfect predictor of what they could actually see. Although we cannot match this ability with our own depauperate vision, our instruments and our inductive ability as scientist observers allow us a glimpse into a world that we humans cannot otherwise know.

How Pigments and Molecular Structures Create the Color of Parrot Feathers

Now that we know more or less how parrots see, what do they do with this information? We will assume that Budgerigars represent all parrots, which not a bad assumption given how conserved the basic visual system is over all orders of birds. To understand a particular visual system, we must find out not only what the viewers behold, but why. After all, if we see not with our eyes but with our brains, the uses to which visual information is put define our vision as much as the organs that collect and process the information.

Let us begin by discussing one property of parrots that makes them so fascinating to humans: the stunning array of colors that adorn the feathers of parrots. We can also ask to what uses such fancy accoutrements might be put.

Parrot feathers are typical bird feathers, made of the protein keratin. They are formed when this keratin is spun and woven as it is exuded from special organs in the skin, the follicles. Feathers come in a wide variety of colors. Without question, feathers completely and totally outdo hair in the color department, and even put to shame most scales. We have already learned, however, that color is a property of brains, not of external objects such as feathers. That Cartesian reality is our first clue that feather colors exist first and foremost as *signals*, to be read by beings with brains. Having only a few thermally sensitive colors, such as white, grey, and black, would suffice to help organisms gain or lose heat. Likewise, a limited array of background-matching or confusing colors would do to help organisms hide

FIGURE 31 A strikingly colored Red-capped Parrot, *Purpureicephalus spurius*, near Mandurah, Australia. Photo © Georgina Steytler.

from the view of their predators and prey. And yet, the feathers of parrots exhibit a dazzling range of conspicuous colors, arrayed in attention-getting patterns, often with the most astonishing effects (figure 31). The colorful feathers of most parrots scream loudly one inescapable message: "Notice me!"

So how are these colors produced? Remarkably, the vast array of colors produced by biological systems are caused by only two basic processes. One is *structural*. Physical structures on the surface or just within skin, hair, feathers, or scales may interact with light in such a way as to reflect certain wavelengths while letting other pass through. Larger structures reflect both longer and shorter wavelengths, producing whites, reds, and yellows, while smaller ones reflect only the shorter wavelengths, like greens, blues, and even ultraviolet.

The wavelengths that are reflected are those we perceive as the color of the animal. Structural colors are responsible for many of the brightest colors seen in the animal world, like the flashing throats of hummingbirds and the iridescent wing spots of butterflies. Parrots use structural colors to produce or augment some of the colors seen in their feathers. You can detect structural colors in feathers by comparing the color of a blue, green, or red parrot feather when you view the feather with the light shining from above you versus when you hold the feather between your eyes and the light source. Feathers with colors produced by structural mechanisms will appear to change color in these two positions, appearing bright and colorful when your eye and the light source are on the same side of the feather and duller when the feather is between you and the light source. This change occurs because small structures inside the feather are reflecting certain wavelengths back toward the light source while letting others pass through.

The second process producing color, one that is more familiar to most of us, is *pigments*. These are special types of molecules within cells that absorb certain wavelengths of light and reflect others. One set of pigments that is ubiquitous among animals is the family of melanins. They produce dark colors, primarily the blacks and browns, as visible to humans (box 4). Melanins also interact with structures inside feathers to produce the blue found in some of the large macaws and in captive-bred budgerigars.

Another set of pigments found in most organisms is the carotenoids. Carotenoids have dual functions in animals. In most organisms, including parrots, they serve as antioxidants, which control the negative effects of oxidative chemicals produced by cells as a by-product of producing energy from food. Parrots are no different from humans or any other animal in their bodily relationships with carotenoids; just like us, they must consume a full complement of carotenoids from their food to remain healthy (box 4).

Many organisms also use carotenoids to color themselves, since these pigments reflect light strongly in the red-orange-yellow part of the spectrum. Many of the most colorful bird species sequester some portion of these dietary carotenoids in their feathers. Imagine the biologists' surprise, then, to discover that parrots have no carotenoid pigments in their feathers—none at all. In fact, the pigments causing red, orange, and yellow plumage in parrots are found nowhere else in the animal kingdom. So unique is this class of biochemicals, it is named after the parrots, being called the *psittacofulvins* by some scientists, and the *parrodienes* by others (box 4). More remarkable still, parrots have no psittacofulvins circulating in their blood, so they do not appear to be used for their antioxidant capacity. Another interesting discovery is that psittacofulvins cannot come from the parrots' plant food, because these compounds are found (so far) in parrots and nothing else. Therefore, the current thinking is that psittacofulvins are manufactured *de novo* by the parrots themselves, probably in the vicinity of the feather follicles.

A clue to what psittacofulvins could be doing in the feathers has been provided by Elena Pini, Aldo Bertelli, Riccardo Stradi, and Mario Falchi at the University of Milan. They discovered that this unusual class of compounds has powerful antioxidant, anti-inflammatory, and anti-tumoral properties. The purpose that these pigments serve in parrot feathers may

BOX 4   THE PHYSICS AND BIOLOGY OF HOW ANIMALS MAKE COLOR

MELANIN AND STRUCTURAL COLOR

The color of birds arises from an interplay of two features of their feathers: pigment molecules that (by definition) interact with light; and structural features that affect how light travels into and out of scales, hairs, and feathers. Parrots use two different classes of pigments, plus structural colors, to enrobe themselves in their dazzling displays.

The melanins are a class of pigments shared by parrots with other vertebrates. Melanins are complex molecules that are made up of repeating segments (polymers) containing nitrogen. These molecules interact with light by reacting selectively to different wavelengths, including the ultraviolet (UV) part of the spectrum. Pigments derived from melanin produce an array of colors from black to brown in vertebrate skin, hair, feathers, and other structures such as eyes.

For melanin to make reflected light appear blue requires the addition of a specific structure in the feathers reflecting that light. This phenomenon is known as *structural color*. In this case the melanin absorbs some of these wavelengths in the light, while a particular structure of the feather surface causes only certain other wavelengths to be reflected. The result is that only a small subset of the spectrum that hits the feather is reflected back to our eyes, corresponding to the blue color we perceive.

The structural property that creates blue feathers lies inside the ramus of a feather barb. (The barb is the first "branch" of the feather arising from the central shaft or rachis. The barb in turn has a mini-shaft or ramus, from which branch the yet smaller barbules.) At the core of each barb's ramus is a spongy center, named in honor of its discoverer, Dutch biologist Jan Dyck, who first described this part of the feather in great detail and associated this structure with the color of blue feathers in parrots. The center of the barb is described as spongy because, like a sponge, it is composed of tissue interspaced with empty spaces. In the case of feather rami, the filled parts are rods of keratin, and the empty spaces contain air. This spongy structure is responsible for the production of structural color.

Studies by Richard Prum and his colleagues Jan Dyck, Staffan Andersson, Rodolfo Torres, and Scott Williamson verified and elaborated on the proposed mechanism for the colored light reflecting from the feather barbs. In the spongy Dyck matrix, there are uniform and regular rods of keratin aligned in a certain structure so that the rods alternate with air-filled spaces exactly the diameter of the rods. The diameter of the rods, and therefore the spaces, determines the wavelength of light that is reflected. As a rule, the diameter of this space is one-tenth of the wavelength that it emits when the spongy matrix scatters full-spectrum light absorbed by the feather barb. The light

(continued on next page)

enters the barb and bounces off the central melanin granules lining the core of the ramus containing the spongy matrix. It is then diffused and scattered as it bounces around the spaces and is redirected when it hits the keratin rods, by a specific physical process called coherent scattering. Only certain wavelengths of light are reflected back outward for eyes to intercept, and these wavelengths are determined by the inter-rod spacing. For years, other hypotheses abounded about the process of scattering that could produce selected wavelengths of reflected light from the feather. Finally, the studies of Prum and his colleagues verified that the tiny rods of keratin had sufficiently regular structure to cause coherent scattering.

OTHER PIGMENTS

Some vertebrates have other pigments in addition to melanin to color themselves. A common pigment reflects light in the yellow-orange-red part of the spectrum. These are molecules known as carotenoids, which bear this name as a historical accident; the first carotenoid known to science was isolated from carrots. Plants, and plants alone, produce hundreds of different variants of these compounds. Most animals need these compounds for a variety of essential functions, so they must obtain carotenoids from their diet. Scientists have found a bounty of carotenoid compounds coursing through the parrots' blood circulatory system, including lutein, zeaxanthin, β-cryptoxanthin, and their metabolic derivatives anhydrolutein and dehydrolutein, in a wide survey of forty-four red-feather-bearing parrot species in all parts of the parrot phylogenetic tree (chapter 1). The dietary sources of these carotenoids are hardly mysterious. Seeds and nuts contain an abundance of carotenoids, including lutein, zeaxanthin, β-cryptoxanthin, carotene, and β-carotene, and more, as do the fleshy fruits and plant parts that parrots exist on in the wild. Surprisingly, none of these carotenoids seem to be used to produce the reds and yellows seen in parrot feathers.

Instead, parrots use another pigment type to produce their yellow-to-red colors. This class of pigment, not yet known to occur in any other type of organism, is called the psittacofulvins, or alternatively parrodienes. Only recently have the secrets of the structure of these parrot pigments been revealed. The psittacofulvins are a class of lipid-soluble polyenal lipochromes, five of which are known to date: tetradecahexenal, hexadecapheptenal, octadecooctenal, eicosanonenal (the names reflect their molecular structure), and a fifth as yet not fully studied.

FLUORESCENCE AND SPECIAL PROPERTIES OF ULTRAVIOLET WAVELENGTHS

Fluorescence occurs when a pigment absorbs UV-length light waves and reflects them back out at longer wavelengths that are within the human-visible spectrum. Although humans need to take a black light into the dark to see its effects when some surfaces reflect visible light back at us, full sunlight is rich in UV wavelengths. We humans cannot see them, but we can become painfully aware of them when our skin burns after too much sun. All my black light tells me is that some surfaces absorb UV light while others may reflect back UV or visible wavelengths of light. If a surface returns UV light, I cannot see it, and if a surface absorbs UV rather than reflecting it, I cannot detect its darkness in the UV spectrum. I therefore cannot see the colors that birds can see on the parts of their bodies exposed to UV light and interacting with it. These colors are visible to birds because their ultraviolet-sensing cones can detect them.

Some scientists have proposed that UV wavelengths have special advantages for use in mate signaling in birds. Others question whether this generalization holds, given that UV light perception also functions in foraging, recognizing foods and, perhaps avoiding predators. UV light does, however, exhibit properties that potentially could make these wavelengths a handy signal in mate choice. For example, these wavelengths scatter more easily than longer wavelengths, making them work best as signals at short distances. The putative advantage to using short-distance communication for courtship lies in not revealing oneself so readily to predators while one's attention is diverted from survival (chapter 5). Another possible advantage to short wavelengths is that they are favored in polarization arising from the backscattering of light, but we have as yet no evidence that birds can even detect polarized light, much less use it for signaling. A third advantage of communicating in the UV channel, for birds, is that such signals contrast well with UV-absorbing chlorophyll in the ubiquitous background of plant parts. These "special" properties of ultraviolet light continue to be debated, as more bird taxa are studied and more of the hypotheses for "specialness" are tested.

to protect the wearer from the harmful effects of ultraviolet radiation. It remains a mystery why parrots do not recruit these beneficial chemicals for wider use throughout their bodies.

But the psittacofulvins, like the carotenoids, are not in feathers only to provide basic protection from the elements. Were that true, most likely one pigment would be best at that job and parrots would be all the same few colors. The splendid diversity of parrot colors is clearly there for use as signals to other parrots. To learn what these functions might be, we have to consider how the parrots view themselves.

Why Parrots are Colorful

How can we even begin to imagine how parrots see their world? I try to grapple with my Cartesian dilemma without much success—how can my feeble senses possibly allow me to experience, and therefore to understand, the universe in its full and true glory? In an act of

FIGURE 32 A pair of Red-winged Parrots, *Aprosmictus erythopterus,* illustrating the plumage differences between the two sexes. In addition to the differences that we can see, there are differences in the ultraviolet portion of the spectrum, which parrots can perceive. The male on the right has a brilliant skullcap that reflects ultraviolet; in the eyes of his mate, on the left, this contrasts strongly with the green feathers elsewhere on his body. Photo © Ian Montgomery.

impulsive curiosity, one night I ventured into my aviary with a "black light"—a lamp that glows in the ultraviolet (UV) range. I figured we had survived the sixties without all that much bodily damage, so what harm could it do to flash a black light here and there at unsuspecting parrots? Standing there in the dark with my bewildered pets, I felt as if I had slipped into another dimension, an alternate universe. What I thought I knew, I did not know. What I thought was one color, was more. What I thought was a body with no pattern, was a striking matrix of contrasting patches (figure 32). And I was not even seeing the parrots the way they see each other. I was getting only a miserable hint of the spectrum beheld by those with tetrachromatic vision. The black light had not created anything in my mind like the images that appear in the minds of parrots. It allowed me only a shadowy glimpse into another world, as if I were a visitor looking out on a new landscape just at sunset, when in the dim light I could only guess at the colors of the scenery.

The ease of using a black light to illuminate museum specimens allowed scientists some decades ago to discover that feathers of many species of parrots fluoresce under UV light (box 4). Among those, male Budgerigars display strikingly contrasting patterns of reflect-

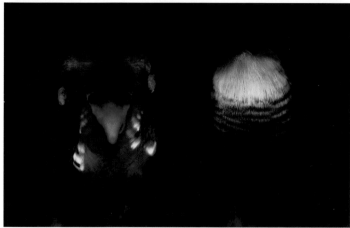

FIGURE 33 A male Budgerigar, *Melopsittacus undulatus*, seen in human-visible light (above) and in ultraviolet light visible to Budgerigars (bottom). Pictures are from a study by Kathryn Arnold and colleagues that showed that female budgerigars preferred males with unobscured UV-reflecting patches over males whose patches were obscured by sunblock. Photos © Justin Marshall.

ance and absorption of UV light. In particular, the unstriped crowns of adults fluoresce bright yellow, and the area of blue spangles and black dots against a yellow background on the throat fluoresces as glowing yellow spots against a dark UV-absorbing background (figure 33). Thus, the differently colored regions in the human-visible spectrum are overlaid with distinct patterns of interaction with UV light in the human-invisible spectrum. How this light is processed in a bird's brain as a color, a human cannot ever know. We can however use instruments to detect what we cannot see and measure the wavelengths of light reflecting off the feathers embedded with different pigments. We can also present the birds with different colors, or modify their colors as presented to another bird, and take note of their reactions, as a first approximation to what the birds are seeing that we humans cannot.

In a parallel series of ingenious experiments on Budgerigars, two different research groups offered trained budgies a set of choices of companions under different regimens of filtered light. Sophie Pearn and her colleagues Andrew Bennett and Innes Cuthill used special light filters placed between experimental birds, while Kathryn Arnold, Ian Owens, and Justin Marshall dabbed sunscreen (which blocks UV light) onto specific areas of color on the budgies. Their experimental setups were similar, allowing a focal bird, typically a female, to choose from a set of companions offered under different light regimens and indicate her preference by spending more or less time in front of companions under different light treatments. In the experiments, the scientists offered both male and female companions to the focal bird, to test whether the highly gregarious budgies were merely meeting their social needs or whether they were flirting with potential mates.

In all of the experiments, without exception, female Budgerigars used light reflected from their companions in the UV spectrum to establish a preference for potential mates. When UV light was blocked, focal females showed no preference for different males or for male versus female companionship. When the full light spectrum was available, however, female budgies clearly showed a preference for unmodified male budgies. These UV adornments were apparently required to make male Budgerigars more studly to females.

The evolution of striking color and other accoutrements to serve solely in mate choice is not news. Charles Darwin was initially mystified at such ornamentation because it seemed at best irrelevant to survival, and at worst could clearly impair the bearer. Then Darwin realized that the whimsies of mates could overrule the wisdom of natural selection (which worked only to improve the ability of offspring to thrive under everyday life-and-death scenarios posed by the environment). The panorama of spectacular colors of parrots would seem to have no other purpose than to say "notice me" and therefore to function primarily in mate choice.

What is new in the experiments of Pearn, Arnold, and others is the discovery that birds (including parrots) use a much fuller spectrum of light wavelengths to signal to one another than do mammals, including humans. Franziska Hausmann teamed up with Arnold and other colleagues to survey 108 species of Australian birds, including 51 species of parrots from 25 genera. The considerable majority (78 percent) of species of Australian birds possess plumage that reflects UV light, and in most, the UV-reflecting regions are used in courtship displays. Of the 51 species of parrots surveyed, one-third reflected UV light in some portion of their plumage. Of these species, nearly 90 percent reflected UV light in a part of the plumage known to be important in courtship, whereas fewer (two-thirds) reflected UV in plumage not known to be used in courtship. This difference, while slight, suggests that UV wavelengths in particular are used by parrots to signal to their mates, or, more importantly, to their potential mates. Hausmann and colleagues also tested for fluorescence (box 4) and for the presence of areas that provided high contrast between neighboring patches of fluorescence and UV-reflecting feathers. They found that in 93 percent of species and nearly all genera, the fluorescing feathers are used in courtship display, whereas 62 percent use fluorescence in regions not known to be important in mate signaling.

FIGURE 34 A pair of Blue-fronted Amazons, *Amazona aestiva*, in the wild. Although the two sexes appear similar in the visible spectrum, there are striking plumage differences in the ultraviolet that are visible to these birds. Photo © Mike Bowles and Loretta Erickson.

An extension of the story of parrot color vision and our human inability to admit to our deficiencies is a study by Susana Santos and her colleagues. In examining the feather colors of Blue-fronted Amazons (figure 34), they found that males and females are strikingly different in coloration, specifically in the UV portion of the spectrum. Their findings provide yet another form of strong evidence that UV-active plumage is dominant in courtship signaling in parrots. In their study, they note the fallacy of our classification of 75 percent of parrots as sexually monomorphic (chapter 5)—that is, *we* cannot see the difference between males and females. Clearly, our assumption that we humans can perceive reality objectively is false; what we considered an established fact about parrots is not true when parrots are viewing each other.

Why UV-active plumage should be used differentially for courtship is controversial (box 4). The correlation between UV activity in certain feathers and their use in courtship signaling, however, is convincing in the studies of parrots. Parrots are tetrachromats through and through, using their visual capabilities certainly for sexual signaling with color. We know this only because curious scientists could not rest, until somehow they discovered what they should not have been able to know.

VOCAL COMMUNICATION

A Natural History of the Sounds Wild Parrots Make

My travels to Australia and South America affected forever how I view parrots. Because of my own species' penchant for causing extinction (chapter 7), I grew up not experiencing parrots in the wild. My reaction, as a grown-up ecologist, was one of childlike wonder and awe at the sights and sounds of wild parrots living free. One sunny, mild afternoon, in the austral winter of Queensland, I sat on a hillside north of Brisbane, gazing over a forested valley floor below. The valley spanned several kilometers between two low ridges, trending respectively to the right and left of me. Suddenly, to my right, a flock of Sulphur-crested Cockatoos *Cacatua galerita* came into view (figure 35). With powerful and measured wing strokes, they traveled together across the small valley, skimming over the treetops below, becoming so many brilliantly white specks in the distance. While they flew, they cried out and were answered by cockatoos on the other side of the valley. Although far away, their calls were still loud and easily discernable, both to me and to their companions. Their raucous screams echoed off the mountainside behind me and filled the winter day with the undeniable presence of parrots. With what seemed like only a dozen effortless wing strokes, the traveling cockatoos united with those on the valley's other side, and the joined group disappeared beneath the canopy.

Clearly the cockatoos were shouting something to their companions across the valley. My first impression was an obvious one: parrots are very loud because they communicate with one another across great distances. Part of that impression contained a subtle assumption. I had concluded that the noises that the birds were making, and that were returned by birds at a distance when they heard those noises, were intentional signals. To a human observer, who happens to be a member of another intelligent, social, and vocally communicating species, the obvious conclusion from my observations was that the birds were purposefully talking to one another, so that they could find each other across a distance and meet up, for whatever reasons they might have (chapters 4, 5, and 6).

Yet, to a Cartesian observer, assigning purpose and meaning to the communication would be problematic. In chapter 4, we explore the still-controversial topics of language and cognition and whether a nonhuman animal has those capabilities. In the meantime, can we at least attribute meaning to the vocalizations as a noncontroversial act of communication that reunited a flock of cockatoos across the valley floor they call home? I shall first attempt to answer this question from the perspective of a natural historian, one trained in the fields of ecology, ethology, and evolutionary biology. A naturalist would begin by describing all the sounds that the parrots make in the wild and then attempt to associate those sounds with contexts that are ecological or behaviorally meaningful (to the parrots).

FIGURE 35 A pair of Sulphur-crested Cockatoos, *Cacatua galerita,* calling from a tree in Australia. Photo © Mary Bomford.

An early paper, published in 1966 in *Australian Natural History* by Australian naturalists John Le Gay Brereton and Robert Pidgeon, provides a perfect starting point for our exploration of parrot communication. "The Language of the Eastern Rosella" begins with a review of why animals (including humans) need to communicate, that is, primarily for sexual reproduction (chapter 5) but also for assembly and coordination of individuals in species that are social, as are parrots (chapter 6). They describe the 24 vocalizations that they recorded in Eastern Rosellas *Platycercus eximius* and examine the ecological or behavioral contexts associated with the vocalizations, thus assigning probable meaning or purpose to the sounds that parrots intentionally make (table 1).

Brereton and Pidgeon proceed with the common-sense assumption that parrots have reasons to communicate with each other—that they need to be aware of their surroundings and to act with intent to survive and to reproduce. Evolution would favor such capabilities in mobile animals that must gather and process information so as to find food, shelter, and mates.

TABLE 1   Examples of vocalizations of parrots and their functions

Category of call and species	Name of vocalization	Comments
Contact calls		
Eastern Rosella[1]	Pipping calls	Contact and flight calls
Short-billed Black Cockatoo[2]	Wy-lah	Contact and flight calls
	Interrogative call	Modified wy-lah
	Whistle call	Females only, in conjunction with wy-lah
Brown-headed Parrot[3]	Double, triple chip	Contact calls
	Kreek call	Response to double chip, always in flight
	Zzweet call	Function unknown
Grey-headed Parrot[4]	Tzu-wee	Long-distance contact call
	Chirp, chatter	Short-distance
Yellow-naped Amazon[5]	Wah-wah	Contact call, modified into dialects with vocal learning
Orange-fronted Parakeet[6]	Chee	Contact call, modified with vocal learning
	Zip, peach	Flock cohesion during flight
Kea[7]	Kee-ah	Contact call, modified with vocal learning
Other functions		
Short-billed Black Cockatoo[2]	Squeak call, ah-ah	Males only, courtship
	Squeak chatter	Males only, territoriality
	Squawk call	Alarm call, agonistic call
Brown-headed Parrot[3]	Chreeo call	Given during foraging and feeding
Grey-headed Parrot[4]	Duet song	Courtship/pair bonding, male and female
	Kraak call	Social function
	Zeek-zeek	Juveniles only, solicitation
Yellow-naped Amazon[5]	Duet song	Courtship/pair bonding, male and female
Orange-fronted Parakeet[6]	Warbling, growl, squeak, churr	Unknown social functions, often given during loafing/resting

[1] Brereton and Pidgeon (1966)
[2] Saunders (1983)
[3] Taylor and Perrin (2005)
[4] Symes and Perrin (2004)
[5] Wright (1996); Wright and Dahlin (2007); Wright and Wilkinson (2001); Wright, Dahlin and Salinas-Melgoza (2008)
[6] Bradbury, Cortopassi, and Clemmons (2001); Vehrencamp et al. (2003); Cortopassi and Bradbury (2006)
[7] Bond and Diamond (2005)

FIGURE 36 A flock of Budgerigars, *Melopsittacus undulatus,* perched near Murchison, Australia. Budgerigars are also called Warbling Grass Parakeets because of the complex warble song sung by males during courtship. Photo © Georgina Steytler.

With this background, let us now proceed to explore the natural history studies of parrot communication. We can then visit the experimental studies testing hypotheses raised by the study of wild parrots' sounds to understand the physiological (how) and evolutionary (why) reasons for the sounds they make.

Sound Perception in Parrots

Since Brereton and Pidgeon's 1966 paper, the first modern study of communication in wild parrots, no fewer than thirty other studies have documented natural vocalizations of parrots in the wild, categorizing them and associating them with ecologically and behaviorally meaningful contexts (table 1). Biologists who study songbirds often draw the distinction between a *call*, which is a relatively short vocalization that references a specific consequence or intention, versus a *song*, which is a longer, less specific, often melodious vocalization (figure 36). In songbirds, songs are quite often associated with territoriality (beautiful, perhaps, but all business, rather like a national anthem) or with wooing a mate and bonding the pair (as are many human songs; see epilogue). In this classification, most parrot vocalizations fall under the category of calls, although some, particularly the duets produced by mated pairs in some species, could be considered songs.

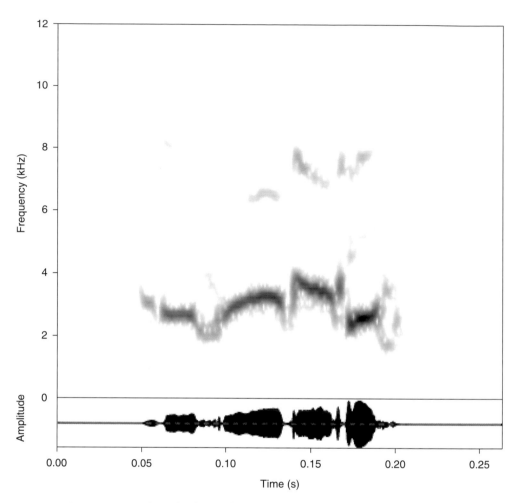

FIGURE 37 Spectrogram and amplitude waveform of a contact call from a Budgerigar, *Melopsittacus undulatus*. The upper spectrogram represents changes in frequencies with time. The lower waveform illustrates changes in the overall energy of the call with time. A typical parrot contact call is short in duration and may be both frequency and amplitude modulated. Figure by Marcelo Araya Salas.

The repertoires of most wild parrots studied so far have between one and two dozen distinctly different calls that have shades of a variety of meanings useful to the parrots. The biologists not only listened to parrot calls with their human ears (not in themselves the most objective instrument, as we will see) but also recorded and processed the calls with mechanical instruments to produce visual illustrations called spectrograms (figure 37).

Vocalizations of wild parrots seem to fall into several broad, functional categories, each of which can contain two or more distinct types of calls. These categories include: (1) begging calls, given by nestlings and fledglings to solicit food from their parents; (2) social cohesion calls, which include calls that maintain contact between mates, family members, and flock members while hidden from view in foliage or in flight, and calls that signal the intention to take flight or to engage in other types of activities that involve the coordination of individuals; (3) courtship and mate-bonding calls, including songs and duets, that are used by a paired male and female (chapter 5); (4) aggression calls, which are used primarily during the breeding season when a pair maintains a nesting territory and excludes conspecifics; and (5) alarm and distress calls, which are used whenever a parrot is frightened or harmed, such as in encounters with predators.

The contact call is the most common vocalization parrots make, and all parrots use it (figure 37). Each species of parrot has at least one contact call; most have several for use in different situations. By the term *contact call* we imply that this vocalization is associated with two individuals interacting. Usually the contact call is given between two individual parrots that know each other and have an established relationship. The call may be given when the two individuals are separated and cannot see each other. It is also often given when two individuals have been separated for a certain amount of time and then come back into contact with each other, such as a male returning with food to feed his mate (the female does all the incubating of eggs; chapter 6).

A typical contact call is rather short. For example, the contact call of Budgerigars is about 100–300 ms (milliseconds) in duration (figure 37). In some species, including the Budgerigar, energy in the contact call is focused in a narrow range of frequencies and sounds pure in tone or pitch. In other species, the contact call's energy may spread across several narrow bands of different frequencies; such a pattern is referred to as *harmonics* and has a noisier or rougher timbre. For most species, the energy in contact calls (in fact all of their calls) typically occurs between 2 and 5 kHz (kilohertz, 1,000 cycles per second). The narrow-bandwidth frequencies typically change during the call; that is, contact calls are usually *frequency modulated*. Likewise, the energy of different frequencies may vary during the call, so that the contact call is also *amplitude modulated*. Scientists have established that the collective characteristics of most contact calls enhance the call's capability to carry important information that two individuals need to communicate.

Sometimes the same type of contact call is exchanged between two individuals. In other cases, depending on the species and circumstances, an initial contact call is answered with a different acknowledgement contact call. Brown-headed Parrots *Poicephalus cryptoxanthus* return a "double chip" contact call with a "kreek" call, but only when the birds are in flight. The Brown-headed Parrots also give a "zzweet" call (table 1). The zzweet call is their most common call and so may be an additional contact call. Studying parrots in the wild has its strengths and its weaknesses. Only when behavior is observed in the wild can its true behavioral and ecological context be determined. On the other hand, it is difficult to assign causality to patterns in observational data. Because the zzweet call is so common, and given in a wide variety of contexts, its meaning cannot easily be correlated with particular activities. Often enough, only laboratory studies can probe precisely enough to reveal causality with certainty, and we soon turn to those to understand parrot vocalizations.

Like the zzweet call, other contact calls are given only when the parrots are in flight (table 1). Some parrots use a particular call only when they are about to take flight, such as the

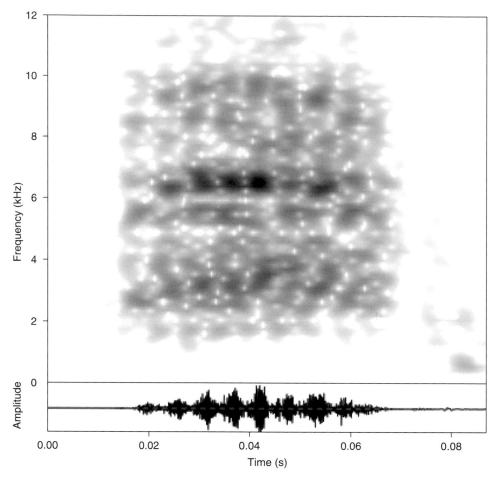

FIGURE 38 Spectrogram (top) and waveform (bottom) of an alarm call from a Budgerigar, *Melopsittacus undulatus*, illustrating the broadband nature of these calls, which typically lack the frequency and amplitude modulation seen in contact calls. Figure by Marcelo Araya Salas.

"peach" call of Orange-fronted Parakeets *Eupsittula canicularis*. St. Lucia Parrots *Amazona versicolor* give the "p-chow" call only in flight. Likewise, Monk Parakeets *Myiopsitta monachus* and Cape Parrots *Poicephalus robustus* produce certain vocalizations only while in flight. Other calls are given only in foraging flocks, such as the "chreeo" call of Brown-headed Parrots. Yet other calls are produced only when the parrots are sitting and loafing between their morning and afternoon feeding expeditions, preening or resting quietly with flock-mates (chapter 6). Thick-billed Parrots *Rhynchopsitta pachyrhyncha* use a single-note "bark" contact call while in flight and a multi-note "laugh" call when perched.

The acoustic properties of parrot calls can give clues as to their function. Alarm, distress, and aggression calls often have properties distinct from contact calls. Usually these calls are broadband and not frequency or amplitude modulated (figure 38). Humans can easily enough interpret meaning from the very nature of alarm calls. Why? Such a broadband signal is harsh, grating, and attention grabbing. It sounds alarming to humans. However, broadband calls do not typically travel as far through vegetation as do tonal calls with a narrow bandwidth. That may be just fine if the intended receiver of the call is close at hand.

In other words, calls that are intended to signal a threat have different properties from calls intended to maintain contact. Contact calls are often intended for specific recipients, and to relay specific information to particular individuals. They may need to travel far in the environment and be unambiguous. Alarm, distress, and aggression calls are general, not intended for any given recipient, that is, a specific individual. The threat (be it a predator or invader of a territory) will be close by, and only individuals nearby will be affected by the information in the call. Such calls need to grab attention but not travel; nor do they need to carry a great deal of specific information.

Thus, the signals that parrots produce have meaning and purpose in their daily lives in the wild, and the properties of the signal will determine how well that purpose is accomplished.

Laboratory Findings

To pursue answers to questions raised by our survey of parrot sounds, we need to now examine studies done in the laboratory. We first explore how and what parrots actually hear, which they tell us by behaving in certain ways in the laboratory experiments. By far most of the laboratory work is done on the little Budgerigar. Budgies are easy to maintain in the lab, and they are obliging about cooperating in experiments. Budgies are trained using operant conditioning to choose different seed hoppers or peck different keys based on the experimental variables offered them, just as they are for experiments on their vision. We can thank in particular Robert Dooling of the University of Maryland and his many students, postdoctoral fellows, and colleagues for our now extensive knowledge of parrot hearing and perception.

The first and most obvious experiment is to offer Budgerigars different frequencies of sound, to discover parameters of their hearing acuity. Doing so allows us to check whether their ability to hear sounds matches the characteristics of sounds that they make as signals.

The ability of budgies to detect and to discriminate sound frequencies (pitch) is greatest in the range of 2–4 kHz, although they can hear pitches both lower and higher, roughly 0.5–6 kHz. A measure of hearing acuity is not simply which frequencies a subject can detect, but how well they can discriminate the difference between two tones or the change of tones in time. We humans are quite interested in this subject, because of our love of music (chapter 8). Students of voice and music must train themselves to recognize absolute pitch, discriminate pitches, produce a vocalization of any given pitch, and read a visual representation of musical notes so that they can accurately reproduce a given pitch with a musical instrument or their voice. Interestingly, humans share with budgerigars the ability to hear best in the range of 2–4 kHz. Thus, unlike color vision, humans and parrots share common ground in hearing frequencies.

Budgies, however, are much better than humans at distinguishing pitches in this frequency range (figure 39). Humans and rats are lackluster at absolute pitch perception, in

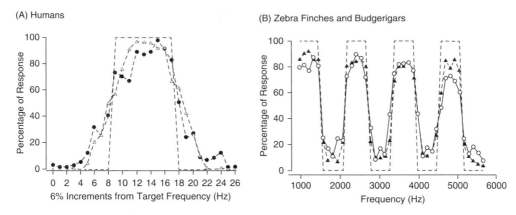

FIGURE 39 Experimental results from Ron Weisman and colleagues on absolute pitch perception in humans, Budgerigars (*Melopsittacus undulatus*), and Zebra Finches (*Taeniopygia guttata*). All three species were tested on their ability to correctly distinguish between rewarded and unrewarded pitches, indicated by the dotted lines on each graph. Humans (left) had weak discrimination abilities, as indicated by the poor match between the dotted square indicating the rewarded frequencies and the solid line indicating their responses. In contrast, both bird species (right) rarely mistook rewarded for unrewarded frequencies, indicating their strong ability to distinguish fine differences in pitch. Redrawn with permission from Weisman et al. (2004).

comparison to budgies and oscine songbirds, as we know from studies by Ronald Weisman and his colleagues. His study is of particular interest in our quest to understand parrots' perception and cognition, as he and his colleagues rejected the notion that a parrot had to be able to "name" a pitch in order to possess absolute pitch perception. All the scientists needed to do was to show that the budgies could accurately perceive and discriminate pure tones by comparing them with memory alone (absolute pitch). They trained the birds (not only budgies but also Zebra Finches *Taeniopygia guttata* and White-crowned Sparrows *Zonotrichia leucophrys*) to alight on perches where certain sounds were played, thereby receiving a reward of seeds. The scientists then used this act of choice as an indicator of perception of the pitch being played in front of the perch.

In an eight-range experiment, tones in the range of 1.0–1.8 kHz predicted a seed reward; tones in the range of 1.8–2.3 kHz predicted no reward; and so on up the scale (figure 39). The goal of the experiment was to learn whether the birds can learn pitches finely enough to distinguish pitches both within and outside of these narrow ranges of frequency—not by comparison of two tones that they hear during a single trial of the experiment but by comparison of a single tone with a memorized one. A bird's cage would have a number of such perch-hopper-microphone apparatuses, and the bird would choose among the various perches depending on which sound was played there. Birds in the experiments earned their living this way, feeding in this manner over a twelve-hour period and each accomplishing over 1,000 choices of tones. Budgerigars and the finches were easily able to discriminate three-tone and eight-tone brackets over the scale of their perceptive ability, scoring nearly perfectly. Humans, in contrast, made many mistakes in the much easier task of discriminating three different sound frequencies (humans were offered the same arrangement, only they pushed buttons and got money for their correct responses).

All of the birds, once trained to understand the task, made virtually no errors, receiving their seed reward for "correct" choices. Thus, not only can the birds perceive absolute pitch, but they do so over fine gradations of pitches. To say that a bird perceives absolute pitch in an experiment like this means that the bird can do all of the following: (1) collect enough information in its ears about the frequency of a sound wave; (2) send that information to the brain; (3) store the exact pitch in memory; (4) compare the experimental pitch of the moment with a memorized pitch from the training; and (5) then use that information to innervate their flight, leg, and neck muscles to fly over to a perch and receive and eat a food reward. That is, the birds needed to be cognizant of the protocol and understand that certain pitches always got them food and other pitches never did. Thus, far more than the birds' hearing acuity was revealed in this experiment, as we continue to explore in chapter 4.

Communicating Acoustically: The Contact Call as an Important Social Signal

The contact call of a Budgerigar is a valuable social signal, and thus makes a particularly potent acoustic stimulus for use in studies of acoustic perception. Recall that the contact call is a tonal call with frequency modulated in the range that fits the Budgerigar's abilities in pitch perception (figure 37). In other words, the budgies' hearing exactly matches the information that they put into an important social signal that they produce.

In the wild, the budgies need this important social signal to be heard above the ubiquitous background noise found in every environment. Examples of environmental noise include a waterfall, rustlings of leaves and other wind-produced noise, and other bird calls in the background. The phenomenon of environmental noise interfering with our detecting important information in acoustic signals is known as *masking*. Budgerigars can use behavior to improve their ability to detect a signal among all the noise, or to produce a signal that can be detected by others. The budgie can simply fly away from the noise source or toward the signal source. Or, a budgie can wait to give a signal until there is a gap in the environmental noise, since few environmental sounds are truly continuous. A budgie can turn its head back and forth on a horizontal plane (azimuth), allowing its two ears to vary in position relative to the sound source. That information then can assist a brain in locating a source or discovering the best pathway between the source and listener for avoiding the irrelevant background sounds.

In addition to these cruder measures to unmask sounds, experimental studies of freely moving Budgerigars confirm that having two ears to hear with vastly improves a budgie's ability to hear, compared to birds with one ear plugged up. Budgie brains use information from the time delay to the far ear to unmask sounds, an ability called *binaural masking release*. Freely moving budgies can achieve an 8–10 dB binaural masking release for sounds within their hearing range of 2–4 kHz. In fact, they possess an ability that scales with that of much larger mammals. Why does size matter? Because the size of the animal's skull determines the slight delay in sound's reaching one ear after the other—the interaural time difference, or ITD. The smaller the ITD, the less the information would be available for use, presumably. The Budgerigar's hearing, however, functions as if its skull were much larger than it actually is. The distance between the two ears of an average budgie is 16 mm, but the improvement in hearing resulting from the interaural time difference is equivalent to a skull width of 28 mm. At this point, how Budgerigars achieve this acoustically effective head size, an improvement of 175 percent (nearly double), is not known, but somehow the contouring of the feathers must play a significant role.

For now, we examine only steps 1 (detection) and 2 (discrimination of pitch as well as signal from noise) and leave the more difficult task of studying step 3 (the signal's meaning) for another topic. A study by Bernard Lohr, Tim Wright, and Robert Dooling tested the first two abilities. They compared the performance of Budgerigars, Zebra Finches and Canaries *Serinus canaria* in detecting and discriminating each other's contact calls against background noise more like that of natural environments (i.e., instead of using easy-to-generate white noise). In all three species, critical thresholds for discrimination of calls were higher than that of simple detection of calls by about 3 dB—that is, extracting information from a sound (pitch discrimination) against a noisy background is more difficult than simply being aware of a sound against that background (detection). Interestingly, a budgie contact call was easier for both Budgerigars and Zebra Finches to detect and to discriminate than were either Zebra Finch or Canary contact calls. Budgerigar contact calls contain relatively pure tones that are frequency modulated, but Lohr and his colleagues proposed that it was the amplitude modulation in budgie contact calls that made them easier to unmask with a background of noise, since even small peaks in amplitude would allow the call to stand out from the background. All told, the easier a sound is to unmask, the longer the distance over which a signal can be detected, discriminated, and presumably understood.

Thus far, all of the sounds we have covered that are used by parrots in the wild are constrained to some degree. These parrots make similar sounds in similar contexts with sufficient consistency that we humans can categorize them and label them with mnemonics like "peach" or "zweet" (table 1). This sort of consistency suggests that the calls might be *innate*, or genetically encoded. And yet practically any time scientists examine these calls closely, they find variation in the call structure, which suggests that individual *learning* might also be important in their acquisition. To understand more about these vocalizations and how they are acquired, we must explore some of the social influences on call development in wild and captive parrots.

Introduction to Vocal Learning in Parrots

The popularity of parrots as captive pets and companion animals surely rests as much on their well-known abilities to reproduce sounds of human speech as on their stunning beauty. As a trained scientist, evolutionary biologist, and natural historian who ponders the qualities of parrots, I had always had the nagging thought that this ability to imitate sounds in the environment, especially the sounds of companions, has to be present in wild parrot populations. In particular, the ability to reproduce sounds must have had some sort of useful function for it to evolve. I was always certain that parrots did not possess this ability as a curious epiphenomenon of their lives in captivity. The first order of business would seem to be to learn the extent of parrots' propensities to copy each other in the wild, and also in captivity under relatively natural social conditions, such as in a flock of free-ranging individuals, kept in spacious quarters and allowed to engage in normal social activities. That is, what do parrots not tutored by humans do?

When we refer to parrots, the terms *mimic* and *imitate* dominate any discussion of vocalizations in which parrots copy the sounds of other individuals or other species, or sounds in the environment. I prefer to use the more general and more neutral (descriptive) terms applied to this phenomenon: *vocal learning* and *copied* sounds. Vocal learners process sounds that they hear and then modify their own vocalizations to match sounds from their environment (biological and otherwise). Without question, vocal learning plays a role in inter-individual communication of some kind or other, given the contexts in which it occurs in those animals that possess this capacity.

Surprisingly, only a small number of animals are capable of vocal learning. Obviously, humans (one species within the order Primates) represent one such taxon. We are not alone, but we are joined by relatively few other taxa. Thus far, vocal learning is well documented in some species in three other orders of mammals: bats (order Chiroptera), whales and dolphins (order Cetacea), and seals and sea lions (order Carnivora); and some species in three orders of birds (class Aves): parrots (order Psittaciformes), hummingbirds (order Apodiformes), and songbirds (order Passeriformes, suborder Oscines). As researchers become more aware of this possibility, more instances of vocal learning are being added to this list of taxa, including, most recently, African elephants, *Loxodonta africana*. The difficulty of establishing vocal learning in nature, using free-ranging animals, should not be underestimated; the problem arises in how to establish that a sound is copied and what model was used.

Tests for vocal learning are most easily done in the laboratory. Enter once again the Budgerigar. The first step is simply to document vocal learning—to demonstrate it in a way that is replicable and that convinces other scientists that such a phenomenon occurs. One study (of many) that is particularly thorough and relevant to our discussion here was done by Susan Farabaugh, Alison Linzenbold, and Robert Dooling on contact calls. The contact call is a good vocalization with which to begin. Not only is the contact call ubiquitous in parrots but, as we learned, its characteristics make it an excellent signal for communicating at a distance.

In the study of Farabaugh and colleagues, six Budgerigars lived in two separate flight cages. Each flight cage contained three males able to interact at will. The birds in each flight cage could hear those in the other, but they could interact socially only with the birds in the same cage. The scientists recorded several dozen variations on the basic Budgerigar contact call, as could be distinguished by eye in sonograms, that were used by these budgies at the start of the experiment. These call "variants" were based on a simple foundation structure common to all Budgerigar contact calls: 100–300 ms duration and strongly frequency modulated in the 2–4 kHz band. Variations consisted of different patterns in frequency and amplitude modulation. At the beginning of the experiment, the budgies recruited from different sources used all the call types on the list. Each budgie's contact call was recorded before he was introduced to the new group and then again after eight weeks of free social interaction.

After eight weeks, the three budgies within each cage had converged on a small number of shared call types. This convergence was reciprocal, in that each budgie learned the call types of his companions, and the group then settled on a couple of shared calls. For example, Enrico entered the group with a type E contact call and Nico with a type A. At the end of eight weeks, Enrico had developed an exact rendition of Nico's call, and vice versa. Moreover, all of the birds in that cage were improvising a new call, a sort of combination of types A and E. Although the birds could clearly hear contact calls from the other cage, the birds in each cage focused on group-specific sharing and did not pick up or converge on the dominant call types of birds in the other cages.

Other studies By Elizabeth Brittan-Powell and Robert Dooling confirm that baby budgies learn their family's contact call as soon as they fledge. Before fledging, while still in the nest, the primary vocalization of the babies is their begging call, for soliciting food from parents. The structure of the begging call contains repeated elements that are incorporated in the first contact calls of the fledglings at approximately 35 days old. Contact calls of the young birds then develop further and are refined, depending on the social dynamics of each of these new flock members as they gain entry into the social structure of the flock. Males are more likely to copy the calls of adult male tutors, whereas females learn their contact calls from any individual, regardless of age or sex.

When the Budgerigars become adults, the copying of contact calls takes on specific adult functions. Studies by Arla Hile, Georg Streidter, and colleagues showed that courting males converge quickly on the contact call of the females that they are attempting to woo, whereas the females do not so reciprocate. Experiments with deafened males show that if a male suitor cannot master the contact call of the female that he is courting, she is not as likely to bond with him, or alternatively stay faithful to him if the pair is force-bonded by experimenters.

In another study, by Marin Moravec, Georg Striedter, and Nancy Burley, a group of budgies of both sexes was assembled by the experimenters and allowed to live in a reasonably natural setting equipped with nesting boxes to encourage pair formation and breeding. The scientists found that pairing ensued quickly, with mates selected based on the similarity of the males' and females' contact calls. Remarkably, contact-call similarity predicted good male parenting. Females were always good parents, but males showed more variation in the quality of parenting they provided. Budgerigar males do no incubation, so their role, although important, is less tied to a particular set of eggs (chapter 6). A bonded male feeds the incubating female, who should be reluctant to leave her eggs because, in the group-style breeding that is characteristic of budgies, marauding pairs can destroy eggs and take over unguarded nests. The male is also needed to help feed the young. An attentive father allows a female, indeed the pair, to raise more young than either parent could alone (chapter 5). Somehow, for some reason, this attentiveness scales directly with the similarity of the male's and female's contact calls, even before the two birds begin their courting process and the male intentionally copies those of his intended mate. In this study, chicks of attentive fathers had lower mortality rates than those of males with lesser parenting skills.

Another interesting finding of Moravec and colleagues was that the contact calls of a bonded pair converge only during the courtship, bonding, and mating phase and later diverge during incubation and chick rearing. This pattern suggests a cost for call convergence. Certainly it takes some effort to listen to calls, pay attention to them, memorize them, and practice them; call convergence is not instantaneous. This cost could be in energy—calling takes physiological effort as well as mental effort—or it could be in a social currency or a trade-off with another important activity.

Authors of studies on vocal learning in captive Budgerigars point to the importance of rapid pair-bonding in wild Budgerigars, suggesting an important function for contact-call similarity. This species is nomadic and breeds opportunistically, whenever and wherever the weather is good and local resources sufficiently abundant within interior Australia. Their mortality rate is also likely to be very high, like that of the Green-rumped Parrotlet *Forpus passerinus* (figure 40), a small Neotropical species that has been extensively studied by Steve Beissinger, Scott Stoleson, Karl Berg and their colleagues (box 6). The high mortality of these smaller parrots means that individuals may frequently need to find new mates (chapters 5 and 6). For these reasons, pair-bonding should still be as fast as possible in the wild. Even bonded pairs would need to respond quickly to resource opportunities. Contact-call convergence seems to speed things along and, most importantly, increase a breeding pair's success in parenting.

Some Functions of Vocal Learning in Wild Parrots

Studying vocal learning in captive parrots, or indeed studying just about anything in captive parrots, is a great deal more straightforward than studying parrots in the wild. One of the first studies on vocal learning in wild parrots was conducted on the most vocally talented of all parrot species: the Grey Parrot. In that study, published in 1993, Alick Cruickshank, Jean-Pierre Gautier, and Claude Chappuis produced sonograms from recordings of a variety of Grey Parrot vocalizations and their putative models. They found convincing evidence that the wild birds were copying eleven different vocalizations of nine different species of birds and one species of bat, making up 27 percent of the vocalizations that the researchers recorded the parrots producing. The copied vocalizations seemed to be woven into the duets performed by bonded pairs of Grey Parrots, a finding suggestive of studies of call use by Budgerigars in the laboratory.

Since the study by Cruickshank et al., behavioral ecologists have begun to work in earnest to discover what wild parrots do with their ability to copy sounds. Cruickshank et al. discussed in their paper why no one had ever thought to look for vocal learning (which they called *mimicry*) in the wild. The first reason they offered was simply how difficult parrots are to study in the wild. Although wild parrots are conspicuous and loud, so that their presence is easily detected, they stay high up in the canopy, moving in and out of the foliage, and they

FIGURE 40 A family of Green-rumped Parrotlets, *Forpus passerinus*, emerges from a nest in Venezuela. Photo © Eduardo Lopez.

do not stay in one place for long. They fly kilometers each day. These challenges make sounds that a researcher is certain came from a parrot difficult to record, especially where the parrots coexist with the species or other sound sources providing the model for the learned vocalizations. It also makes parrots hard to trap and mark so that one can follow known individuals around and record all the different call variants they make. Cruickshank et al. also proposed that no one had found parrots in the wild copying sounds of any kind simply because no one had looked for this phenomenon, perhaps assuming that the behavior of captive parrots is an amusing artifact. Alternatively, precisely because most wild parrots copy each other's vocalizations, as we learned from laboratory studies, vocal learning would be difficult to detect *de novo* in wild parrots.

Subsequent studies of wild parrots have demonstrated that vocal learning occurs routinely in populations of wild parrots and even highlighted likely functions for vocal learning. A group of behavioral ecologists including Timothy Wright, Jack Bradbury, Sandra Vehrencamp, Thorsten Balsby, and other colleagues, studied contact calls and other vocalizations of

Orange-fronted Parakeets and Yellow-naped Amazons *Amazona auropalliata* in the Guanacaste region of Costa Rica, with an eye toward understanding the role of vocal learning in wild parrots. Their studies have revealed far more than the occurrence of vocal learning in the wild. Their findings about vocal learning have led them to discover more about how wild parrot flocks are structured, how parrots disperse from their natal haunts, and how parrots interact in large social groupings. Interestingly, these studies have found little evidence of mimicry of other species, as found in the Grey Parrot. Instead, copying of the vocalizations of others appears to be very common, but occurs exclusively among members of the same species, as seen with the captive Budgerigars.

In the first of these studies, Timothy Wright followed and recorded Yellow-naped Amazons in populations scattered throughout the wet forests of Costa Rica and Nicaragua. The "wa-wa" contact call was used by wild Yellow-naped Amazons, much as has been found for other species, to maintain social contact and coordination between members of a pair or among members of a flock engaged in a group activity (table 1). One such group activity that Wright exploited was their nightly roosting behavior. As is common in many social species, Yellow-naped Amazons gather in communal roosts to spend the night. The birds in the roosts are all regulars for that roosting spot, but they are not closely connected with each other socially, other than in where they roost. The main unit of flock structure is still the bonded pair and their offspring (chapter 6).

These roosts are postulated to be information-gathering centers. Scientists hypothesize that members of a roost use each other to discern good foraging possibilities for the next day, in a dynamic that Bernd Heinrich describes as "resource parasitism" when discussing his Common Ravens *Corvus corax*. The birds roost together not to help each other find food but to help themselves find food by following other individuals that appear to know where they are going first thing in the morning. Evidence for such "leader birds" is now being established in studies of Orange-fronted Parakeets by Thorsten Balsby and Jack Bradbury.

The roosts of Yellow-naped Amazons do not have fixed memberships over the very long term, in that genetic relatedness of the individuals within a roost is not correlated with the spatial distribution of the roosts themselves in the parrot species studied so far. Roost members however do share a common dialect of the wa-wa contact call. Wright found that he could distinguish contact calls from different regions, namely Nicaragua bordering Costa Rica, northern Costa Rica, and southern Costa Rica, both by ear and by examination of a sonogram of the calls. Roosts within these regions shared a common contact-call structure. Wright and his colleagues have determined that the dialects have remained stable over eleven years of their study so far, although surely individuals have turned over in these populations. In roosts on the borders between dialects, the parrots are actually bilingual, able to produce both types of contact call.

Although no one has yet followed known individuals with radio-tracking equipment over long periods, Wright deduced from these findings and other observations how parrot movements change over their lifetimes (chapter 6). Young Yellow-naped Amazons leave their parents and birthplace some time within their first year or so. Eventually, these juveniles settle

FIGURE 41 A pair of Orange-fronted Parakeets, *Eupsittula canicularis* (formerly in the genus *Aratinga*). Studies of this species in Costa Rica show it has flexible group membership and extensive matching of vocalizations among group members. Photo © Hans D. Dossenbach.

down in a new area and join the local flocks. There the juveniles mix for several years while they seek their life-long mates (more detail in Chapter 6). We already learned that pairing in Budgerigars may be hastened when courting individuals share similarities in their learned contact calls. The implication for Yellow-naped Amazons is clear. The newly arrived, dispersing juveniles may work their way into acceptance in the local flock by picking up the local vernacular (more below). The dialects of contact calls are distinct and discrete, suggesting that once the birds settle into a region, if not a single roost, they remain there, probably for the rest of their lives. At least they do not move far, or else the dialect boundaries of Yellow-naped Amazons would not exist, and contact-call structures would grade into one another, making a continuum of gradually changing variants.

This alternative scenario precisely describes vocal learning in another Costa Rican parrot, the Orange-fronted Parakeet (figure 41). Orange-fronted Parakeets do not develop distinct, discrete dialects that are geographically delimited. Rather, across Costa Rica from north to south there is a continuous gradation of contact-call structure. This pattern leads us to two features of the biology of Orange-fronted Parakeets. First, individuals of this species also engage in vocal learning, with members of closely knit social groups copying each other's contact calls, in the wild under natural conditions. Second, the dispersal and social structure of Orange-fronted Parakeets are clearly different from those of the Yellow-naped Amazons even though the two species co-occur in the same eco-regions. Orange-fronted

Parakeets travel and forage during the day in small flocks of unrelated individuals. Unlike Yellow-naped Amazons, Orange-fronted Parakeets roost in small dispersed roosts that seem to form wherever the foraging flocks alight for the night, rather than in semi-permanent locations like the Yellow-naped Amazons. Orange-fronted Parakeets do show site fidelity, however, with home ranges of around 9 km square. Interestingly, the gradations in the geographic clines of contact-call variation are roughly on the order of home-range size.

Bradbury, Vehrencamp, Cortopassi, Balsby, and colleagues consider the Orange-fronted Parakeet an example of a *fission-fusion* society. This term refers to the fluidity of the daily foraging flocks of the parakeets (chapter 4). The membership of such flocks is not stable over long periods of time, because family groups mix in and out of different flocks, depending on how they meet up at food sources. This fluidity in flock membership probably explains the absence of well-structured vocal dialects in this species.

Recent studies of Australian parrots find the same patterns as we encountered in the parrots of Costa Rica. Both Australian Ringnecks *Barnardius zonarius* in Western Australia and the Galah *Eolophus roseicapillus* use vocal learning to modify their contact calls to converge on the contact calls of response to conspecific flock members. Myron Baker showed that, like those of the Yellow-naped Amazon, the contact calls of Australian Ringnecks converged rapidly in a given area to form dialects among individuals with different genetic backgrounds. A special twist of his study was that two genetically distinct populations of Australian Ringnecks diverged during climate change in the late Pliocene or early Pleistocene (chapter 1); the genetic distinctions were revealed by the physical appearance of hybrids between subpopulations coming back into contact with one another. Galahs, in contrast, appear to have more of a fission-fusion structure to their populations, similar to the Orange-fronted Parakeets of Costa Rica. Remarkably, a study by Judith Scarl and Jack Bradbury showed that Galahs converged contact calls within literally minutes of interacting with unknown birds, which they encountered as recorded calls in call-back experiments set up by the investigators. This kind of rapid response could benefit mixing groups of Galahs in a fluid society such as in fission-fusion populations.

Next, we explore what is known of yet another species that might shed more light on the social function of call convergence in all parrots.

Vocal Learning for Social Cohesion

In New Zealand, the Kea *Nestor notabilis* (figure 42) earns its common name from the sound of its contact call: "kee-ahh." Alan Bond and Judy Diamond have studied Keas in the wild for many years; we will encounter their work again in chapter 6. Their study also revealed geographic variations in the contact call in both adults and juveniles. They asked the basic question of whether this variation was a result of chance drift in call structure or, alternatively, local cultures resulting from vocal learning. Their recordings revealed the perplexing result that the contact calls' variations differed between adults and juveniles along the same cline.

The answer to this puzzle may lie in the social structure of Keas. Young Keas disperse soon after they gain independence from their parents, and they join "gangs," in which they

FIGURE 42 A Kea, *Nestor notabilis,* vocalizing near Arthur's Pass, New Zealand. Juveniles and adults in this species show different patterns of vocal variation, suggesting different patterns of social association. Photo © Andrius Pašukonis.

band together until they find mates (chapter 6). These groups of juvenile Keas then develop their own dialects, different from those of their parents. The juvenile groups are apparently fission-fusion also, as indicated by the clinal (gradual geographic) variation in their call structure, arising from the same processes that we observed in the Orange-fronted Parakeet. Adults also have clinal variation, but their calls were acoustically distinct from those of juveniles and varied in different acoustic dimensions among the sampled populations.

The significance of age-structured vocal learning in Keas is probably revealed by the studies of Budgerigars that we reviewed, as well as studies of Bottlenose Dolphins. Being able to join a group is critical for young, vulnerable, inexperienced individuals. Only by joining forces can they find enough food, test and learn the right kinds of foods, and avoid predators. Although their parents begin to provide them with this information, the parents and offspring must nevertheless part ways early on, before the juvenile parrots have learned everything they need to know. Meanwhile, the juveniles find solace in each other, in playing and in group learning while they develop adult skills (chapter 6). During this time they copy each other's contact calls. This call matching is as if they had a code word to announce themselves and claim a right to join up with an established group.

Perhaps by now you are thinking, "Wait! This sounds familiar—it reminds me of high school." Human ecologists and linguists study local colloquial use of language by subgroups in society. These subgroups could be clubs, with their secret words, or regions, such as the San Fernando Valley of California, or neighborhoods of downtown Los Angeles or New York City or London, or entire cultures. When entire cultures or their regional subsets share commonalities that seem only about cultural identity, anthropologists refer to this pattern as *ethnic marking*. In a variety of studies, Peter Richerson, Robert Boyd, Charles Efferson, and colleagues discuss how ethnic markers arise by social learning (as opposed to being innate). Ethnic markers are informative to group members for predicting what they need to know to function in the group and, among other things, gain access to resources. Mastering a local dialect allows quick entry into an established group, and an established group contains experienced members who are good at finding food and avoiding predators.

Parrots may not be so different from humans. Keas have an age-specific, socially transmitted trait that may function to gain entry into social groups. Like humans, parrots are long-lived and must learn most of what they need to know to survive and reproduce, as opposed to relying on instinct alone. Being part of a group and sharing information is key for social beings to succeed in life. One idea for how vocal copying in both humans and animals might play a role in this process is the "password hypothesis," which suggests that copied sounds act as a password of group membership. If you haven't learned the password you can't gain admittance to the group. An arduous study by Alejandro Salinas-Melgoza and Timothy Wright tested this hypothesis by translocating Yellow-Naped Amazons from one dialect region to another in Costa Rica. The results were surprising. Of the ten birds they were able to translocate and subsequently follow with radio tracking, four birds managed to leave the release site and fly back to their old home (more than 30 km). Five others remained in the new dialect; but instead of learning the local dialect they found the other translocated birds and formed a kind of "immigrant enclave." One bird who stayed at the release site, however, did socialize with local birds, and six weeks later it started producing the calls of its new dialect. Intriguingly, this bird was one of only two juveniles in the study, suggesting that the juveniles were either better able to learn new calls than the adults, or, alternatively, more willing to join new social groups. In either case, the results provided little support for the password hypothesis, since the juvenile was able to join the local social groups well before it learned the local dialect.

Individually Specific Labels and How Parrots Use Them

Several of the studies of the vocalizations of parrots have discovered that individuals in one way or another use a contact call that is distinguishably different from other individuals in the flock. For example, in the study by Cortopassi and Bradbury, pairs of Orange-fronted Parakeets used a smaller number of contact variants than those available to the pair based on the variants used by the population at large. Surprisingly, perhaps, unlike Budgerigar pairs, each member of a pair used a single variant most of the time and not the same variant as its mate. Thus, Cortopassi and Bradbury proposed that Orange-fronted Parakeets have individually specific calls, as do some other species with fission-fusion societies. Scientists studying cetaceans (whales and dolphins) first labeled the unique sound that only one individual in a social group makes a *signature* call. Researchers have gone so far as to consider the signature call of social species to be a "name," that is, a label for a specific individual in a social group that members of the group know somehow refers to a specific individual in that group.

Just demonstrating that individuals vary in the quality of sounds that they produce, even when making a specific call, does not demonstrate that such individuality means anything. Signature calls do not function as names until either that individual or a member of its social group discerns the unique acoustic properties of that call and somehow responds appropriately to it in a social context.

Studies on captive Spectacled Parrotlets *Forpus conspicillatus* show that they do just that. Ralf Wanker and his colleagues made a scientifically convincing case for their use of individually specific labels for different members in their social group. That is to say, Spectacled Parrotlets call each other by name, apparently, depending on what restrictions one applies to the definition of "name." The acoustical parameters that Spectacled Parrotlets use to vary their contact calls are the same as those of the other parrots that we have reviewed. Parrotlet contact calls vary in a great many of their acoustic characteristics; thus, contact-call variants provide sufficient information to distinguish a reasonable number of the individually specific labels needed in an average parrotlet social grouping, in their normal, noisy, wild environments.

In this study, the scientists recorded the dominant contact calls that one individual parrotlet uttered in the presence of certain companions. The individual bird calling was the "focal" or "tested" bird. Let us say that in one example experiment the tested bird was called Eddi (by the scientists). The tested bird was exposed to "stimulus" birds, that is, companions that he or she interacted with and reacted to. In an example experiment, the experimenters would then record Eddi's dominant contact call in the presence of his mate Renee, in the presence of his son Uvo, and in the presence of his other son Ustinov. In each case Eddi uttered slightly different contact calls, individualized for the partner he was with and repeatable, similarly to male Budgerigars using a specific contact-call variant in the presence of a mate.

To complete this particular experiment, Wanker and colleagues then played Eddi's vocalizations back to each of the stimulus birds. The scientists found that each stimulus bird reacted more strongly to the contact-call variant Eddi uttered in his or her presence than a variant Eddi used in the presence of another bird. In other words, Renee was more attracted to Eddi's special contact call for her than to his contact calls for their sons.

Wanker and colleagues repeated this protocol for a number of social groups within the flock, and the results of their experiments were straightforward. Adults were better at distinguishing the sets of contact calls (those made in the tested bird's presence versus not) than were juveniles. In addition to experience, the closeness of the social relationship made a difference. The male and female of a bonded pair responded more to calls made in each other's presence than did siblings calling to each other. Pairs and siblings both responded more to their cohorts' calls made in their presence than to those made in the presence of less socially bonded group members.

The authors of this study therefore concluded that the parrotlets create and use names for each other. In interpreting their results, they used the terminology of *refer to* and *referen-*

tial. These terms mean at minimum that a signal (in this case vocal) is attached to an object, in this case another bird. Importantly, a referential signal is a *symbol* of that object, because the signal is not the object itself; the signal *refers to* that object. A name would necessarily be a referential signal.

Usually animal communication is not so abstract. Many calls or vocalizations fall into the category called *motivational*. This type of call signals the motivational state of the signaler, and use of this term is without controversy. A motivational signal could mean, "I'm about to fly, even though I'm not flying now," which is how biologists interpret the flight call of parrots. More directly, an alarm call signifies that the individual making the call is in a state of fear, which it shares with flock members presumably to warn them of an impending danger. Or a begging call is made by a hungry chick to communicate to its parents, "Feed me!!!"

Although perhaps surprising, the results of Wanker et al. were to be expected from the previous studies of Budgerigars and Orange-fronted Parakeets. If a female Budgerigar is more likely to bond with a male that utters a specific contact-call variant for her and moreover has a stronger bond that allows more successful parenting, then surely that female must somehow respond to the sound of that call. Moreover, her response must be meaningful in a social context such that raising more offspring successfully is the result.

The study of Wanker et al. thus contributes two new bits of information building on what we know about individually specific contact calls. First, the scientists demonstrated this attraction as a behavioral response to the individually specific contact call alone, as opposed to the natural combination of the call and the flock member uttering it (who might emit many other cues). Second, the scientists extended the use of individualized contact calls to other classes of social bonds than that between mates. They showed that the intensity of the response to the call attenuated with the degree of social bond, providing evidence that the individualized contact call does perform a social function and is not simply a random variant in the call parameters.

Collating the results of all studies of vocal learning in parrots, we see that vocal learning for identity and cohesion purposes may occur at any social level. For some species, such as the Yellow-naped Amazon, vocal learning to establish identity focuses primarily on the level of large groups, such as regional subpopulations. For other species, such as Orange-fronted Parakeets, Spectacled Parrotlets, and Budgerigars, establishment of identity is more focused on less inclusive social groupings, such as that of the immediate family or pair. These studies of communication among parrots lead us to ponder just what parrots understand about each other and the world they live in, a topic that we expand more fully in the next chapter. We will revisit the phenomenon of vocal learning there, to learn about the brains of parrots and all that they do with them.

FOUR

The Thinking Parrot

The Brains of Parrots and How they Use Them

COGNITIVE ETHOLOGY

Je pense donc je suis.[1]

—RENÉ DESCARTES, 1637

CONTENTS

Introduction to the Study of Learning and Cognition 117

The Brains of Birds 118

    Vocal Learning as a Model for How Brains Accomplish Learning 118

    Why Vocal Learning? Why Not? 122

    The Structure of Parrot Brains and How Birds Do What Mammals Do 125

Cognition in Parrots 128

    Introduction to Cognitive Ethology and Alex the Parrot 128

Our Education of Alex, and His of Us: Testing the Cognitive Abilities of Grey Parrots 130

Evolution of Cognition in Parrots and Larger Patterns across Animals 135

Mental Distress in Captive Parrots 138

    Freestone's Story 138

    Abnormal Repetitive Behaviors 139

    Parrots and Their Signs of Mental Distress 141

INTRODUCTION TO THE STUDY OF LEARNING AND COGNITION

The studies of hearing and vocal learning that we discussed in chapter 3 have led curious scientists on a journey deep inside the parrot brain. Perhaps because we humans depend so greatly on vocal learning, many studies have pursued the question of how vertebrate brains accomplish it. These scientists argue that vocal learning is a good model for how the brain accomplishes learning as a general phenomenon. In turn, learning is the basis for intelligence and problem-solving, which fall under the conceptual umbrella of *cognition*, the topic of this chapter.

We begin by learning what scientists know about the brains of birds, specifically those of parrots, and how they function to enable vocal learning and other mental abilities. Then we proceed to learn about what parrots actually think, as told to us by parrots trained to

1. "I think, therefore I am." Descartes also expressed this in Latin, as *Cogito ergo sum.*

FIGURE 43 Alex, a Grey Parrot, *Psittacus erithacus,* that was the subject of many of the groundbreaking studies of avian cognition conducted by Irene Pepperberg and colleagues. Photo © Arlene Levin-Rowe and the Alex Foundation.

communicate with humans in human language, because parrots are skilled vocal learners, perhaps second only to humans (figure 43). Along the way, we explore the topics of intelligence, complex communication, and social learning, and why these traits seem to evolve together.

This chapter wraps up with another area related to thinking parrots. We learn that parrots may think too much to remain happy in barren, unengaging environments. Perhaps their capacity for insanity speaks the most persuasively for parrots' capacity for intelligence and consciousness. These topics are controversial but essential, not only for understanding parrots but for our own enlightenment as living beings sharing our planet with others.

THE BRAINS OF BIRDS

Vocal Learning as a Model for How Brains Accomplish Learning

Much of what we know about parrot brains has come from intense study of one particular species of parrot, the Budgerigar *Melopsittacus undulatus,* in investigations of how exactly birds accomplish vocal learning.

Vocal learning is a good model for sleuthing the inner workings of vertebrate brains. A brain is all about gleaning information from the environment, processing this information to extract useful meaning and to decide on an appropriate response, and then sending instructions to muscles to carry out that decision. We could choose any sensory-locomotor processing pathway, but vocal learning is a good place to begin our journey through the parrot brain because we know so much about the behavior of vocal learning (chapter 3). Tracing the pathways information follows to transform sound into action permitted neurophysiologists to realize that the brains of birds were not as different from those of mammals as originally thought. Breaking this mental block (no pun intended) allowed neurophysiologists not only to have a greater appreciation of the intelligence of birds, but also to learn more about cognitive functions in all vertebrates.

Vocal learning begins with hearing the sounds that will eventually be copied. Within the inner ear is the cochlea, a bony structure that receives and translates information from pressure waves in the atmosphere to the sensory hair cells that cover a fleshy membrane called the basilar papilla. Neurons connected to these hair cells are then activated by the sound pressure waves as they are translated into mechanical pressure waves that cause the basilar papilla to vibrate. Specific frequencies of sound cause specific parts of the basilar papilla to vibrate, and specific hair cells to fire, a phenomenon known as a tonotopic map. The signal from these hair cells flows along a cascade of firing neural synapses, whereby vesicles of special chemicals (neurotransmitters) are released between each pair of neurons tracing this pathway. The information in the form of this signal travels from the periphery (the ear) into the central nervous system, to the posterior hindbrain (rhombencephalon), where, in birds, a structure called the cochlear nucleus resides. The incoming auditory information is then further processed by other nuclei in the hindbrain before it is collated in the midbrain (figure 44).

Once in the midbrain (mesencephalon), the information lands in a coordinating center known as the dorsal lateral nucleus. To be "understood" and given meaning, the auditory information has then to travel to the forebrain (cerebrum or telencephalon), a journey that begins in the thalamus, where all sensory input is gathered and organized before it is sent to the cognitive centers in the forebrain. In parrots and songbirds, this thalamic nucleus is a structure called the ovoidalis. After reaching this gateway into the forebrain, the information then moves on (in birds) to a consortium of cognitive processing nuclei in the forebrain, which collaborate to derive meaning from this sound stimulus.

So far, sound information travels the same path in vocal-learning and in non-vocal-learning birds, and the same path in parrots and in songbirds, the two vocal-learning birds most often studied. Non-vocal learners are nevertheless auditory learners; they process sounds the same way and learn to derive meaning from them just as vocal learners do, even if they cannot reproduce the heard sounds themselves. For vocalizations to incorporate learned sounds, two additional general neural pathways must be present. First, the auditory information that is interpreted in the forebrain must be stored as an auditory memory, as it is in all auditory learners. Second, this information must be eventually retrieved from memory and sent somewhere to innervate the vocal motor neurons, allowing the copied sound to be produced by the animal's vocal apparatus, a feat accomplished only by vocal learners, by definition.

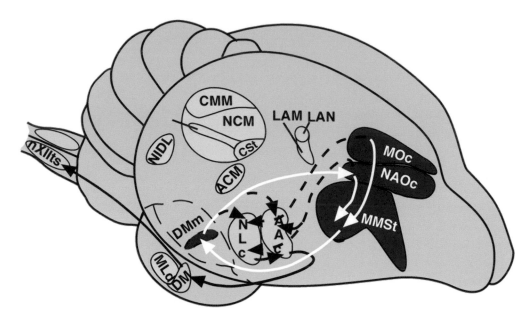

FIGURE 44 A schematic of the brain of the Budgerigar, *Melopsittacus undulatus*, illustrating brain areas involved in vocal learning. Colored areas with abbreviations are specific brain nuclei within three interactive pathway: blue indicates areas devoted to auditory perception; yellow, areas devoted to motor control and production of vocalizations; and red, areas involved in learning new vocalizations. Redrawn with permission from Feenders et al. (2008).

When the vocal learner utters this copied sound, it hears its own version as it utters it. The copied-sound stimulus will be processed by the same auditory route as the original sound, but it also will be compared with the auditory memory. The forebrain auditory and cognitive centers then team up to produce a closer rendition of the copied sound, in a feedback loop of neural pathways in the brain. Vocal learners rarely achieve a perfect copy of the sound the first time they attempt it, certainly not without practice in sound copying. This skill likely increases with the individual's age and experience (at least up to a point), as we saw with Budgerigars copying each other's contact calls.

The vocal learner thus engages in an iterative process, whereby later copies of the sound are compared with the auditory memory and improved until a good match is achieved. This learning process is also a developmental process, during which baby vocal learners get better at copying sounds. These copied sounds, after all, are essential to the communication system of vocal learners. Many studies have been performed in humans to learn how children acquire and master language. Studies on Monk Parakeets *Myiopsitta monachus* and Grey Parrots show that parrots use their tongues, as do humans, to produce the fine gradations of frequency and amplitude modulation (chapter 3) required to copy sounds precisely, and studies of Grey Parrots reveal that parrots achieve fluency in copying sounds similarly to human children.

FIGURE 45 A pair of Budgerigars, *Melopsittacus undulatus,* in flight in Australia. Budgerigars have been the primary model for laboratory studies aimed at understanding how the parrot brain accomplishes vocal learning. Photo © Jim Bendon.

In parrots, no fewer than nine vocal nuclei participate in vocal learning and communication, distributed throughout the three brain regions but mostly concentrated in the nidopallium, mesopallium, and arcopallium of the telencephalon (figure 44). That is to say, processes essential to vocal learning occur deep in the thinking part of the brain. The oscine songbirds use different parts of the telencephalon for their auditory-vocal feedback pathways than do parrots, but other parts are similar or exactly the same. Although brain centers for mammalian vocal learning differ in yet other respects from those responsible for avian vocal learning, the same general neuroarchitecture is used by vocal-learning mammals too.

This observation has prompted some neuroevolutionary biologists to conclude that the major pathways for information input and processing, including vocal learning, are highly conserved. In other words, the basic neuroarchitecture of the vertebrate brain has not changed significantly during amniote evolution for the past 300 million years, when the synapsids (which gave rise to modern mammals) and the diapsids (which gave rise to the reptiles, the dinosaurs, and their modern descendants, the birds) last shared a common ancestor (see chapter 1). All sensory information and motor coordination occur in the same general brain regions of all amniotes, whether they are vocal learners or not. An implication of this fact is that all birds and mammals have the pre-existing machinery to be vocal learners.

We could then ask: So, why does vocal learning make such a spotty and rare appearance in such a huge diversity of organisms? Does being rare make vocal learning really special, as we humans consider it to be? Or, does its rarity imply that vocal learning is *not* special and that there are other ways to accomplish the same goals, in an evolutionary sense? We need to learn more about vocal learning, the brain, and thinking to answer that question (figure 45).

Why Vocal Learning? Why Not?

To begin, let us review the taxa with vocal learning. Birds capable of vocal learning (chapter 3) are: parrots (Psittaciformes), oscine songbirds (Passeriformes), and humming-birds (Apodiformes)—which are separated by at least 90 million years of evolution from their last common ancestor. No other close relatives of these three avian clades have the ability to modify their vocalizations based on heard information. Likewise, vocal learning appears in only a few branches of the mammalian phylogenetic tree, in five clades (so far) separated from a common ancestor on roughly the same time scale as birds. Within each order and among the various taxa, the degree of skill in and extent of vocal learning vary considerably. Most of these vocal learners copy only *conspecific* sounds, that is, sounds made by individuals of the same species—and not many of those sounds, at that. Copiers of sounds made by other species, or anything (living or not) in the environment, and copying of many sounds, seem to be limited to some oscine songbirds, parrots, and humans.

Vocal learning, as it turns out, is a nifty neural-genetic trick that requires some remark-ably simple changes in the gene regulation and neuroarchitecture that, again, are shared by all birds or mammals. It is possible that any lineage can be transformed into vocal learners by a relatively few mutations (substitutions) in a couple of multipurpose, universal genes involved in coordinating sensory and motor functions and shared by all amniotes. Thus, the neural mechanisms permitting vocal learning are simple and could arise quickly from non-vocal-learning ancestors under pressure from natural selection. A first, obvious conclusion for an evolutionary biologist then would be that vocal learning evolved independently in the eight clades (so far) containing species capable of vocal learning. This observation leads to another. The rarity of vocal learning in species of communicating and social vertebrates speaks not to the difficulty of evolving such an ability—quite the opposite, as vocal learning has been reinvented at least eight times. Rather, its rarity may well imply that other solutions, such as the use of innate and unchangeable vocalizations, are equally suitable for meeting certain ecological and evolutionary challenges. Scientists point to the costs of vocal learning, such as the possibility of mistakes, the time needed to learn the copied sounds, and greater investment in expensive brain tissue. For all of these reasons, innate vocalizations serve most species quite well, showing few limits in their utility.

Nevertheless, the very rarity of vocal learning also prompts us to try to understand why it might have evolved in place of other solutions that are perhaps equally good, or perhaps not. Because parrots are exceptional vocal learners, we need to consider why it might have evolved in parrots so that we understand more about them.

Erich Jarvis and his colleagues have proposed a laundry list of hypotheses on why vocal learning evolves and, by implication, why it doesn't. Jarvis lists general functions that are accomplished by vocal learning and that might favor its evolution: (1) individual identifica-tion, (2) mate attraction, (3) territory defense, (4) semantic communication (i.e. vocalizations

TABLE 2 Functions of copied sounds produced with vocal learning ability and examples in parrots

	Function[1]	Examples in parrots documented so far[2]
1	Individual signature	Budgerigar; Spectacled Parakeet; Orange-fronted Parakeet
2	Mate attraction/bonding	Budgerigar; Yellow-crowned Amazon; Orange-fronted Parakeet
3	Territoriality	Yellow-crowned Amazon
4	Symbolic/referential communication	Spectacled Parakeet; Grey Parrot
5	Syntactical communication	Grey Parrot
6	Sound propagation	All parrot examples
7	Social cohesion (beyond the pair)	Yellow-crowned Amazon; Spectacled Parakeet; Kea; Galah; Australian Ringneck
8	Abstract communication and thinking	Grey Parrot

[1] Adapted from Jarvis (2006).

[2] Scientific names: Budgerigar *Melopsittacus undulatus*; Spectacled Parakeet *Forpus conspicillatus*; Orange-fronted Parakeet *Aratinga canicularis*; Yellow-crowned Amazon *Amazona ochrocephala auropalliata*; Grey Parrot *Psittacus erithacus*; Kea *Nestor notabilis*; Galah *Eolophus roseicapillus*; Australian Ringneck *Platycercus zonarius*.

as labels), (5) complex syntax (vocalizations that are organized strings of labels, involving procedural memory), and (6) rapid accommodation of sound propagation in different acoustic environments. To his list, I would add from studies of parrots (chapter 3): (7) social cohesion or group identity, as distinct from his (1); and (8) abstract conceptualization, which he may imply in his (4) or (5), or both (table 2).

We covered in chapter 3 how parrots use their learned and copied vocalizations for individual recognition in the form of a signature contact call that an individual gives either itself or another individual in the group, to which it is bonded (function 1). We also learned that parrots use their ability to copy vocalizations for ethnic marking, which could permit rapid entry into social groups, in turn assuring access to resources and protection from predators (function 8). Other organisms, however, accomplish these goals with highly sophisticated innate vocalizations, such as those used by cooperatively social canids, arguing against functions 1 and 8 as primary evolutionary drivers for vocal learning. Parrots also inject vocal learning into attracting mates and defending territories (figure 46), but so many species do equally well without vocal learning, which suggests that functions 2 and 3 are also unconvincing as primary drivers of its evolution. Parrots clearly use their copied sounds for function 4, if we allow the "names" of Spectacled Parrotlets *Forpus conspicillatus* to pass as symbolic, referential labels. Parrots could use their fixed, innate vocalizations for purposes that also serve as labels. For example, a number of birds and primates that have specific alarm calls for predators on the ground (snakes or carnivores) versus predators in the air (raptors). Likewise, parrots use vocal learning for social cohesion (function 7), as do most other species with vocal learning, yet innate vocalizations also do the trick for social cohesion in social canids such as the dhole. Thus nothing in this list of known functions of vocal learning uniquely explains its evolution.

FIGURE 46 A pair of Sulphur-crested Cockatoos, *Cacatua galerita,* examine a nesting hollow in a eucalyptus tree in Australia. Finding and defending a suitable nest cavity requires close coordination between members of a pair. Photo © Bent Pedersen.

After reviewing the innate vocalizations that do most if not all of these jobs well, and also discussing the costs of vocal learning, Jarvis proposes that a combination of factors must be required to favor vocal learning over innate vocalizations. In particular, this combination of useful functions triggers feedbacks that can cause strong natural selection for vocal learning over alternate modes of vocal communication. I agree with his proposition. Parrots and other vocal learners have taught us that the vocalizations that are most likely to incorporate learned sounds nearly all have traits that make the vocalizer easy to locate. Therefore, I propose that vocal learning in parrots evolved first from a combination of needing to propagate socially important information through acoustically challenging environments (functions 1, 2, 3, and 7: individual and group identification, mate attraction, and territory defense). Mate attraction and selection, for example, raises the ante to promote selection on any other trait to increase the probability that an individual's genes will be passed to the next generation—that is, mate attraction is absolutely fundamental to the currency of fitness (chapter 5). Similarly, in social species, group cohesion signals also carry high stakes in fitness.

Were we just interested in vocal learning, we could stop here, and we would have discussed this topic in chapter 3. The functions and properties of human language, however, raise the possibility that vocal learners can then co-opt vocal learning, once it evolves, into functions 4, 5, and especially 8 (semantic, syntactical or abstract communication), thus reinforcing its evolution even more. Parrots—of all other vocal learners—are potential candidates for using their abilities to copy sounds to extend their abilities to communicate beyond simple identification purposes, as have humans. Let us continue to explore this possibility.

The Structure of Parrot Brains and How Birds Do What Mammals Do

As we begin to differentiate specific functions within the general mental tasks of gathering, processing, and acting on information, we discover the different parts of the vertebrate brain that are dedicated to those tasks. The more complex or inclusive the task—the more that task dominates in the lifestyle of that organism—the more neural tissue we might expect the brain to devote to accomplishing it. Hence, scientists measure the volume of the brain parts that seem to be devoted to certain tasks to test hypotheses about those tasks. Scientists exploring these questions in birds include Erich Jarvis and Andrew Iwaniuk and their numerous colleagues. Here we consider their findings for parrot brains.

A simple and rather easily tested hypothesis is the proposal that the cerebellum of birds is relatively so much larger than that of most mammals because most birds fly and most mammals do not. Certainly, the cerebellum and entire hindbrain and midbrain dominate the overall brain volume in birds and stand out in morphological complexity, compared to most mammals. The hindbrain, including the cerebellum, and the midbrain are involved in sensory-motor coordination. Flight is especially complex as a form of movement because it occurs in three spatial dimensions rather than two, and because air molecules are much less densely packed than are water molecules. Therefore, flight as a form of locomotion requires more exacting sensory-motor coordination than do walking, climbing, and swimming. The relatively enormous cerebellum of birds surely reflects these neural challenges and the difficulty and complexity of accomplishing flight (figure 47).

In parallel reasoning, we might expect the thinking parts of the brain to be relatively larger in animals with more complicated behavior, such as problem-solving ability, vocal learning, and other kinds of cognitive processes. Because a bird's telencephalon is rather measly in comparison to its cerebellum, and smooth in appearance, relative to mammals, one hypothesis has been that birds lack the capacity for all those mental qualities possessed by mammals, in particular humans. But two lines of evidence argue against this hypothesis. The first, we already have covered: the case of vocal learning. In researching vocal learning in parrots and passerines, neurobiologists have uncovered many homologies between bird and mammal brains, more than were suspected. They have also discovered numerous examples of independent evolution of functions (analogies) shared by birds and mammals. These analogous functions are accomplished by recruiting the same general brain regions possessed by all amniotes but deriving different specific neurological solutions, as we discovered for vocal learning.

Another line of evidence speaks to the trade-offs that are universally faced by all organisms and by natural selection in achieving solutions to environmental challenges. An example most relevant to the evolution of bird brains is that the demands of flight preclude having

FIGURE 47 A flock of Short-billed Black Cockatoos, *Calyptorhynchus latirostris,* take flight in Yanchep National Park, Australia. Photo © Georgina Steytler.

a big, heavy head. Natural selection would then work away at making neuroarchitecture more efficient, instead of just bulky, as increasingly more complex behaviors are also favored by natural selection. This line of evidence shows no evolutionary limitations of neuroarchitecture in birds that would prevent birds from acquiring any of the most complex behaviors and mental functions seen in mammals.

In a study of 180 species of parrots, Iwaniuk, Karen Dean, and John Nelson estimated the relative sizes of different parts of parrot brains to test (in part) the hypothesis that more complex behaviors, including capacity for cognition, require more neural tissue. They compared brains of these parrot species with other species of birds and with primates. First they investigated whether the gross morphology of brains is under biomechanical and developmental constraints, limiting how much the brain can be remodeled through evolution to achieve different types of tasks. One idea is that the basic architecture of the vertebrate brain is conserved, because of overriding constraints on the essential life functions (needing to detect the environment and to direct muscles) or on the way that an adult brain develops from an embryo.

Their evidence favored the alternate "mosaic hypothesis," which proposes that the brain can be redesigned to meet various environmental challenges. Nothing inherent in neuroarchitecture restricts natural selection in changing the morphology of at least some regions of the brain relative to others, depending on what tasks need to be accomplished. Iwaniuk and

FIGURE 48 A Palm Cockatoo, *Probosciger aterrimus,* perched at a nest hollow in Cape York, Australia. Parrots are the largest-headed of birds, proportional to their body size. Photo © Steve Murphy and Brian Venables.

Peter Hurd described five *cerebrotypes* of bird brains, by comparing the major orders of birds. Parrots and passerine birds represented one of the five cerebrotypes, characterized by an emphasis on complex cognitive abilities, compared to other cerebrotypes focused more predominantly on specialized modes of predation or locomotion.

Iwaniuk and his colleagues then turned their attention to aspects of the architecture of parrot brains that differ from those of other birds and that would be consistent with their known cognitive abilities. They discovered that parrot brains are proportionately larger than those of bird species in every other order but the Passeriformes. The telencephalon (the thinking part of the brain) of parrots and of passerines makes up most of this differential in relative brain size. Iwaniuk and his colleagues found that the expansion of the nidopallium and mesopallium contributed the most to the relatively large brains and telencephala of parrots.

These scientists also note the parallels between evolution of large brain size in the parrots and in the primates (figure 48). In particular, they point to the similar proportionality of brain parts in birds and primates, with parrots on the large-brain end of the continuum in birds, and with hominid primates (humans and other great apes) on the large-brain end of the continuum in mammals. Further studies may add the cetaceans (whales, dolphins, and their relatives) to the list of species with complex cognitive behaviors and sociality that show this trend of expanding certain neurological substrates in the brain.

The next task, then, is to explore what parrots in particular might be doing with so much of their unusually large brains devoted to thinking components, relative to other birds. Could the parrots be similar to unusually large-brained mammals, such as ourselves? As it happens, because of their skill in vocal learning, we can teach parrots how to use human language and tell us themselves.

COGNITION IN PARROTS

Introduction to Cognitive Ethology and Alex the Parrot

The field of cognitive ethology was launched in the 1970s by the late Donald Griffin of Rockefeller University and a number of his colleagues. This branch of ethology seeks to understand what animals (including humans) know, how they know it, and what kinds of environments and circumstances might favor animals possessing flexible and learned behavior. That is, why did cognition evolve? This question encompasses, first, our discovering what cognition is, in all the different animals species with flexible behavior—and, second, putting those behaviors into an ecological and evolutionary context so that we may have a greater understanding of them.

The ethological approach to animal behavior is distinct from that of other behavioral sciences, specifically in its focus on an animal's behavior in the ecological setting in which it evolved. In this way, ethology is closely allied to the fields of ecology and evolutionary biology, which share with ethology a strong natural history foundation. Humans are included in this ethological worldview, but they must take their humble place among all of the other creatures in the biosphere and follow the same principles of ecology and evolution as all of the others do.

One argument that still holds ground in some fields of scholarship is that vocal language, with formal syntax, is a prerequisite for advanced cognition—a kind of Doolittle argument. One thing about this argument is certain: If we could talk to other animals, we could just ask them what they think and why they think it. Talking to other animals and having them answer back would be a start in understanding what is going on in a nonhuman mind (figure 49).

Primatologists have been conversing with other primates for some decades using mainly the gestural (nonvocal) American Sign Language. Parrots, however, can—well—talk. One scientist in particular, Irene Pepperberg, has spent three decades teaching Grey Parrots to speak English. Pepperberg systematically taught Grey Parrots English words using their referential and functional contexts, much as human parents teach their children language. As a result, the Grey Parrots that she has engaged in her research—Alex, Alo, Kyaaro, Griffin, and Arthur (Wart)—have been able to communicate with Pepperberg and other researchers in proper human discourse. Pepperberg's research program has also spun off, with scientists in other laboratories following a similar approach to understanding parrot cognition.

FIGURE 49 A pair of Palm Cockatoos, *Probosciger aterrimus,* at a nest. The bird on the right is using a stick as a tool to drum on the hollow stump. Photo © Christina Zdenek.

In the mid-1970s, Pepperberg walked into a random pet shop in the area of Chicago, Illinois, and selected one of eight recently weaned, domestically raised, juvenile Grey Parrots to begin a most remarkable odyssey. Alex, the chosen one, has changed forever how we think about birds. He lived and worked with Pepperberg until he reached the age of 31. While I was writing the first draft of this chapter, the world received the stunning news of his unexpected and premature death on September 7, 2007. His loss to humanity, to science, and to Irene Pepperberg and his other human companions cannot be measured.

Alex had a career much like any adult human. He lived in Pepperberg's laboratory, and for some of the hours between 8 A.M. and 5 P.M. each day he worked for his living, by conversing with his human instructors and taking batteries of tests that they designed for him. His rewards included being able to play with the objects he was tested with, provided that he supplied a "correct" answer, and he was otherwise attended to faithfully by his human caretakers. Should he get bored with the tests and want to play on his gym, he had only to say "wanna go gym" and the laboratory assistant would offer an arm to transport him to his desired location. His parrot presence filled up the laboratory and dominated his relationships with all who knew him.

Like any adult human's career, Alex's work occurred in earnest. His tutoring and his accomplishments were not done for his or our amusement, as entertained as we might be by a parrot who talks to us and solves riddles in a genuine human language. Pepperberg's research goals were conceived during the inception of cognitive ethology, out of the heady quest for nonhuman thought. She would teach a parrot to use a human code, for the precise and sole aim of getting a bird to tell us what he thinks. Although her research methodology draws heavily from those in the fields of linguistics and human psychology, the study of Alex is not only about human language or about how parrots, or humans, communicate, though we learn much about those areas from her research. Her work is summarized in her fascinating book, *The Alex Studies: Cognitive and Communicative Abilities of Grey Parrots*, and presented in a legion of peer-reviewed papers.

Scientists and laypersons alike are astonished to hear a parrot speak his thoughts clearly in a human code. The studies summarized above tell us that his speech is accomplished using the same brain elements that vocal-learning mammals, including humans, use and the same cognitive elements that humans use, with the especially large brains that they share with parrots. The fables of Dr. Doolittle and King Solomon, among many others in human cultures around the world, came alive in a modest, unremarkable grey bird, in the context of carefully controlled and meticulously executed modern science.

Our Education of Alex and His of Us: Testing the Cognitive Abilities of Grey Parrots

The Approach

Pepperberg and her colleagues drew on the literature in human psychology and child development, primarily, to identify types of cognitive skills to test. These skills could be ranked from simplest to most complex, from easiest to most difficult, from acquired early (i.e. at a young age or developmental state in humans) to acquired later, from first order of abstraction to abstraction on multiple levels. Throughout these batteries of increasingly difficult tests, Pepperberg and her colleagues first had to crystallize the exact aspects of cognition that were being tested, pinning down minutiae of the characteristics to be probed for and their potential neural mechanisms (most of which were not actually known). They distilled these essences of criteria from exacting interpretations of the vast literature in linguistics, psychology, and other behavioral sciences, most of which were intended only for the study of humans.

Then the scientists had to design a true and unbiased test of the characteristic that they had isolated and about which they were hypothesizing. After they isolated a concept to test and defined it precisely enough, they created a test for that concept that could engage Alex and, most importantly, that could *falsify* whether Alex possessed that attribute of cognition. In other words, Alex had to have the potential to fail each test, so that his not failing pointed to his being capable of that attribute of cognition. Alternative and simpler explanations also had to be winnowed out by these tests. The parrot's responses were subjected to analysis using sophisticated statistics, because Alex made "mistakes." Among the first of many simpler hypotheses to eliminate is Alex's using his extensive vocabulary simply to get the right answer by chance, as I mentioned, or next, to get the right answer using cues from the experimenters or the environment, rather than actually thinking the problem through.

TABLE 3 Criteria for aspects of cognition tested in Grey Parrots. Adapted from Pepperberg (1999, 2006a)

Cognitive ability	Comments
Symbolic labels	Identification of objects, actions, and states of being/attributes
Category concept	Organizing similar objects actions or attributes (examples: color, size, material)
Comparison of categories	Using two categories to rank or identify objects, actions or attributes
Counting	Ranking numbers of objects, to varying degrees of precision ("1, 2, many" or "1, 2, 3, 4, 5, . . . n").
Algorithms and logic	Recursive and conjunctive tasks, following set rules and logical bifurcations (*and*, *or*)
Intention and planning	A goal that is mentally (consciously) formulated in advance
Referential and abstract communication	Using flexible language in communication and group problem-solving

As we review each of the cognitive abilities for which Alex and his fellow Grey Parrots were probed, follow along with table 3, which provides a summary of the basic tests.

Association Using Symbolic Labels

The first step in teaching Alex was to build a vocabulary of labels. A label would be a vocally produced, auditory signal that was a symbol representing some aspect of Alex's environment. A label could be attached to an object, living or otherwise. It also could be attached to an attribute of an object, such as its color, or to an action, such as "going," or to a motivation or state of being, such as "I want." To be symbolic, a label would be a sound that is exclusively and consistently used for one particular association and not for others.

The first step in acquiring a mode of communication is to associate a symbolic label with something. Pepperberg discovered that, just as for human children, labels are acquired by parrots more quickly if they are more relevant to the learner. That is, words in a language are learned faster the more functional they are, the more in-context they are, and the closer the social interaction between tutor and student. We therefore refer to a label as firstly and foremost symbolic and associative, and secondly, relevant to function and context. At minimum, a label is a signal linked to something in the environment that has relevance to the two communicating individuals, or else they would not be sending and receiving signals about this something.

The Concept of Categories

Alex easily acquired the understanding and appropriate use of dozens of labels for items, actions, and states of being, provided that such referents had functional or contextual relevance to him. Once a vocabulary was built, Alex could then be queried, to probe for more derived mental abilities than just association. The first of these possibilities to be tested was Alex's ability to abstract a *category concept*. Examples of such abstraction would be color, shape, or material. Color is a state of being, or attribute, of an object, at least as our brains interpret the wavelengths of reflected light (chapter 3). Alex could easily acquire correct labels for color by association, such as a green pumpkin seed, a rose cork, and so on. Once Alex could correctly describe an object's color, shape, or material, Pepperberg and her assistants began to ask him if he understood the concept of color, shape, and material. They could inquire by asking "What [object is] green?" or later by asking "What color [is this object]?"

In tests carefully designed to eliminate other explanations for his answers, Alex quickly mastered the ability to form categorical concepts and correctly communicate them to his human companions. In an unplanned demonstration of this ability, Alex revealed his curiosity about his own color when he was shown his image in a mirror. Here we are not addressing whether he knew that the image was of himself, only that he saw an object in the mirror that was a color for which he had not been taught a label. He queried the humans, "What color?" and got the answer "grey." After that, he could correctly identify grey versions of his repertoire of corks, keys, squares, triangles, hides, and so on.

Comparison Using Categories

Next, Alex's cognitive skills were tested for arguably yet more derived abstractions. Those scholars who ponder such skills in humans reason that logic increases in difficulty with additional steps in abstraction. First, a set of category concepts is established in the subject's repertoire. Next, the subject is tested for comparisons, which require that two forms of categorization be kept in mind simultaneously and then compared for yet another level of attribute relative to the categorization itself. Whether their reasoning fits how brains actually work to accomplish derived abstractions is not usually tested. That is, whether the derivation is more difficult to accomplish with the brain's neuroarchitecture than a simpler task such as association is typically unknown for most of these tasks. To be sure, we can imagine that extra steps of logic would be required to test these derivations, such as whether two objects are the same or different with respect to some category; in which of multiple categories do two objects differ; and whether the property of absence versus presence can be detected and communicated. Studies of development of cognition in human children suggest that these tasks are more difficult, because they take longer for children to master than simpler tasks involving fewer steps of reasoning.

Alex had no difficulty in identifying, and more importantly communicating, what property of two objects was different, when queried.

INTERVIEWER TO ALEX: "What different?" (showing him two keys, one yellow and one green).

ALEX TO INTERVIEWER: "Color."

INTERVIEWER TO ALEX: "What color bigger?"

ALEX TO INTERVIEWER (CORRECTLY): "Green."

In this exchange, you can see that Alex had to achieve several logical derivations of concept categories. The first was "color" as an attribute of objects. He had to go beyond just telling the interviewer what color a single key was, which would be simple 1:1 association. Rather he had to compare two objects with different colors and identify the category itself (color), demonstrating that he understood the abstraction of a category concept. Next he was asked to apply yet another category concept, relative size, and his response signified that he understood the value of the object in that category, that is, he understood "bigger." These kinds of tests were done from all angles of the problem, to ensure that Alex could not be using some simpler algorithm than was intended by the test. Because Alex had labels for five colors, four shapes, and four materials, the number of permutations possible in his answers was so large that testing his responses against chance (lucky correct answers) was easy. The odds against lucky guesses were so small that the experimenters could easily show using statistical conventions accepted by other scientists that chance alone could not explain Alex's performance on the tests.

Absence and Object Permanence

Mastering the concept of absence versus presence was similarly considered to involve extra steps of reasoning, again paralleling how normal human children acquire this concept. The first level would be quite familiar to any parrot owner: rejection. Rejection is the desire for the absence of something, which my own pet parrots practice with gusto on a daily basis. The next levels involve noticing that something is missing (nonexistence) or malfunctioning (absence of process), or that something has been dropped or has disappeared (has gone from presence to absence). To assess whether the subject has noticed that something is absent, linguists and psychologists require that the subject discuss the absence, rather than just notice it. A concept related to these levels of understanding absence is known as *object permanence* in psychological circles. Supposedly noticing, thinking about, and communicating about an absent object takes a greater level of processing than simply noticing only what is in an individual's environment at the time. A final stage of noticing absence is supposedly being able to distinguish truth from deception, a criterion that seems to be to be highly specific to those social species in which group members deliberately engage in deceit. Humans clearly fit the bill, but overall this trait has a spotty distribution among animal species in the specific sense of communicating deliberate lies (with or without conscious awareness). That this level should be included in interspecific comparisons of cognition probably reflects the paradigm of human-only studies and not that of comparative studies of communication.

Alex's tests for absence grew naturally out of those for other comparisons. He was taught the label *none,* and then he was given his usual comparison tests, in which one of the correct answers was "none [of the above]." Alex was able to accomplish this task also, as evaluated by increasingly complicated protocols, in which their complexity was proportional to the degree of abstraction required of Alex.

Counting

Next came tests of whether Alex could count and do mathematical computations, either by literally counting or having some cruder method (such as a shortcut) for perceiving quantity. In a short video on Alex's web site, you may see him looking at a plate of his objects, containing four blue blocks (square), two green blocks, and two green balls (round). In this test, his objects are all one material (wood) but two colors and two shapes. Within that category variation, the objects vary in number. Alex is asked "How many blue block?" to which he correctly replies "four." In other words, Pepperberg designed a test that she intended to be more than just asking Alex to count. She was trying to determine in part whether he could simultaneously process other tasks to count correctly. She was also trying to eliminate his use of pattern estimation to assess cruder types of quantification such as whether there are simply more blocks than balls or more blue objects than green.

Algorithms and Logical Associations

Adding so-called *recursive* and *conjunctive* tasks ratchets up the cognitive skills required to perform correctly on increasingly more exacting tests. A recursive task involves knowing a set of rules, or an algorithm, and applying these rules to solving novel problems. An example of a recursive task is the one described in the last paragraph, in which he was required to sift through category comparisons before being able to count correctly. A companion concept is *conjunction*, in which a comparison or a task such as counting requires an *and* condition, as in "blue and block." Alex clearly mastered this aspect of cognition, as revealed by the examples of Alex's tests that I have provided here.

Intention and Planning

Lastly, Alex was probed for perhaps the most abstract of all concepts, that of *intention*. What is intention? At one level, we might assume that Alex had intentionally given all of the answers he uttered, as evidenced by the high percentage of his correct answers—even his mistakes were highly revealing of either Alex's own abilities or the nature of the testing itself. I find that ethologists and psychologists alike struggle to articulate an objective, precise definition of intention and intentionality. Is simple goal-oriented behavior intentional? That definition would presumably be too broad, because intention should surely include a certain level of cognition and consciousness.

The definition of *intention* proposed by these scholars that most fits my preconception is "a goal that is mentally formulated." Such a goal has cognitive existence before it is attained and, importantly, before the goal-seeking behavior is initiated. Some might say that we know whether a behavior is intentional if we can ask the doer of the behavior what he thought about before he initiated the goal-attaining behavior that we observe. Cognitive ethologists often look for other indications of planning before imputing intentionality, but the entire point of Pepperberg's study was to be able simply to ask and get a comprehensible reply.

Evolution of Cognition in Parrots and Larger Patterns Across Animals

In my view, the approach taken by Pepperberg and her colleagues is ingenious as a method to probe the cognitive abilities of species other than humans. The paradigm under which most cognitive ethologists work, however, takes another approach, one that is not so anthropocentric in this one way. Cognitive ethologists seek to elucidate what animals can and cannot do cognitively regardless of a particular species' form of communication. In this approach, their task is to devise tests of cognition that do not rely on the attributes of or use of human language. Their approach also centers their search for animal cognition on the relevant ecological and evolutionary context for any given species, and most importantly not on cognitive attributes of humans.

Thus I turn now to asking how parrots apply their considerable cognitive skills in living their lives in the wild. With the knowledge gained by the remarkable findings of Pepperberg and her colleagues, we now know that we should be looking for such skills in wild parrots.

Although studies of Grey Parrots in the wild document that they copy sounds, the best evidence so far is in the form of their copying songs of other birds and bats that would be convincing to match using sonograms. Their use of environmental sounds that are more broadband and less frequency and amplitude modulated would create challenges for matching with a sonogram. As a result, scientifically rigorous evidence could be difficult to provide for the kind of behavior we see commonly in captive Grey Parrots. Yet, despite such challenges, on top of the usual difficulties of studying parrots in the wild, we must consider that Grey Parrots use copied sounds in the wild as they spontaneously do in captivity. The possibility exists therefore that wild Grey Parrots create labels that are mutually understood by close associates in small, stable social groupings, and further that the labels may be used in syntactical communication. Such communication might relay information that has high value for survival and reproduction, for example in predator detection and location of important resources. Christine Dahlin and Timothy Wright found that Yellow-crowned Parrots use syntax in their courtship duets, so the step to using syntax in more flexible communications is plausible in wild parrots.

Although animals without vocal learning can match the average vocal learner's vocabulary using other acoustical tricks such as biphonation (the dhole, for example), the use of vocal learning could considerably up the ante in complexity and flexibility of communication (see epilogue). Many social species transfer information from generation to generation through various kinds of social learning, a phenomenon referred to as *culture* by most evolutionary biologists. Some scholars parse the term culture into two terms, assigning *tradition* to mean social learning that does not accumulate over generations and reserving *culture*

FIGURE 50 A flock of Yellow-crested Cockatoos, *Cacatua galerita parvula,* evades a Brahminy Kite, *Haliastur indus,* on the island of Komodo in Indonesia. Photo © Mehd Halaouate.

to mean social learning that accumulates over generations and multiple domains of behavior, as occurs in humans and perhaps only in humans. Thus far, evidence supports that the social learning of parrots, and indeed most other non-human animals from cattle to primates, results in non-accumulating traditions. Nonetheless, developing a more a flexible and potentially complex communication code has sufficient potential value in terms of fitness that investigating what Grey Parrots do with their copied sounds in the wild is warranted.

Whatever their use of copied sounds in the wild, the composite picture that many studies paint of parrot cognition is one that is a common theme found in other animals that are social and intelligent. In one study, Cynthia Schuck-Paim, Wladimir Alonso, and Eduardo Ottoni demonstrated that the sizes of brains of 100 species of Neotropical parrots are directly correlated with the environmental variability of the region where any given species occurs. In other words, brainier parrots occur where the environment is more uncertain, as an evolutionary adaptation. Making a living from resources that are locally abundant but unpredictable in time and space is a considerable challenge that natural selection has met frequently with a certain combination of traits in multiple taxa. Intelligence and problem-solving

FIGURE 51 A male Salmon-crested Cockatoo, *Cacatua moluccensis,* investigating a nest hollow in the wild. Photo © Mehd Halaouate.

are perfect tools for meeting challenges in an environment that is not forthcoming with its riches. Sociality contributes because the group can be more than the sum of its parts. In particular, members of social groups share information on where rare and widely scattered resources can be found and how to evade predators (figure 50). This information can be shared among members in a concurrent group, such as a roost, and it can be passed along from generation to generation in the form of tradition or culture. Such social animals therefore need abilities for flexible communication as well as intelligence. Living a relatively long time then becomes valuable as a means of storing this socially learned information, and longevity is permitted because adults have a better chance at it through evasion of predators and calamities (chapter 6). The more information that is required for success with a particular lifestyle in a particular environment, the more living longer and sharing that information with offspring, relatives, and other group members should be favored by natural selection.

Parrots share all of these traits with other species faced with the same challenges, such as corvid birds (ravens, crows, and their relatives), whales and dolphins, some canids, elephants, and primates, that we know of so far. As handy as instinct can be when it is needed, flexible behavior generated by intelligence and problem-solving is more efficient and power-ful than behavior that is hardwired and rote.[2]

MENTAL DISTRESS IN CAPTIVE PARROTS

Freestone's Story

Freestone was a Salmon-crested Cockatoo *Cacatua moluccensis* (figure 51) whose life inter-sected with mine nearly twenty years ago. He had been captured in the wild, in the Maluku (Moluccan) Islands of Indonesia, probably Seram. He was then legally exported from Indo-nesia and imported into the United States in the heyday of international trade in wild-caught parrots (chapter 7), probably sometime in the late 1980s. Once out of U.S. quarantine, he was sold in the commercial pet trade to a well-meaning family of ordinary pet owners, much as if he were a goldfish. A man and his son tried hard to make him a family pet. They under-standably knew little about parrots in general and wild cockatoos in particular. The Salmon-crested Cockatoo is the largest of the white cockatoos and arguably the loudest. As typical parrots, they are highly social, consorting in often large flocks composed of mated pairs and their offspring (chapter 6). Salmon-crested Cockatoos bond with a single mate with whom they remain for life (chapter 5). As large, forest-dwelling parrots, Salmon-crested Cockatoos fly miles each day, searching for food and communal roosts (chapter 6).

Given all these factors, it is unsurprising that Freestone had a hard time of it in captivity. As is usual for a pet parrot, he was socially isolated. His human companions could not offer him what he needed. He was kept in a cage that was small by rain forest–dwelling, free-fly-ing cockatoo standards. He had nothing to do. He was given an excess of water and food, monotonous food at that, in bowls he could reach without even having to move. One day when he was let out of his cage, he put his idle beak to work dismantling the family's leather couch. That project was the last straw. The father and son realized that their bird was unhappy and that they could not fill his emptiness. In an act of responsibility motivated by a mixture of annoyance, guilt, and compassion, they took the bird to a local pet-store owner who had several Salmon-crested Cockatoos in her home aviary. She recognized Freestone's plight and offered him the promise of a better home.

I became involved in Freestone's life when I agreed to foster him with me for a while, until my friend could arrange better housing for him. I too tried to meet Freestone's needs, but it was difficult. Freestone took no pleasure in my company. He was unreceptive to the better and more varied diet I offered him. He barely ate anything, showing interest only in a part of a cob of corn and a couple of peanuts each day. Although he was free to fly and move about my house, he did not choose to go far. When he was let out each morning, he mostly sat on a corner of the top of his cage. He readily returned to his cage whenever I threatened

---

2. David Sloan Wilson argues in his book *Darwin's Cathedral* that open-ended processes can be more efficient than fixed processes in biological and ecological contexts and are therefore often favored by natural selection. Cognition, vocal learning, and social learning are examples of open-ended processes, in contrast to instinct.

to pay attention to him, so putting him away for the night was not difficult. He did not play with his parrot toys. He spent a little of his time at my house remodeling my furniture, but mostly he simply sat on that one corner of his cage. At dawn and at dusk each day, his forest music made the windows of my entire house rattle. When Freestone was "calling," my other parrots could be seen opening and closing their beaks, but none of their own screams and calls made it through the gargantuan sounds made by Freestone. Sometimes when I was out jogging around the neighborhood, I could hear Freestone from blocks away.

After a few weeks, Freestone began to wail at about 2 A.M. each night. His nightly descant was soulful and plaintive, nothing like his crepuscular choruses. Freestone acted and sounded miserable to me. His vocal protestations, I suspected, were attempts to contact companions, as he would do in the wild. Between his depressed behavior and these futile communications, his torment seemed palpable. After a week of interrupted sleep, I asked my friend if she could please take him back and offer him what I could not. Even if he were in a small cage while waiting for larger quarters, at least he might be able to see her other Salmon-crested Cockatoos and take some comfort.

For a while longer, he waited in a large wrought-iron cage in her busy pet store. He had one large perch going across the cage, within which hung some toys. He could not really spread his wings inside. While so housed, he would spend his day in an oddly rhythmic dance. His head would trace a three-dimensional figure-eight, first arching far back as if he had cerebral palsy. He would then swing his head to his left, then down to his front, bowing low on his perch. He would repeat the same circular motion to his right. As his head looped back and forth, he would turn slightly in each direction, shuffling on the perch with each foot. He occupied himself this way for long stretches of time. Another friend noticed him doing this when she was visiting the store and later asked me whether this behavior was normal. I said no, it was not.

Eventually Freestone moved to better quarters, slowly progressing toward our goal of finding him a compatible companion of his own species. I had always hoped that he would eventually enjoy some measure of contentment in his captive life. He did not. Before we could help him, the angels took pity on Freestone. They folded him into their arms one night, when he was housed next to parrots of other species during an outbreak of Pacheco's virus.

Abnormal Repetitive Behaviors

Wild animals kept long-term in captivity often develop repetitive patterns of behaviors that seem to fulfill no function and that are not observed in free-ranging individuals of those species in the wild. In ethological terms, *stereotypy* refers to invariant and repeated chains of behavior or speech. Animals kept in cages often develop stereotypic behaviors, hence they are often called *cage stereotypies*. Behavioral stereotypies are also observed in domesticated animals kept in close and artificial quarters in various forms of agriculture, when they are raised for human food, clothing, or amusement.

Georgia Mason and her colleagues have studied such behaviors in zoos and other captive settings where formerly wild animals are permanently housed. Stereotypies are specific examples of a class of behavioral phenomena now put under the broader umbrella of *abnormal repetitive behaviors* or ARBs for short. These behaviors are "abnormal" in the most neutral sense of not being normally observed in wild, free-ranging animals. That is, they differ quantitatively or qualitatively and can be statistically distinguished from behaviors routinely performed by wild animals. Also, they are abnormal in the sense that they often have no readily identifiable function even in the captive environment. These abnormal behaviors can sometimes be goal-oriented and resemble naturally occurring, functional behaviors, but even so, they are somehow inappropriate for the context. Feather plucking is an example of this other kind of ARB in parrots. ARBs are also abnormal in a third sense, in that they can be associated with other mental or physical dysfunctions, as we next explore.

The existence of ARBs is an obvious artifact of captivity, because they are always inappropriate for the context even when they derive from expected, goal-oriented behaviors. A human observer can recognize the abnormality of these behaviors, even if the person is uninformed about the natural behavior of the species to which the performer of ARBs belongs. Without much further information, one can easily venture the conclusion that the ARBs are somehow related to the performing individuals' unhappiness. Our human capacity for empathy, however, could lead us to this possibly incorrect and anthropomorphic deduction (see epilogue). Was I inappropriately assigning Freestone's behavior to a feeling I know that I would have, should I have to live under the same conditions as he did?

Studies of zoo animals demonstrate physical evidence of poor welfare in individuals that perform ARBs in captive settings, proportionate to the number and intensity of ARBs observed. This evidence is provided, for example, by higher levels of the excreted products of cortisol or corticosteroids, vertebrate hormones known to be associated with various forms of stress in human and animal studies, compared to levels observed under husbandry conditions where individuals do not perform ARBs. Other measurable correlates are attributes such as lowered rate of reproduction and higher rates of mortality and morbidity in captivity where animals exhibit ARBs, compared to animals in circumstances where ARBs are not seen. An important component of this evidence therefore points to captive husbandry as the instigator of ARBs. In large meta-analyses, Mason and her colleagues found ARBs to be consistently more common under certain types of captive husbandry. For example, animals living in unenriched and unstimulating environments exhibit more ARBs and greater degrees of stress indicators than those living in environments that are meant to stimulate the natural behaviors of that species.

Short of assigning the quality of *unhappiness* to the animal's state of mind, these trends confirm our suspicion that captivity cannot meet all of the requirements of formerly wild animals. Animals do not thrive when kept in barren, unenriched environments, where they are forced to be physically and mentally inactive and socially isolated, in a manner uncharacteristic of wild counterparts. Studies further show that in most instances, the ARB is a symptom of the animal's chronic stress. For all of these reasons, the conclusion of ethologists studying captive wild animals is that their caretakers should have zero tolerance of ARBs. ARBs are a symptom of a state of poor welfare in captive animals. Their arising in a captive setting is a red flag that the husbandry of that animal is inadequate, even cruel. For zookeepers, who often care for endangered species and who focus on public education, poor welfare of animals and the resulting abnormal behaviors are detrimental to any conceivable goal that the caretaker could have for captive husbandry. Owners of undomesticated animals, such as are parrots, also should find the correlates of their performing ARBs to be counter to every possible motivation for having a pet.

Among primates, the tendency to exhibit ARBs in captivity appears proportional to the intelligence, the level of cognition, and the social and self-awareness characteristic of that species, with 40 percent of great apes developing ARBs in captive conditions compared to around 10 percent for monkeys and prosimians. Parrots, as the bird parallel of hominid primates, are prone to various kinds of inappropriate behaviors in the apparently deprived environments in which they are typically kept. Many captive parrots live lives like that of Freestone. They are kept in social isolation, often quite deliberately so that they will develop into "good" pets. Even the most enriched of captive, pet environments will inevitably be spare and arguably barren for a parrot. Pet owners and aviculturists face the same pragmatic limitations as zookeepers. To simulate a wild environment, and then only in a paltry imitation, one would need an outdoor aviary the size of an airplane hanger so that larger parrots could actually fly. Even the most spacious aviary does not allow a bird the size of a Salmon-crested Cockatoo the distance to travel more than a few wing beats.

Parrots and Their Signs of Mental Distress

Many pet and other captive parrots develop two types of ARBs. The first type comprises the official stereotypies, such as Freestone performed. Such manifestations of behavior are centered typically in the basal ganglia, in the cortico-striatal circuit loop. Orange-winged Parrots *Amazona amazonica* kept in banks of cages in a research colony engaged in corner flipping. In this stereotypy, the parrots paced an invariant route from a high perch, grabbing onto the side of the cage and next its top, and then somersaulting back down to the perch. Caged carnivores, including polar bears, foxes, wolves and domestic dogs, perform similar circuits in their cages or kennels.

A second type of ARB commonly exhibited by parrots comprises inappropriate repetitions of normal, goal-oriented behaviors. The most common such behavior in parrots is feather picking, which is a manifestation of excessive and inappropriate grooming. Grooming is a normal, variable, and appropriate goal-oriented activity, necessary for proper maintenance of feathers. Parrots spend a considerable amount of time in the wild grooming themselves or their companions. In abnormal grooming in captivity, the parrots may destroy feathers or pluck feathers out by the shaft. For each feather so removed repeatedly, the follicle can become irreversibly damaged, and the bird become permanently bald in those areas. Cockatoos seem to be prone to the most extreme forms of over-grooming, in which the birds begin to pick away at their own flesh, creating gruesome wounds. Self-mutilating parrots can die of these injuries.

Cheryl Meehan, Joy Mench, Joseph Garner, and their colleagues have studied behavioral abnormalities in captive Orange-winged Parrots kept as a large research colony at the

University of California, Davis. These parrots as a population showed a significant incidence of ARB's in an experiment designed to study this abnormal behavior. Both kinds of ARBs occurred in the population. Fourteen percent of the birds exhibited stereotypies (cage flipping, route tracing, and sham chewing), and at least 10 percent, feather picking. Their research has revealed evidence that these two types of abnormalities arise as expected from similar deficiencies of husbandry, but they differ in cause and development.

This evidence for different etiologies for feather picking and cage stereotypies arises from various experimental results. Feather picking seems to have a genetic origin, in that the statistical incidence differed significantly among families. Stereotypies showed no familial relationships. In contrast, stereotypies were correlated with the parrots' performance on a "gambling" test that was designed to diagnose humans with autism and schizophrenia. This test is designed to reveal the behavior of stereotypy, which is to repeat a behavior without any influence of feedback as to whether the behavior is functional.

In the gambling test protocol, the subject is taught to try to predict which of two choices will yield a reward. The choices presented to parrots were two identical food cups that contained especially desired food treats. In the experimental protocol, experimenters adjusted the reward protocol to be biased against whichever cup had had more rewards in the previous set of trials. Overall, the reward rate of each side still approximated a random draw, and the best strategy was to consider both sides equally likely to yield a reward. In the human version of such games, the subjects are asked to discover the best strategy and tell the experimenters what rules they think govern the dispensing of rewards. In the parrot version, for which such communication would be impossible, the parrots were trained in such a way that they learned to anticipate the reward in any given cup. By this pre-training, parrots then revealed by their behavior (going over to the cup on a given side) which rules they thought the cups were following.

For parrots tested in this way, individuals engaging in higher-than-average rates of stereotypy (up to 80 percent of their waking hours) indeed performed more poorly on the gambling test. These parrots revealed their greater tendency to continue to repeat behaviors that did not serve a function or were more unrewarding than other behaviors. Because of this analogy with human medicine, and all that is known about human disorders, the scientists could conclude that dysfunctions of the basal ganglia are the likely source of stereotypy in captive parrots. Moreover, in humans, disorders of the basal ganglia are known to be associated with chronic stress, frustration, and suffering. Although we cannot easily ask the parrots to discuss their feelings with us, we can infer their possible states of mind from other data. For example, in the study by Garner and his colleagues, parrots housed next to the door where caretakers entered were more likely to develop feather picking than those housed in parts of the room where parrots perhaps felt less vulnerable. Parrots housed as singletons were more likely to develop cage stereotypies than those housed with even a same-sex, platonic companion; and singly caged parrots housed in isolation were more likely to develop cage stereotypies than those housed closer to their neighbors.

Once parrots, and other animals housed in deprived captive conditions, develop ARBs, these behaviors may well persist despite improvements in husbandry. This persistence is

FIGURE 52 A family of Blue-fronted Amazons, *Amazona aestiva,* consisting of a mated pair and two recently fledged offspring. Chicks will remain in close contact with their parents for several months after fledging. Photo © Mike Bowles and Loretta Erickson.

more common with cage stereotypies than with variable, abnormal goal-oriented behaviors such as over-grooming. Thus, the persistence of cage stereotypies that were pre-existing is not good evidence for currently poor husbandry. However, the feather-picking Orange-winged Parrots improved significantly in feather condition when the deprivation-treatment birds were given the same environmental amenities as the birds in the enriched treatment. In other words, once their environments were enriched and they were made to feel secure, feather-picking parrots often lost that habit completely.

Many parrot owners believe that the single life is best for causing their pet parrot to develop a bond with its human caretakers. Experiments on the Orange-winged Parrots, however, did not support that motivation for keeping parrots in isolation from conspecifics. Parrots kept singly were less likely to explore and play with toys. They exhibited higher levels of neophobia—fear of novelty and the unfamiliar. Parrots kept in pairs were more likely to engage with their human caretakers than were parrots kept singly. Parrots living in an enriched and normal social environment, paired up with individuals of their own species, were overall much more able to cope with the stresses of captive husbandry. In the controlled experiment performed by Meehan, Garner, and Mench, nearly 60 percent of the singly housed parrots developed cage stereotypies by the end of twelve months, while none of the parrots living with conspecifics did. In my albeit anecdotal experience with parrots over the years that I have kept under my own captive husbandry, I have found that only companions of the same species truly fill each others' needs. Furthermore, my companion Cockatiels and Budgerigars stayed friendly with me even after pairing up with conspecific mates.

This chapter has included much in the way of evidence that parrots are intelligent, sentient, highly social beings (figure 52). In the following chapters, we will discover how rich are their lives in the wild. Surely an enriched physical, psychological, and social environment is the least we can provide for them in captivity. After all, the decision to remove them from the wild was ours, not theirs.

PART THREE

The Lives of Parrots

Mating, Life History, and Populations

Sex and Marriage

The Mating Systems of Parrots

BEHAVIORAL ECOLOGY

The course of true love never did run smooth.

—*A Midsummer Night's Dream*, Act 1, Scene 2

CONTENTS

Introduction to Mating Systems and How They Evolve   147

Monogamy   148

   Patterns of Monogamy in Parrots   148

   Does Social Monogamy Mean True Genetic Monogamy?   151

   Why Life-Long Monogamy Means Being Really Choosy about Your Mate   155

Divorce in Parrots   158

Other Mating Systems   162

   Alternatives to Monogamy   162

   Polygyny in Action: The Lek Mating System of the Kakapo   163

   Polyandrous Mating Systems   166

   Other Possibilities   174

INTRODUCTION TO MATING SYSTEMS AND HOW THEY EVOLVE

Budgerigars (*Melopsittacus undulatus*) were among the first parrots to be kept as caged pet birds, and they remain among the most popular yet today. Now commonly called "budgies" or "parakeets," they were known as "love birds" in the nineteenth century. Today another type of small parrot, species in the genus *Agapornis*, formally sports the vernacular name of lovebird. In French, they are known as *inséparables*. Anyone watching a bonded pair of parrots can understand these monikers. Mates spend hours cuddling closely and allopreening, that is, preening each other (figure 53). A male will also feed his mate, a behavior that undoubtedly is borrowed from parental care but now is used to strengthen the pair's bond, as do other ritualized behaviors often specific to particular types of parrots. If they become separated, both male and female cry out with attention-getting, urgent sounds (chapter 3).

In this chapter we learn that the majority of parrots appear to be loyal life-long to one chosen mate, displaying a level of fidelity that humans value but often do not achieve. We also discover that there are exceptions to this life-long commitment to one mate among the parrots, as scientists discover an increasing number of alternate mating systems. These

FIGURE 53 A pair of the endangered Red-browed Amazons, *Amazona rhodocorytha,* allopreening in Linhares, Brazil. Mated pairs are the fundamental social unit in most parrot species. Photo © Carlos Yamashita.

various marital arrangements can be explained within the same rubric of evolutionary and ecological principles in parrots and humans alike. The common denominator is that members of each sex pursue a mating strategy that maximizes that individual's ability to survive and reproduce—what is called *fitness* in the scientific literature. We will discover that among parrots the fitness interests of the two sexes often converge, perhaps more often than they do in humans. Nevertheless, in other settings, parrot males and females may come into conflict in their respective pursuits to leave the largest possible number of copies of their genes in the next generation. This chapter explores the ins and outs of marital harmony and discord in parrots.

MONOGAMY

Patterns of Monogamy in Parrots

Fidelity to one mate, and true love, have been so extolled throughout the centuries in song, poetry, and religion that one might consider *Homo sapiens* to be the unique proprietor of these ideal qualities among the entire animal kingdom. Alas, we humans do not achieve this

FIGURE 54 A pair of Blue-winged Macaws, *Primolius maracana*, engaged in mutual preening to maintain their long-term pair bond. Photo © Luiz Claudio Marigo.

exalted state, officially called *monogamy*, nearly as well as most parrots.[1] Although relatively few parrot species have been studied in detail so far, most of these practice life-long monogamy that is both *social* (they consort socially with only one partner) and *genetic* (they produce offspring with only that partner and don't participate in extra-pair matings). Parrot biologists are increasingly overcoming the difficulties of following individual parrots in the wild throughout their long lifetimes. These studies are fleshing out the story of parrot marriage in nature, especially now that molecular methods easily reveal the secrets of how many males actually fertilize the offspring of each female.

Small and large parrots of all regions form pair bonds in the wild (figure 54) that last beyond rearing one brood, despite their spending the nonbreeding season in large social flocks. This consistency across the parrots is surprising, given that mating systems reflect ecological and demographic conditions and so often evolve quickly and may differ among even closely related species.

It is a general pattern across living organisms that smaller species have shorter life expectancies than larger species (chapter 6). We might therefore expect the smaller parrots to change mates more often for this reason alone, as they are more likely to lose mates to

1. Humans are classified as *facultatively polygynous* by anthropologists who study hominid behavior, meaning that we are sometimes monogamous and sometimes polygynous (males mating with multiple females).

FIGURE 55 A pair of Yellow-tailed Black Cockatoos, *Calyptorhynchus funereus*, feeds together on cones in Australia. This species has been shown to maintain socially monogamous pairs over many years. Photo © Steve Martin.

death. The turnover of mates between breeding seasons does appear to be higher in the small parrots. Only 60–70 percent of the adult breeders of the 23-gram Green-rumped Parrotlet survive between consecutive breeding seasons, meaning that up to 40 percent could be faced with finding new mates each year. Nevertheless, the rate of divorce (that is, leaving a mate that is still alive to find another) is estimated to be only 1–2 percent of pairs per year in wild Green-rumped Parrotlets. Other species of small parrots demonstrated to maintain pair bonds beyond one season include Orange-chinned Parakeets *Brotogeris jugularis*, Canary-winged Parakeets *B. versicolurus*, and Orange-fronted Parakeets *Eupsittula canicularis* in the Neotropics; the lovebirds themselves (the species of *Agapornis*) in Africa and Asia; and Budgerigars in Australia.

Among the large-bodied species of parrots studied in the wild, the great majority of pairs followed through more than one breeding season stay together until the death of a mate. This pattern has been documented definitively in the wild in the well-studied Australian cockatoos, including the Major Mitchell's Cockatoo *Cacatua leadbeateri* and the Yellow-tailed Black Cockatoo *Calyptorhynchus funereus* (figure 55), and in the equally well-studied Puerto

Rican Parrot *Amazona vittata*. Nevertheless, some of the detailed studies of Australian cockatoos in the wild show a background divorce rate as high as 15 percent, with the rate among younger individuals approaching 25 percent, as G. T. Smith of Australia's Commonwealth Scientific and Industrial Research Organisation discovered in his extensive work on the Western Corella *Cacatua pastinator*. Why parrots sometimes divorce should tell us more about why they usually do not, and we turn our attention to that topic shortly.

When a mate dies, and there are no chicks yet, the widowed bird abandons the nest. For most species of parrot, that action makes perfect sense. The female incubates the eggs full-time, and the male forages for both himself and her. When a mate dies after the chicks have hatched, the outcome of that season's reproduction is less certain. Widowed Major Mitchell's Cockatoo, Puerto Rican Parrot, and Green-rumped Parrotlet hens were observed to rear a brood to fledging by themselves, but a male who lost a mate usually (but not always) abandoned his babies. Even if the survivor became bonded to another bird that season, the new pair did not attempt a nest of their own in the remainder of that breeding season. The forming of a solid pair bond takes a long time in parrots (chapter 6). Thus, a newly formed pair is not likely to nest until the next breeding season at the earliest.

In a few instances, birds in these studies divorced a mate and re-paired during the breeding season. The individual stories are interesting. One divorce occurred when a pair of Major Mitchell's Cockatoos broke up after being goaded by a Galah *Eolophus roseicapillus* that had been raised as a foster child in a nest of the former species. The strategy of nest parasitism (purposefully laying eggs in another pair's nest) is very rare in parrots. However, in this particular case it did occur, and the Galah odd fellow ended up being raised by the Major Mitchell's Cockatoos as foster parents. Perhaps unsurprisingly, he seemed confused about his species identity when he grew up, and so he interfered with a normal pairing of two Major Mitchell's Cockatoos. A general lesson to be taken for understanding parrot divorce is that in some cases it may be driven by the determination of unmated individuals to change their status, particularly if there is a shortage of one sex, as we learn to be true of some parrots in the wild.

Given a preponderance of life-long, stable monogamy in species of parrots for which breeding behavior is known, we might first ask whether social monogamy means genetic monogamy in parrots, as has been explored in other socially monogamous birds. Then we can consider what "normal" divorce would be for parrots.

Does Social Monogamy Mean True Genetic Monogamy?

The Burrowing Parrot *Cyanoliseus patagonus* of Patagonia is a noisy, gregarious bird that was more widespread in Argentina before persecution by humans. Burrowing Parrots nest in large colonies, an unusual habit for parrots. They earned their vernacular name for their equally unusual habit of tunneling their nests into the sides of limestone or sandstone cliffs (figure 56), undoubtedly a handy means of avoiding nest predators (chapter 6). Because of the limited supply of suitable cliffs, this species often forms dense colonies, with many nests close together. This observation led Juan Masello, Petra Quillfeldt, and Thomas Lubjuhn to

FIGURE 56 A flock of Burrowing Parrots, *Cyanoliseus patagonus*, approaching a nest colony on a cliff face in Argentina. Photo © Carlos Yamashita.

ask whether the increased opportunity afforded by close proximity might lead the Burrowing Parrot to have higher levels of infidelity.

The researchers followed nearly fifty families over the course of two years, determining the genetic identity of over 250 birds, making their study the most thorough examination of parrot paternity in the wild yet done (figure 57). They found, much to their surprise, that not a single chick was a *nullius filius,* the product of an extra-pair copulation, also nicknamed EPC by scientists.

They did however find 2 chicks, out of the 166 examined, to be related to neither of the two birds raising them. With this finding, we may have genetic evidence of egg dumping

FIGURE 57 A pair of Burrowing Parrots, *Cyanoliseus patagonus,* at the opening of their nesting burrow in a cliff face. Despite the opportunities afforded by colonial nesting, this species has been shown to be genetically as well as socially monogamous. Photo © Carlos Yamashita.

and possible intraspecific nest parasitism occurring at a low level in one species of parrot. We know too little about this nest sharing and fostering of another pair's young in the wild to say much about why it happens or whether parrots try to prevent it from happening, as do parents in many species of birds that are routinely parasitized.

The study by Masello and colleagues was the first to investigate genetic monogamy in parrots on such a large scale of the population as a whole. A recent population study of parrots also done on an impressive scale is that of the Green-rumped Parrotlet *Forpus passerinus* (figure 58), in extensive scientific work by Steven Beissinger and colleagues over many years (more to come in chapter 6). In this socially monogamous species with a low divorce rate but high annual death rate, up to 7 percent of the chicks in a season are fathered by a male that is not their mother's mate. Up to 14 percent of families (offspring of a single pair) contain at least one chick with different paternity from the rest. Because Green-rumped Parrotlets do not nest in colonies, EPCs do not appear to be less likely in solitary-nesting parrots. Could they be more likely? Perhaps this rate of genetic infidelity is tied to the particular life history of an unusually small parrot, or perhaps it is more typical than we yet know.

FIGURE 58 A Green-rumped Parrotlet, *Forpus passerinus*, in Venezuela. This relatively short-lived parrot maintains socially monogamous pairs but scientists have recorded some infidelity. Photo © Eduardo Lopez.

Most of what we know about genetic monogamy in the animal kingdom comes from species other than parrots, but its application to parrots makes perfect sense. On this broader comparative scale, genetic and long-term monogamy is related to life span. When adults can expect to enjoy long lives and the young require care of both their parents, the degree of genetic monogamy tracks the life expectancy of individuals in that species or even within different populations of the same species. For example, in wild Canaries *Serinus canaria,* genetic monogamy is more likely on islands, where predators are absent and life expectancy is longer, than in populations of the same species on the mainland. This study on canaries favors the hypothesis that Green-rumped Parrotlets have a discernable rate of infidelity because of their life history and their lower life expectancy relative to larger parrots.

Parrots share this prospect of a relatively long life, the necessity for a male to assist in rearing his offspring, and genetic monogamy with other large birds. These include most geese and swans (Anseriformes), albatrosses, fulmars, shearwaters, and petrels (Procellariiformes), and some raptors such as the Black Vulture *Coragyps atratus* and the Little Owl *Athene noctua* (Falconiformes). The list of genetically monogamous birds is not very long, even in a taxon with a high frequency of social monogamy. Although monogamy is the norm in birds, the term typically applies for one breeding season only and in many species is more social than genetic. Life-long monogamy and true fidelity as practiced by parrots are rare even in birds.

Why Life-Long Monogamy Means Being Really Choosy about Your Mate

Given the high standard of fidelity to one mate held by most parrots studied so far, we can now consider its causes and consequences. First, we must ask why fidelity and life-long devotion happen in parrots in the first place, given that polygyny, with its philandering males, is the evolutionary default to which all other mating systems are compared. Second, we explore the choosiness in selecting a mate that is a consequence of the number of mates had by each individual in any particular mating system. That is, parrots face the ubiquitous trade-off between how many mates you intend to have and how selective you must be about each one. In a facultatively polygynous species (humans), Cyrano de Bergerac pled passionately to convince the discerning Roxanne that he was the most worthy of her favors. As we are about to see, the biological elements involved in the causes and consequences of monogamy are inseparably intertwined.

Pairing up in the first place is a significant investment of time and energy, all the more so the choosier an individual is about a mate. Those expecting to pair with only one mate are the most finicky of all, as they are literally putting all of their eggs in one basket. We find several interrelated and self-reinforcing characteristics that lead to the evolution of extended monogamy in parrots.

It all starts with the necessity of needing both parents to care for the young, which are born in an *altricial* (extremely undeveloped) state. The female knows that all the offspring are hers, because she laid the eggs. Males, on the other hand, have the ancient problem that they cannot be quite so sure. If a male is going to invest in care of one female's young, they had all better be his. Some scientists have proposed that some aspects of the strong pair bond serve as *mate guarding*, rather like a form of chastity insurance.

More importantly, if you are going to have only one mate in your long lifetime, and one partner to raise the few offspring that you will ever have, that mate had better be the best one you can possibly find, genetically and behaviorally. Therefore, choosiness is intertwined with the slow reproductive rate of parrots, as it is with the procellariiform birds, which is in turn part and parcel of long life expectancy. Parrots wait a long time to start reproducing, at least in the larger species (more in chapter 6). In part, this delay is caused by how long members of an incipient pair take to bond with one another before they attempt to raise young together. Small parrots, such as the Green-rumped Parrotlet, not surprisingly get right down to business, setting up housekeeping in the next breeding season after their birth, because they have relatively short life spans. Their haste should be expected to make waste, in the currency of possibly regrettable decisions about who to mate with.

In the much longer-lived and larger cockatoos, pairing up also starts toward the end of the birds' first year, sometime after fledging and weaning, when young-of-the-year gather in large flocks and become independent of their parents in all ways. Biologists were surprised to find out how soon wild parrots start pairing up and how long they consort as a pair before starting a family. Male parrots apparently need a little longer than females before they can become sufficiently good parents, by about a year. We learn more about these various life histories in chapter 6, but for now, we consider that pair-bonding in the wild begins when the females are relatively young—often in their second year—and continues for up to two or three years before a pair settles down and starts a family.

We know much about the bonding behaviors of parrots, because those are so easy to observe, both in the wild and in captivity. What we know much less about is the basis for this choosiness, that is, why a parrot chooses one individual over another. This question is particularly salient in the wild, where options are many in the large flocks that most parrots gather in between breeding seasons. In most mating systems, the differences in appearance and behavior between the sexes provide us with some clues. For example, the female in a polygynous system is the choosier of the two sexes. She picks her beau based on his ability to acquire good resources for her in competition with other males, or on his ability to transmit good genes to her offspring, which may be indicated by showy or exaggerated traits. Hence, females tend to throw in their lot with the larger or showier males. What clues do parrots give us?

As aviculturists know all too well, most parrots are frustratingly monomorphic and require surgery or DNA testing for us to determine their sex. Although we now know that parrots are more dimorphic in color than humans can see (chapter 3), some of the most strikingly dimorphic species of parrots, for example the Eclectus Parrot *Eclectus roratus*, which is dimorphic in color in the human-visible spectrum, and the Kakapo *Strigops habroptilus*, which is dimorphic in size, with males 30–40 percent heavier than females, are (guess what?) not monogamous (more on these fascinating cases below). Most scientific papers and treatises on parrots remark on the social monogamy of parrots and the lack of outward morphological differences between male and female parrots and leave it at that.

Biologists who have looked hard enough at the essentially monomorphic species of parrots have found some subtle differences between sexes and between individuals within a sex that may provide us those clues as to how a parrot decides to accept another as a mate (figure 59). First, we consider differences in size between male and female that might be instructive as a clue to mate choosiness and fidelity in each sex. In Burrowing Parrots, Major Mitchell's Cockatoos, Keas *Nestor notabilis* and Kakas *N. meridionalis,* males are somewhat heavier than females, about 5 percent—up to 15 percent in the Kaka. The males of shining parrots in *Prosopeia* and the *Platycercus* rosellas are also reported anecdotally to be somewhat larger than females. In Keas, Kakas, the Palm Cockatoos *Probosciger aterrimus,* and Ground Parrots *Pezoporus wallicus,* the size of the bill is markedly dimorphic (males have longer bills than females), and in Keas, it is disproportionately so, relative to the dimorphism in weight. For now we speculate that the presence of slightly larger males could betray some need to win the favors of a finicky female against the intrusions of other males. As a metric of a female's choice, a larger beak on a male suitor may predict a better forager for her and her offspring. These possibilities are as yet hypotheses, and the general pattern of very similar

FIGURE 59 A pair of Blue-winged Macaws, *Propyrrhura maracana,* displaying the red ventral patch that may indicate the physical condition of its owner, as it does in Burrowing Parrots, *Cyanoliseus patagonus.* Photo © Luiz Claudio Marigo.

morphology suggests that parrots of both sexes are equally choosy, as one would expect in a highly monogamous organism.

Masello and Quillfeldt found fascinating, and unexpected, evidence of the role played by a color patch in mate choice in Burrowing Parrots. This potential color criterion is monomorphic, as befits a genetically monogamous species. The scientists discovered that within bonded pairs of Burrowing Parrots at the breeding colony, members of a pair tended to be in similar body condition. Specifically, their relative weight and the size of their red abdominal patches varied together among the pairs in the colony. Parrots in better condition were heavier for their overall length, signifying that they were getting plenty to eat and had plenty of reserves to rear young. As a crucial link in the evidence chain, the young of parents in better condition when nesting started indeed fledged heavier nestlings. The red feather patch on the abdomen was larger and brighter in birds with better body condition, thereby serving as a signal of the potential quality of the bearer as a mate and parent (figure 60). No doubt by judging the size the red feather patch, male and female Burrowing Parrots in better condition tended to choose one another, like pairing with like.

The most obvious conclusion to draw from this study (and many others, of other types of organisms) is that a parrot chooses its mate based on clues as to the quality of a potential suitor and predictors of that other individual's ability as a parent. For example, female Budgerigars chose males by the characteristics of their contact calls in a way that predicted their

FIGURE 60 A pair of Burrowing Parrots, *Cyanoliseus patagonus,* showing the red ventral patch that provides "honest" information on that bird's condition. Photo © Mauricio Failla.

success as parents (chapter 3). This marvelous study on Burrowing Parrots is the first to demonstrate the tangible reasons for this choosiness in the wild. We find that the parrots sport a flag of their quality, "honestly" revealing to another parrot whether their prospective mate is worthy—unlike Cyrano, who hid in the shrubbery after dark to woo Roxanne. Or perhaps not? Perhaps Cyrano's eloquence and devotion to Roxanne were better indicators of his quality as a mate to the coy woman than was his oversized nose.

Divorce in Parrots

With this growing evidence of how parrots choose mates, we can begin to understand the grounds for parrot divorce. To that end, another anecdote of a rare divorce in monogamous parrots is particularly telling about what life is like for wild parrots. For many species of parrots, suitable nest sites are so scarce that pairs compete fiercely for them. In their book on the Puerto Rican Parrot (figure 61), Noel Snyder, James Wiley, and Cameron Kepler describe a dramatic fight between two pairs over a nest site known as the North Fork. The male of one pair vanished after the fight and was presumed dead, and the female of the other pair was blinded in one eye during the altercation. Although the victorious pair with the one-eyed female immediately took possession of the spoils, only a few days later the male abandoned

FIGURE 61 A pair of Puerto Rican Parrots, *Amazona vittata*. Pairs will fight heatedly for possession of a nest cavity. One member of this pair is wearing a radio collar as part of an ongoing study of the breeding ecology of this endangered species. Photo © Tanya Martinez.

her for the widowed female who had fought them so hard for that precious nest hollow. The biologists assumed that the disappearance of the one-eyed female could only be due to her death, but later they were surprised to find her in another valley, paired with another male— so surprised in fact that they had to check several forms of evidence before they believed that it was the same bird.

As with the one-eyed, injured female Puerto Rican Parrot, the condition of a mate may suddenly change for the worse. Heartless as it may be to dump a mate, wild parrots are governed by the hard rules of natural selection. At stake is their lifetime production of carriers of their DNA (otherwise known as offspring).

We might also now understand the parrot equivalent of the drama that plays out at every high school prom. As the most attractive individuals find each other quickly and pair up on the dance floor, the less desired are faced with a choice: settling for one of the other leftovers and joining the dance right away, versus investing time and effort in a possibly futile attempt at winning one of the beauties. Such a process is labeled *assortative mating* by scientists, who find it just about everywhere they look, in virtually every kind of organism, from single-celled Protista to plants and animals. After all, choosiness is at the core of one strategy for sexual reproduction, but so are trade-offs and expediency.

In other words, some of the individuals did not get the best mate they could, and they have a long time ahead of them to suffer the consequences of making do, such as the fickle male Puerto Rican Parrot. Some traits that might lessen an individual's quality could be

FIGURE 62 A family group of Cockatiels, *Nymphicus hollandicus*, near a water hole in Australia. A study of captive Cockatiels found that compatible pairs raised more offspring and were less likely to divorce than those that did not engage in affiliative behaviors. Photo © Jim Bendon.

life-long, as was the loss of one eye in the female Puerto Rican Parrot. Alternatively, a trait such as body condition or the red patch of Burrowing Parrots could be transient, either age-related or dependent on seasonal and annual variations in resources. Could any of these elements be the recipe for future divorces? In fact, biologists have found that in birds including parrots, as with humans, those individuals marrying young do have a higher likelihood of divorce later, the more so the longer they are expected to live. In other words, they have more time available to correct an error of foolish youth or expediency, and the consequences of not doing so have a longer time to build and grow.

A study of mate compatibility in Cockatiels *Nymphicus hollandicus* (figure 62) by Tracey Spoon, James Millam, and Donald Owings found some more clues about grounds for divorce in parrots. Their study simulated a natural situation as best as could be done in a controlled laboratory setting. Cockatiels from individual cages were introduced under varying protocols into flight cages with room and facilities for about ten pairs. Spoon et al. then observed how pairs formed and set up housekeeping. They found that bonding pairs differed from uncommitted birds by their closer proximity to one another, greater synchrony of behaviors (the male and female did the same things at the same time, such as eating, preening, and resting), more allopreening, more sex, and fewer conflicts with each other. As would be expected in a species exhibiting life-long monogamy, Cockatiel males and females did not differ much in their behaviors or roles in the bonding process. The slight differences that the scientists observed were that males initiated courtship more often than females and that the males were crankier toward other Cockatiels than were the females.

As pairs formed, Spoon et al. followed them through raising young and beyond. They found that pairs exhibiting more affiliative behaviors were more likely to stay together as a pair throughout the study. Importantly, these compatible pairs raised more chicks to fledging. Conversely, members of pairs exhibiting fewer affiliative behaviors were more likely to check out unpaired birds on the side, culminating in copulation with particular individuals that were not their mates (EPCs). Eventually, members of a pair parted ways when one of them engaged in extrapair copulations (which both sexes did, about equally). In each case of divorce, the newly single individual formed a more compatible bond with his or her former extramarital lover.

These interesting results tell us that somehow members of a pair are assessing compatibility and that the outcome of a favorable assessment is more affiliative behavior, or in the vernacular, public displays of affection. Under the experimental conditions, the relatively short-lived Cockatiels are not inclined to dally in forming pairs, yet not all first attempts are successful. If only we knew what characters each pair is assessing and, more importantly, why that matters. The importance of compatibility is clear, however. The more stable the pair bond, the more offspring are reared.

We are thereby presented with some quandaries that behavioral ecologists have spent considerable time debating. What can we infer about parrot monogamy and divorce? The portions of the theoretical debate most relevant to parrots suggest that we need to consider a list of costs and benefits. We have been discussing what biologists call the *better-choice option*, the benefits of which are self-evident. Never mind that an individual parrot is himself or herself unworthy, he or she still tries for the best possible future partner. Opportunities to do better may be rich if a parrot is continually confronted with easy comparisons. This shopping around is possible for parrots, as it is in birds with routine divorce, when pairs congregate in large social flocks during the off-season between annual chick-rearings (chapter 6). But the restless parrot must weigh that possibly better option against the cost of divorce, or conversely the benefit of fidelity, either of which can be considerable. Given that parrots rarely avail themselves of this seasonal opportunity to remarry, either the cost of divorce or the benefit of fidelity, or both, must be high.

As for costs, divorce means more courtship, which takes time and risks injury. Injury can happen if there is competition for the most desirable mates (fighting over Helen of Troy). Worse, a suitor risks death at the talons of predators, because courting, wooing, and jousting with other males are distracting, to say the least.

On the other hand, staying married has tangible benefits besides simply avoiding the costs of new courtship. Parrots may reap the same benefits from fidelity as another perennial and truly monogamous group of birds, geese. Studies of geese determined that a pair got increasingly better at rearing young the longer the pair were together. The male and female became practiced in working as a team to raise their offspring. If for some reason (usually death) a goose had to start over, even with an experienced partner, they were not as good at parenting as before, until several breeding seasons together. Researchers have discovered in several species of parrots that switching to a new mate costs at least one breeding season outright. Nevertheless, biologists have hypothesized that divorce should be more common, not less, in the longer-lived species. Skipping a breeding season represents a lower percentage of a long lifetime than of a short lifetime. Therefore, larger and longer-lived species should incur a lower cost to divorce—that is, unless staying together longer makes them better and better parents. In parrots, we indeed have evidence of the opposite trend: the shorter-lived parrots are more likely to divorce than the longer-lived, although the number of species examined remains small. We should therefore pursue testing the hypothesis in parrots that experienced pairs get better at raising young, the longer they are together.

Once quantified into a rigorous model, all these costs and benefits of marriage relevant to parrots predict that seldom or never divorcing should be much better than always divorcing. Always divorcing is what most birds actually do—they are monogamous for at most one breeding season. Thus the model says that most of the time, parrots should be life-long monogamists and never sequential monogamists.

That strong fidelity in parrots could be favored by natural selection is evident (perhaps ironically) when parrots find new mates after the death of a partner. Male parrots have been seen to court a female while she is still raising the young of her recently deceased mate and to help as if they were his own. Perhaps demonstrating that he is an excellent father is actually an aphrodisiac to the female. We have seen evidence in parrots that a strong and long-lasting bond with another individual has the prospect a high payoff, equally for both sexes and all the higher the longer the partnership. This payoff may be so generous that both the cost of fidelity in the relatively rare chance of widowhood and the benefit of getting slightly better sperm elsewhere through assortative pairing are negligible. Hence, divorce rates in monogamous parrots are low, and philandering is rare to nonexistent.

OTHER MATING SYSTEMS

Alternatives to Monogamy

The arguments to explain why parrots show such high fidelity are convincing enough that we may wonder why any parrot should not be monogamous, especially if they are large and long-lived. Yet, given the wide range of environments, habitats, and other ecological conditions under which adaptable parrots live, we should not expect so many varied species to have the same mating system. Mating systems directly reflect the immediate ecological setting in which a population is found. Both natural selection and sexual selection can be especially strong when acting directly on the production of offspring. Therefore, mating systems evolve rapidly. When more parrots are studied in the wild, I expect that we will find more exceptions to monogamy. Until then, some fascinating departures from monogamy in parrots are well worth exploring in detail. They are of interest not only in their own right but also for the lessons that they can teach us about all parrots.

Polygyny in Action: The Lek Mating System of the Kakapo

"But it doesn't look like a parrot!" says the title of an article on the Kakapo (figure 63) by Don Merton and Raewyn Empson. To that we can add, "And it doesn't act like a parrot either!" This oddest of birds, never mind just parrots, has the dubious distinction of perching so close to the edge of extinction that it is also one of the most intensively studied parrots, thanks to the devoted efforts of teams of biologists and wildlife managers in its native New Zealand. The story of its dire plight resides in chapter 7; here we delve into its remarkable mating system.

Perhaps fitting for such a bizarre bird, Kakapo happen to have one of the strangest ways to propagate. In the lek mating system, males hang out and display in a fixed location, and females come find them when they are ready to mate. That brief encounter is the only relationship that the sexes have together—a bond lasting minutes. Kakapo fit the picture of lek polygyny well, as we know it from a wide variety of other species of birds and some mammals. Although the Kakapo is unique among parrots (in just about every aspect), it is completely unremarkable among lekking species. A synopsis of its biology reveals many features that are standard fare in other lekking species, rare though they may be among parrots.

Kakapo are native to the dense, mossy southern beech *Nothofagus* and coniferous *Podocarpus* forests and sub-alpine scrublands in the isolated island-chain nation of New Zealand. Kakapo are the world's largest parrots, weighing up to four kilograms. They are far too heavy to fly, and so the Kakapo is the world's only flightless parrot, although not the only primarily terrestrial one. The diet of Kakapo is unusual among parrots in that they partake of the bounty of the dense forest, including the foliage, fruit, seeds, stems, buds, and roots of plants (chapter 2). Kakapo are solitary, another trait that may well be unique among parrots. Kakapo individuals are widely dispersed, presumably because each one needs a large area (about 50 hectares) to get enough food. Because they do not fly around to forage, more than one is a crowd if any expects to get enough to eat.

A combination of terrestrial living and their miserly folivorous diet then sets the stage for Kakapo to become unparrotlike, if not also unbirdlike. A solitary existence is obviously incompatible with life-long monogamy, in which two sluggish parrots would have to range twice as far for both to get enough to eat. Combine that unfortunate reality with life in a dense forest, where finding a mate every breeding season would be like looking for the proverbial needle in the haystack. Voilà, the recipe for the evolution of lek polygyny is present in just about everything unique about the Kakapo.

Also working in the favor of a lek mating system is the highly seasonal nature of breeding made necessary by the unpredictable fruiting of southern beech and podocarp trees, such as the rimu (*Dacrydium cupressinum*). These trees reproduce irregularly in a vast spasm

FIGURE 63 A Kakapo, *Strigops habroptilus*, named Sinbad. The flightless and highly endangered Kakapo is native to New Zealand, where it has evolved a polygynous mating system in which males display to attract multiple female mates. Photo © Stephen Jaquiery.

of fruit production known as masting (more in chapter 7). Only then do female Kakapo have enough nutritious, energy-rich food to produce a clutch of eggs and feed a brood of enormous, protein-craving young within walking distance of their nests.

Kakapo lekking falls in the category of a loose or "exploded" lek. This description best fits a group of males gathering more closely than they would when not interested in breeding but still not all crowded into a small area as with other birds that lek. Up to fifty males of the otherwise unsociable parrots pack into several square kilometers for several months. At best, the food supply is scant for this density of Kakapo, even in masting years. As a result, the males essentially fast for three to four months, during which time their breeding activities require a higher-than-normal energy output. Feeding in some abundance before moving to the lek grounds is essential.

Once gathered in the appointed location, each male creates an elaborate system of shallow depressions in the ground called bowls, each anywhere from 30 to 60 centimeters in diameter and up to 20 centimeters deep, which are connected by tracks. Apparently, the bowls act as sound reflectors to amplify the already loud, low-frequency booming that the parrots generate with their vastly inflatable thoracic air sacs. These sounds travel up to 5 kilometers to beckon females. The bowls also function as a site for his displaying when a receptive lady parrot appears. The bowls are meticulously prepared and maintained by the males. They are also fiercely defended from any usurping male who would rather skip the trouble of making a bowl system himself or perhaps prefers the location of another male's system. Fights between two males over bowls can continue for days and can end in death for one of the combatants.

Not surprisingly, male Kakapo are significantly larger than the females, around 30–40 percent heavier. This degree of sexual size dimorphism is otherwise unheard-of in parrots but run-of-the-mill in other polygynous species, especially those that lek or otherwise engage in intense competition for females. Female Kakapo, the object of all the fuss from the lekking males, come into breeding condition less often males and attempt to breed only every 2–5 years. The natural food scarcity ensures that females are dispersed widely in forest habitats and so are rare. Females ready to breed in any given year are rarer still. It would be pointless for males to search for females under these conditions, particularly because they are flightless and victims of scarce food supplies themselves. Searching for females— never mind their defense—would be too costly and overall a losing proposition. All the ingredients for the evolution of a lek mating system prevail in the Kakapo.

At first, desperate managers supplemented female Kakapo with food to encourage greater success at rearing young under these conditions, which were difficult at best even before humans arrived in New Zealand (chapter 7). A fascinating outcome of this food supplementation was the discovery of a possible reason for a male-biased sex ratio that had long puzzled scientists. Steven Trewick had discovered a ratio of two males to every female in prehistoric populations using subfossil material deposited before the arrival of the first Polynesian colonists to New Zealand; in the remnant populations extant at the turn of the twentieth century, this ratio had increased to six males for every female. This unusually male-biased sex ratio might be a side effect of a rapid population decline (more in chapter 7) or of a high mortality rate of females, which is common enough in birds, as we soon discover for Eclectus Parrots.

An alternate explanation, however, is intriguingly possible. In polygynous mating systems with large variation in mating success, females may try to manipulate the sex ratio of their offspring. They do so to maximize the return on their investment in the currency of fitness. When conditions permit, they invest in the more costly sex because they can expect a higher fitness return. Male Kakapo are more costly to produce, because males are larger than females. On the other hand, larger males are more successful in battles over bowls against smaller males, and females are known to prefer larger males. Male chicks have to grow faster to reach the larger size by fledging, and of course they need more food than female chicks. If chicks frequently risk starvation because of unreliable and variable food supplies, then for Kakapo mothers to produce expensive male chicks when food might be scarce would be throwing good money after bad, so to speak. Females cannot afford the considerable energetic investment of eggs, incubating them while having to feed themselves and then feeding the more demanding sons, only to lose them if the food suddenly gives out (there is no male to attend her as in monogamous parrots). Conversely, if the female is raising a family during a bonanza mast-fruiting year (which she usually is), then why not take advantage of the abundance to make more expensive sons and gamble that her son will be the big man in the lek? Should she spawn the one male that sires nearly all the chicks in that population for a generation, she has won the natural-selection lottery.

"What?" you say—how could she do that, even if she wanted to? In a number of bird species, females can apparently do just that, including several species of parrots, including Eclectus Parrots, Crimson Rosellas *Platycercus elegans,* and, as it turns out, Kakapo. A team of researchers working on the Kakapo then took advantage of years of supplementary feeding to test this hypothesis further. They found that females given extra food raised significantly more sons than did females raising their broods on natural food. The managers did not intend to provide an elegant test of evolutionary theory. Instead, these results presented them with the horns of a conservation dilemma, to which we return in chapter 7. The conservationists, as opposed to the Kakapo, considered the females the more valuable sex, because they are the ones producing more Kakapo. Most of the males sire no chicks at all (typical of polygynous mating systems). This skewed mating success made most males useless in recovering the population, while the more studly males were swamping the gene pool and reducing genetic diversity to potentially hazardous levels through inbreeding. The skewed mating success of polygyny might not matter in a large population, but it might further reduce a small population through the fatal consequences of inbreeding depression.

Polyandrous Mating Systems

Admittedly, the Kakapo is odd, but the Vasa Parrot *Coracopsis vasa* could give the Kakapo a run for its money in an oddness contest. If monogamous parrots evoke our admiration, and those with lek polygyny our curiosity and amusement, the mating system of the Vasa Parrot and other polyandrous parrots probably strikes us as downright kinky. A number of species of parrots, mostly those living on islands, exhibit some variation of the flipside of polygyny on the mating-system spectrum: polyandry. In these species, females choose intentionally and openly to seek the sexual favors of more than one male, and conversely, males agree or at least acquiesce to this arrangement. Parrots exhibiting polyandry include the Vasa Parrot and its congener the Black Parrot *C. nigra,* both of Madagascar; the Eclectus Parrot of New Guinea and surrounding isles and peninsulas; and the New Caledonian Parakeet *Cyanoramphus saisseti* (recently split from its congener the Red-fronted Parakeet *C. novaezelandiae*) and the Horned Parakeet *Eunymphicus cornutus,* of isles surrounding New Zealand.

These parrots have in common a particular manifestation of polyandry. First, their mating systems are actually polygynandry when examined closely—that is, both males and females mate with multiple individuals. Second, these cases also all seem to be a result of making the best of a bad lot, rather than a finely tuned system of cooperative polyandry such as we observe in other species of birds. In other words, some environmental contingency is making females or the resources that they require for reproduction so scarce that as a rule regular monogamy is not feasible for males. Perhaps this contingency aspect explains why all of the well-described cases of polyandry in parrots are species restricted to islands, either real islands or habitat islands. Let us now examine each case of apparent parrot polyandry to amplify these generalizations and better understand this mating system in parrots.

FIGURE 64 A calling Vasa Parrot, *Coracopsis vasa,* near Morondava, Madagascar. This island-dwelling species has a highly unusual mating system in which females display to attract multiple males as mates. Photo © Konrad Wothe.

The Second-Oddest Parrot

Vasa Parrots (figure 64) and Black Parrots are restricted to islands off the southeast coast of Africa, primarily Madagascar, the Comoros, and the Seychelles. Recall that Madagascar is distinguished among the world's largest islands by its long isolation. The breakup of the giant southern continent of Gondwanaland left it stranded from both Africa and India some 150–165 million years ago (chapter 1). The combination of its intermediate size, great age, and distance from sources of plant and animal colonization left Madagascar with a most remarkable fauna, with the vast majority of its often odd biota found nowhere else in the world. The *Coracopsis* parrots are good ambassadors for Madagascar because they are quite unlike most other parrots. Most of their lives Vasa Parrots sport the dullest of drab, dark plumage. They are proportioned somewhat differently from other parrots and overall characterized by a developmental quickness not typical of parrots, particularly given their large size. Their eggs hatch quickly, and the young grow unusually fast.

The polygynandrous family style of the Vasa Parrot was overlooked both in captivity and in the wild until very recently, although the risqué accoutrements of its mating system were all too easy to notice. Jonathan Ekstrom and his colleagues studied the Vasa Parrot in the Kirindy Forest of western Madagascar. There they found ready evidence of social and genetic polyandry, with an average of five or six unrelated males attended each female's nest and provided all of the food that she and her chicks needed for the entire breeding effort. Females mated openly with up to five attending males and copulated with multiple males frequently. Genetic analysis confirmed that the polyandrous associations were not only social. All the broods in their study were of mixed paternity, and in most of these, brood-mates were the offspring of at least three fathers.

Female Vasa Parrots engaged in some remarkable sex role reversal, but the basic necessity for her to lay and incubate the eggs did not change. As with nearly all parrots, the female did all of the incubation. The female Vasa Parrots did not leave the nest at all to feed during incubation. Unlike many other parrots with female-only incubation, the male Vasa did not enter the nest to feed either the incubating female or later the chicks. Instead, male feeding occurred as part of a peculiar ritualized competition, described below. Female Vasa Parrots were highly territorial, chasing other females out of their territories and males away from their nest cavities. Females did not leave their territories during the entire breeding effort. Because no food was found in the territories, the female depended completely on the males to find and bring them food.

Males were not faithful either, as congenial as they were in attending a female and her chicks with such devotion. Each male consorted with an average of three to four females and spent time around each of their nests, even though the females' territories were large and nests widely separated, presumably as a result of the intense territorial bent of the females.

Females therefore had the dual burden of competing for the male's favors (which included both sperm and food) and fiercely defending their nest sites and local consorts against other females. Female Vasa Parrots are 25 percent heavier than the males, which in reverse is a ratio worthy of the intensely competitive polygynous mating systems. Females also sang loudly to attract males from great distances in the dense forest. Singing commenced in earnest during egg laying and incubation, and it peaked when the chicks hatched and needed to be fed. Females that sang more frequently received more food from attending males. The effort to produce this song is considerable, so female Vasa Parrots incurred the various costs of courtship that usually fall on the males, in addition to the considerable investment of energy in producing the well-endowed eggs. The females in better condition were the ones that both sang more fervently and laid more eggs per clutch. Thus, the quality of a female's song was an honest indicator of her quality as a mate and mother of a male's offspring, with males now being the choosier sex. Neighboring females, perhaps those in less good condition, might perch on the periphery of another female's territory and intercept males attracted to the better female's song.

By the time the quickly growing chicks demanded large amounts of food, the female Vasa Parrot had transformed from her ordinarily dull physical appearance to one that was quite striking. The female Vasa Parrot lost all of her head feathers, and the skin thus exposed turned a bright yellow-to-orange color, quite a contrast against her dark plumage. One can only surmise that her appearance is supposed to attract males and keep them more attentive when she needs their provisions most. This kind of fancy ornamentation is certainly the norm for polygynous males in highly competitive mating systems, where it is the focus of mate choice by members of the opposite sex.

Black Parrots, the other species in the genus *Coracopsis*, may engage in a similar mating system, but one not quite so extreme as the Vasa Parrot's. Because mating systems evolve readily in response to local ecological conditions, study of the Black Parrot's variations on polyandry would be very helpful to understand why both of these species reproduce the way they do.

Interestingly, all the general taxonomic references refer to the two sexes as monomorphic in both species of *Coracopsis*. But if the orange head of Vasa Parrot mothers had not attracted much attention, the striking and unusual genitalia of the Vasa Parrots did not escape notice. Both sexes have an everted cloacal structure, which is large and highly vascularized in the males, transforming from flesh-colored to deep red during and after copulation. This quasi-penis, known as an *intromittant organ* in the few species of birds that have it, functions to produce a long-lasting tie, with copulation lasting up to an hour. After this considerable coitus, a liquid that is probably semen may be seen streaming out of the female's cloaca and, to a lesser extent, the male's. Full copulation does not occur more than once a day in captivity (where only one male is paired with a female), perhaps because such a large volume of semen takes that much time to replenish. In the wild, Ekstrom and colleagues found that the female and her various consorts engaged in both infrequent long copulations and frequent short ones (lasting only a few seconds and occurring many times a day). Around a third of these "quickies" resulted in intromission and presumably the transfer of sperm. A female and her paramours began to copulate up to thirty days before egg-laying commenced and continued through chick-rearing. Not surprisingly, however, most copulation occurred just as egg-laying began.

In stark contrast to the aggressive females, the many males drifting in and out of a female's territory hoping to inseminate her eggs and helping to feed her young were all easy-going fellows, tolerant of each other, though they must presumably have been sexual rivals. As in socially monogamous males, the male Vasa Parrot has no clues to determine the paternity of a female's chicks, except the knowledge that he has joined at least once with the female in that territory (if he has). Hence, like nearly all monogamous and polyandrous males, each male Vasa Parrot helped that female rear her chicks as if they were also his own. Quite in contrast to polygyny, polyandrous societies of cooperative males tend to be egalitarian. In some species of birds, the males may be relatives, but in the Vasa Parrot they are not, making their tolerance of each other all the more puzzling.

Similarly to the Kakapo, the Vasa Parrot's sex life may be truly bizarre for a parrot, but Vasa Parrots share nearly all their unusual physical and behavioral traits with other polyandrous and polygynandrous birds. In other words, the mating system is the common denominator to all of these associated traits in any given species and not its evolutionary history. Males of three well-studied polygynandrous passerines, the Dunnock *Prunella modularis*, Alpine Accentor *P. collaris*, and Smith's Longspur *Calcarius pictus*, also possess relatively enlarged testes and modified cloacas that serve as an intromittant organ, quite unusual for birds. Scientists do not yet agree on the reason that male birds, as a group, lack a penis, particularly when males of all other terrestrial vertebrates have one, including the ancestors of birds. Copulating without a penis must not only work, it must be advantageous, or else male birds would be better endowed.

No matter the reason for a general lack of male equipment in birds, the reason for males' having a functional penis in polyandrous mating systems is clear: sperm competition. The impressive size of the intromittant organ in the male Vasa Parrot is obviously related to delivering the maximum amount of sperm that his large testes can manufacture. The long coitus is designed to keep this massive volume of semen inside the female so that his sperm can get a good head start on where they need to go, in case she should mate that same day with another of her many husbands. The female Vasa Parrot in turn shares many sex role reversal features with females of other polyandrous species and males of polygynous species.

Ekstrom and his colleagues consider the role-reversed mating system of the Vasa Parrot a result of significant food shortage, although they point out that that explanation alone cannot satisfactorily account for all of the characteristics of the Vasa Parrot's behaviors. After all, if multiple males—not just one as in monogamy—are required to help one female rear a brood, then how do the males get away with assisting other females and their families on the side? These associated features are easy to understand in hindsight but beg the question of what set them along the evolutionary path to polyandry in the first place. The next polyandrous parrot that we will cover provides us with a possible solution to this mystery. Shortage of housing in combination with shortage of food may push the Vasa Parrot's mating system one more notch toward committed polyandry than is true of any other parrot. Let us learn more to explore this idea.

The Enigmatic Eclectus Parrot

To get a better sense of the diversity of sex role reversal mating systems that are grouped under polyandry, we now turn to the fascinating research on wild Eclectus Parrots done by Robert Heinsohn, Sarah Legge, and their colleagues. Eclectus Parrots come in an intriguingly intense and unusual dichromatism (figures 65 and 66). Males are unremarkably green (for a parrot), with some touches of contrasting colors, including a red upper beak and accents of blue, red, and purple in the wings, underwings, and tail, polished off with a yellow band at the end of the tail. Females present such a startling contrast that initially they were thought to be a different species. They are predominantly bright red over their bodies, with a black beak where the male's is red, purple-blue under their wings and around their mid-bodies, and yellow under their tails. Of course these colors are only those in the human-visible spectrum. As in numerous other species of parrots, both males and females sport patterns of reflectance in the ultraviolet that are certainly for sending sexual and courtship signals (chapter 3).

Any pronounced differences in appearance or structure between the males and females of a single species beg for an explanation in the mating system. Because specific patterns in sex roles are related to the type of morphological differences between the sexes, our first inclination is to examine how such dramatically contrasting males and females partition their sexual division of labor. Heinsohn and Legge did just that by studying their breeding

FIGURE 65 A female Eclectus Parrot, *Eclectus roratus,* at a nesting hollow in Cape York, Australia. The sex-role reversal of Eclectus Parrots extends to their plumage, which is bright red and blue in females and a more modest green in males. Photo © Steve Murphy and Brian Venables.

FIGURE 66 A male Eclectus Parrot, *Eclectus roratus,* foraging. Several males will attend and feed a single nesting female. Photo © Mehd Halaouate.

biology in a population of Eclectus Parrots in the Iron Range National Park on Cape York Peninsula, Australia. For four years, they followed over twenty nest hollows and their occupants, for a total of over 100 female-years, to determine exactly what their mating system is in the wild. First let us consider the general reproductive behavior of Eclectus Parrots and then return to the role of plumage color for both males and females.

Female Eclectus Parrots set up housekeeping in tree hollows that they guard fiercely and persistently from other females and to which to they are extremely loyal. Eighteen of the twenty-one females that Heinsohn and Legge followed returned to use the same hollow each of the four years. Their attendance at their selected nest tree lasted for up to eleven months, well beyond the time needed to rear the typical one clutch per year. Most females arrived at their nest hollows in July, no matter when they commenced laying eggs, and they stayed close to their nest hollows even after their chicks no longer need their brooding for warmth. Their close attention to the nest cavity was not simply neurosis, especially for one endowed with eggs or offspring. Female Eclectus Parrots have been observed to enter another female's cavity and kill the offspring therein, and females are known to kill each other over nest cavities.

As in most other parrots, female Eclectus Parrots incubate the eggs themselves. Their dogged presence at the nest hollow, in addition to the ordinary demands of sole incubation, required that they and their chicks be fed entirely by the males. Like the Vasa Parrots, the males were free to forage far and wide, while the female took on the entire responsibility of guarding her precious nest hollow from intruders.

Also like the Vasa Parrot, each Eclectus Parrot female was attended to and fed by several males. In these studies a particularly popular female was attended by seven males, but the usual number was around four. One successful mother was observed to mate with more than one male at her nest site, but matings were not as conspicuous in this species as in the Vasa Parrot. Males in attendance could be tolerant of one another, but more often they fought with each other at the nest's entrance, jockeying for position as they came to feed the incubating nest holder.

DNA again reveals more than observing behaviors at the nest. In the study by Heinsohn and Legge, males attending single nest hollows were not related to one another, similarly to Vasa Parrots. In contrast to the Vasa Parrot, paternity was not as widely shared among the attending males. Apparently, one male was the primary mate of that female, and he sired most of her young, usually 100 percent of them in small clutches of two eggs. Rarely did more than two males sire the chicks in a single brood. On the other hand, in their long-term study, Heinsohn and Legge found that in any given year, 11–60 percent of the nestlings had half-sibs in another female's nest. In other words, the males that were not the primary mate of a nesting female did not hesitate to exploit their opportunities elsewhere to propagate their genes.

This study then began to assemble a picture of a mating system driven by a serious scarcity of suitable nest hollows for a nesting pair, combined with a serious scarcity of adult females. In adult Eclectus Parrots in the wild, the sex ratio is two males for every one female. We need then to examine both of those contingencies—how they arise and what are their consequences.

Concerning the scarcity of nesting sites, the essential ingredient is finding that rare hollow that does not let rainwater flow onto the eggs and nestlings. Given that Eclectus Parrots are restricted to tropical rainforests, this is a tall order. Safe and dry hollows are so rare that once she has found one, a female dare not leave it. Thus, she stays put nearly all year, a residency quite unheard-of in other species of parrots. She stays at the nest hollow and displays her ownership, warning the other females to stay away by flashing her conspicuous colors. Her adornment in reds and blues is an intentional contrast to the surrounding green vegetation, signaling to others with tetrachromatic vision that can easily distinguish red from green (chapter 3). Another consequence of this requirement of defending her hollow is that she has to depend on males to bring her and the chicks food.

Males, in contrast, spend most of their time out foraging. After all, an adult female rarely feeds herself, so a male's responsibility in supplying her food vastly surpasses that of other parrots, even the Vasa Parrot. His colors are dominantly green so that he will not be seen by predators when he is rummaging in the vegetation for food. His ultraviolet portions, however, could be quite conspicuous to any animal that can perceive that part of the spectrum. So, he could very well be flashy, as he must be to woo females. To come to his potential rescue, a marvelous aspect of using ultraviolet for courtship is convenient for foraging male Eclectus Parrots. He can literally turn off his courtship colors by staying in the shade. Direct sunlight is the richest source of ultraviolet wavelengths, and by avoiding it, a male Eclectus Parrot can avoid reflecting in the UV part of the spectrum, making himself a relatively dull green. His red coloration also is silent if he is foraging, because most of the red can be seen only from behind and below as he flies.

This discussion introduces the other key variable necessary for explaining the Eclectus Parrot's mating system: the skewed sex ratio and scarcity of females. By puberty, females are half as abundant as males. This pattern is unusual. In other parrots, we see in the next chapter (chapter 6) that the sex ratio of males and females is about the same until females start committed reproduction, when her reproductive demands become more draining than those of males. In Eclectus Parrots, undoubtedly this process also applies, adding to an already short supply of females. Yet, what causes the dearth of female Eclectus Parrots before reproduction even commences? The answer seems to be differential predation.

Those brilliant reds and blues that female Eclectus Parrots sport are meant to be noticed, and along with their ultraviolet hues, females are easy targets for any animal with at least trichromatic color vision. A female with a nest hollow displays her warning signals in a conspicuous spot, to save her and other females the trouble of a confrontation at the nest cavity. However, a predator is just as likely to see her as another female Eclectus Parrot. Fortunately, a female defending a nest hollow can just dive inside, should a predator appear on the scene. Juvenile females do not have a refuge, unfortunately, and they appear to be highly visible to certain predators.

The likely culprits to detect red and blue parrots against a background of vegetation are probably Peregrine Falcons *Falco peregrinus* and Rufous Owls *Ninox rufa*. Unlike mammalian predators, which are usually dichromatic and so cannot distinguish red from green, falcons are tetrachromats like parrots, and owls are trichromatic, surprisingly enough given their bent. In other words, the likely avian predators are fully capable of detecting the intentionally conspicuous signal that female Eclectus Parrots display, making them easy pickings. Until she can find a nest hollow to call her own, a female Eclectus Parrot apparently plays Russian roulette with her choice of plumage color. Why has a more camouflaged coloration not evolved for hiding juvenile parrots, to be turned on by increasing hormones, as happens so often in other species? The answer is likely to be related to her need for those colors in the first place. Parrots take a long time to court and bond with a male (chapter 6), and these colors might be necessary for her to find her first and her primary mate. Also possible is that she begins to search for and defend nest hollows far in advance of puberty, because finding one takes as long as finding a suitable male in monogamous parrots.

Thus, in the wild, free-ranging Eclectus Parrots engage in polygynandry and moderate reversal of sex roles. Their mating system is similar to that of the Vasa Parrot, except that female Eclectus Parrots advertise with color and do not have a song. In both species, their polygynandrous mating systems fall short of the more specialized cooperative polyandry seen in some birds. The males and females of both Eclectus Parrots and Vasa Parrots appear to be making the best of a bad lot; otherwise they would stick with monogamy.

The New Caledonian Parakeet

Yet another island-bound parrot, the New Caledonian Parakeet, appears to have a mating system along the model of the Vasa Parrot and Eclectus Parrot. In the small but hard-won sample of observations collected so far, Jörn Theuerkauf of the Polish Academy of Sciences and his colleagues observed three nests attended by two males who both fed the female incubating eggs in each nest hollow. They did genetic analyses to establish the paternity of the chicks and found that brood-mates always had more than one father if more than one male attended the female at the nest. They also observed, however, what could be called the primary male guarding the female from the other male, by sleeping with her in the nest hollow and fighting with the other male when he entered the nest. These observations affirm our suspicions that monogamy would prevail were it not for a contingency, such as scarcity of females, the hollows in which they nest, food, or any combination thereof. This study also adds to the generalization that island ecosystems may be depauperate relative to those on the mainland, imposing shortages of necessary resources for their parrot residents.

Other Possibilities

Other studies hint at some other intriguing possible mating systems, but as yet formal observations by scientists are mostly sketchy or nonexistent. These variations include not only multiple males but also now multiple females, who share a single nest and the task of incubating eggs from more than one mother. We do not yet know enough about any of these cases to understand fully what is going on and therefore how and why such a mating system evolved, if indeed it did. These possibilities deserve mention here, however, to make clear that monogamy is not universal in parrots and that the evolution of mating systems in the Psittaciformes would be just as labile as it is in other organisms.

The Horned Parakeet, another resident of New Caledonia, was also observed by Theuerkauf and his colleagues. They recorded two instances of two females harmoniously sharing a single nest hollow, with two males apparently not quite so accepting of the other's presence. In both cases, there was evidence that in some years the pairs occupied that same nest hollow sequentially, but in other years the two pairs used it at the same time. At the least, we can come to the unsurprising conclusion that suitable nest hollows are scarce, especially on islands. Otherwise, we are left to wonder whether these observations are of another kind of polygynandry in action, or whether these parakeets are simply monogamous and making do with a bad lot, as best they can, with no particular adaptations for doing so.

In the mainland Neotropics, more intriguing observations have been made by scientists but not pursued yet with formal study. The Golden Parakeet *Guaruba guarouba* may have a cooperative mating system, as might the Sun Parakeet *Aratinga solstitialis*. In the Golden Parakeet, females appeared to lay eggs in a communal nest, attended by multiple males, in the one study done in the wild. Perhaps these consortiums are truly cooperative and not just communal nesting by individual but autonomous pairs, such as in the Monk Parakeet *Myiopsitta monachus* (chapter 6) or possibly the Horned Parakeet. If so, then some of the Neotropical parakeets will exhibit yet another polygynandrous mating system, resembling those of some species of woodpeckers, jacana, and gallinules. In such mating systems, related, cooperating females join their nests. Scientists would begin by testing whether the helpers at the joint nests of the Golden and Sun Parakeets are close relatives.

We have every reason to expect mating systems other than social and genetic monogamy within the group of more than 350 species of parrots. For alternative mating systems not to exist would be more surprising, as we have discussed. If further study supports social and genetic monogamy as the prevailing mating system among such a wide variety of species, then we can conclude that all of these parrots have much in common, despite the differences in their sizes and in their environments. A prevailing theme, a thread woven throughout the chapters of this book, is that parrots seem to represent a particular adaptive syndrome: that of organisms with relatively long lives, low reproductive rate, high investment in a few offspring, close attention to parental care, and other associated traits such as intelligence and social learning (see epilogue). This syndrome also happens to be one that is associated with monogamy, because males are needed as committed fathers and mates to send well-prepared offspring into the world on their own. In the next chapter, we turn to learning more about how parrots live their lives in the wild and explore this hypothesis further.

SIX

From the Cradle to the Grave

The Life Histories of Parrots

EVOLUTIONARY ECOLOGY

Every man's life ends the same way, and it is only the details of how he lived and how he died that distinguishes one man from another.

—ERNEST HEMINGWAY (Quoted in Hotchner 1966, xiv)

CONTENTS

Introduction to the Study of Life Histories   177

Life in the Nest Hollow   178

   Preparing for the Family: Cavity Nesting and How Parrots Go About It   178

   Nest Construction and Coloniality   180

   Eggs and Chick Rearing: Parental Care and Hatching Asynchrony   186

   Brood Reduction: Hedging Bets with Babies   192

      Box 5: The Short-Billed Black Cockatoo: A Sad Tale of Two Populations   197

Mass Recession: Parent–Offspring Conflict?   196

Life out in the World   200

   Fledging the Nest and Becoming Independent   200

      Box 6. The Green-Rumped Parrotlet: A Small Parrot Tells a Big Story   201

   Stories from the Wild: A Peek into the Lives of Some Wild Parrots   203

   Evolution of Mortality and Senescence   211

INTRODUCTION TO THE STUDY OF LIFE HISTORIES

The scientific field that seeks to understand the variation in life histories of all the different species on the planet is well summarized in Ernest Hemingway's purported view on the matter—with a little tweak. To paraphrase, nearly every organism's life ends the same way, with death, but the details of how organisms live out their lives distinguish one species from another. This science of life histories delves into the very meaning of life and death, as did Hemingway, by exploring answers to these questions: How does an organism best leave an evolutionary legacy in the currency of its individual fitness, the offspring that carry its genes into the future? What environments and contingencies make a certain set of choices better for one species and not for another? Why must there always be trade-offs, such as that between self and offspring, or between more offspring and more investment in each offspring? Why is death even necessary? We humans struggle in our daily lives and lifetimes with versions of these questions. They represent ageless dilemmas about life itself, as we began to explore in the last chapter.

The rich lives of wild parrots are being revealed to us by a growing body of scientific data, collected by devoted biologists who are not deterred from pursuing some of the most difficult animals to study in nature. The lives of only a handful of the more than 350 species of parrots have been studied thoroughly in the wild, but the research done so far has representatives from all major groups and from all over the world. Although there are the inevitable exceptions to any rule, fascinating by their oddness, the lives of parrots are remarkably similar regardless of taxon or geographic location.

In this chapter, we delve into the best-studied examples and explore the life of a parrot, from birth onward. The sections of this chapter feature each of the basic life stages that a parrot encounters as its life unfolds. The best-studied parrots take center stage, either in the main text or in boxes that integrate concepts from all parts of the main chapter. Let us now start with the activities of a pair, preparing life for their future offspring.

LIFE IN THE NEST HOLLOW

Preparing for the Family: Cavity Nesting and How Parrots Go About It

Parrots belong to a select fellowship of birds that nest in cavities: holes in a substrate, with solid walls and a small entrance. For most Psittaciformes, these cavities are in trees. Unlike the resourceful and better-equipped woodpeckers (Piciformes), parrots do not make their own cavities, with a few telling exceptions. Being a (so-called) *secondary cavity nester* is a trait that parrots share with most of the other cavity-nesting birds. Surely they prefer to move into an existing cavity because making an entire chamber is likely to be a costly and specialized activity. In the vast majority of species, parrots nest in ready-made tree cavities that they discover and take ownership of (figure 67). Clearly, owning a cavity is beneficial. Cavity-nesting species of all kinds lay on average more eggs per clutch than birds that do not nest in cavities, all else being equal. The relatively short list of parrot species that do not follow this rule is fascinating and enlightening. These exceptions include some species of parrots that construct their nests out of sticks (lovebirds of the genus *Agapornis* and Monk Parakeets *Myiopsitta monachus*); some that dig their nests in the face of cliffs (Lear's Macaws *Anodorhynchus leari*, Burrowing Parrots *Cyanoliseus patagonus*, and Pacific Parakeets *Psittacara strenuus*, formerly *Aratinga strenua*) or termite nests (some *Agapornis* lovebirds and *Eupsittula* parakeets); some that nest in existing rock cavities (Bahamian Parrots *Amazona leucocephala bahamensis*) or otherwise make do on the ground on islands (Rainbow Lorikeets *Trichoglossus haematodus flavicans*); and some that grub out spare depressions on the ground in dense vegetation (Kakapo *Strigops habroptilus* and Ground Parrots *Pezoporus wallicus*).

Nesting in a cavity must have great benefits for parrots for virtually all of them to make the evolutionary commitment to nesting in pre-formed holes. One likely benefit is protection from predators. Predation on vulnerable eggs and young is a grim reality faced by all parrots, including predation by humans removing chicks to sell into the trade in wild-caught birds (chapter 7). From an evolutionary standpoint, the question is how much greater that predation would be if the eggs were not protected and hidden in a nest cavity. Nest cavities

FIGURE 67 A parent Red-crowned Amazon, *Amazona viridigenalis,* feeding a chick in a tree-cavity nest. Photo © Mike Bowles and Loretta Erickson.

also clearly protect young from the elements, a necessity illustrated by the all-too-frequent loss of parrot offspring when the cavities fail to keep out rain and the eggs or chicks die of hypothermia, or drown.

Yet, studies of parrots in the wild suggest that this commitment also has costs. Chief of these is that suitable nest sites are evidently in short supply, judging by how seriously pairs fight for and defend them (chapter 5). The key word is *suitable*—not just any old hole in a tree will do. For the many large species of parrots, the cavity must be large, meaning that the tree also has to be large, and therefore old. The life expectancy of these perfect hollows is therefore short, because old trees continue their senescence and fall apart. And in this age of ubiquitous human activities elderly trees are found only in increasingly rare old-growth forests, where trees are left alone to die a natural death. More often than not, given all of these contingencies, holes in trees meeting the criteria for good parrot nests are few and far between.

Cavities form in trees by a variety of means. Those used by parrots are formed mainly by wear and tear on trees, and less often by former occupants that excavated the hole in the first place. Branches may grow too heavy and crack slightly at their bases, allowing fungus and bacteria to enter the tree's tissue. Termites and beetles also bore into living trees or work on dead, partially amputated branches. Once these inroads into the tree's integrity begin, rain-water scours out the cavity's interior. Now it is just right to be remodeled a bit by an ambitious parrot, perhaps to make it roomier or to improve its ability to shield the chicks from the elements.

Usually the male of a pair scouts out the cavity (figure 68). He explores and pokes around trees, testing the architecture of promising caverns in the tree's interior. Once a pair decides on a likely domicile, the female takes over the task of remodeling. She sits in the cavity, nestling where she will be laying eggs, and from that vantage point she chews away at the walls. As she molds and sculpts the natal chamber of her offspring, she generates their bedding of wood chips and debris. This process may take days, or in some species months, before the birthing room is ready to receive her eggs. When this housekeeping occupies the pair for a long time, the activity has other functions than preparing the nest chamber. Prolonged fussing with the nest cavity serves to solidify the bond between the male and female, or signal to other prospecting pairs that this territory is spoken for (chapter 5).

The aviculturists among you may rightfully wince when considering how parrots in the wild prepare their natural nest boxes. Imagine supplying your breeding pairs with a rotten log, and the chicks hatching in what is essentially a compost heap. Are wild parrots subject to some magic of life in Nature that allows them to rear chicks in an environment would be completely unacceptable in a fastidious aviary? Not at all. Wild cockatoo chicks in particular succumb frequently to some unidentified malady that is marked by sweet, putrid smell coming from the dying nestlings. Any aviculturist knows what that smell means: yeast and opportunistic bacteria. These microbes apparently cause an affliction in the wild as much as in the careless aviary, lending another disadvantage to nesting in secondary cavities.

Parrots that use existing tree cavities often defend their nesting areas as territories from which they fiercely expel intruding pairs. Parrot biologists have tried, with mixed success, to determine how many suitable existing cavities are not occupied in a given season. This information could shed light on whether nest sites are in limited supply or whether nesting pairs are more spread out than expected by chance, perhaps because they defend territories that encompass several suitable cavities and chase other pairs away. What has been found, though, is that occupied nest sites for parrots tend to be a bit aggregated in the environment, perhaps because trees with the best hollows occur in the same areas. Alternatively, parrots may be using the presence of other pairs as a quick estimate of habitat quality and choosing to settle in locations that other pairs have already deemed satisfactory.

Nest Construction and Coloniality

While answers to those questions are still forthcoming for most species, in some species of parrots, nesting pairs are conspicuously gregarious, obviously choosing togetherness over territoriality. The three taxa of parrots that build their own nest hollows are those with the most pronounced togetherness. The Monk Parakeet and some of the *Agapornis* lovebirds build their nests out of sticks and scraps of vegetation. The Burrowing Parrot does as its

FIGURE 68 A male Scarlet Macaw, *Ara macao,* investigating a nest hollow near Punta Islita, Costa Rica. Photo © Steve Milpacher.

common name implies, burrowing in the sides of cliffs (chapter 5). Jessica Eberhard of Louisiana State University was the first to notice that nest building and nesting in colonies go hand in hand in birds, and that association is certainly true in parrots. Once the birds can construct their own nests, as opposed to making do with existing cavities, they gain the advantages of being able to do what they want with their nests. We introduced the Burrowing Parrot in chapter 5 and will discuss them elsewhere in this chapter, so here we can delve into the other two examples.

The Monk Parakeet, Purveyor of Thorns

Monk Parakeets live in dry scrub forest and open woodlands in an area centered mostly in Argentina, spilling over into Brazil, Bolivia, Paraguay, Uruguay, and Chile. We know from other studies of wild parrots that suitable nesting trees for most parrots need to be large, and so they may be in short supply in woodlands, savannas, and scrub vegetation. Their choice of habitat would then leave Monk Parakeets with a shortage of promising natural cavities in trees. These logistics have convinced the biologists studying Monk Parakeets in their native lands, Enrique Bucher, Mónica Martella, and Joaquín Navarro, that this species turned to nest construction to make up for the deficit of suitable tree hollows.

In Argentina, Monk Parakeets favor twigs of the thorniest shrubs for the foundation material of their nests (figure 69). They prefer a sturdy tree or other platform for their loose and inherently unstable aggregation of sticks. Perhaps tellingly, their favored nest platforms in this age are *Eucalyptus* trees (introduced into South America from their native Australia), and telephone poles and other human constructions, a habit that often leads them to be considered pests. When suitable thorny material is supplied to flocks of Monk Parakeets in zoos and aviaries, the parrots fall upon the sticks like starving predators. Like all parrots, they employ their powerful beaks and opposable toes to great advantage and rip off twigs as fast as they can, carrying them in their beaks to the intended destination. In their native habitat, the constructed conglomerations of armored twigs are formidable, and it is not difficult to imagine that they deter many kinds of predators from entering the nest chambers.

Although impressively massive nests with dozens of chambers are seen occasionally, the average stick nest in the wilds of Argentina has slightly under two chambers. In other words, far from all nests are compound, and in some places most nests have only one chamber, although occasionally compound nests house four pairs, and rare nests are seen with up to a dozen pairs. In different locations in Argentina, solitary nests (stick nests with one chamber and used by only one pair) make up from one-third to nearly three-fourths of the nests in the population. In the famous feral population of Monk Parakeets established in Hyde Park, near the University of Chicago (chapter 7), solitary nests made up just under half the nests in that population. Thus, in most areas, compound nests are a minority.

Because they build their nests out of the thorniest sticks they can find, we have to consider predation the top hypothesis for bringing Monk Parakeets together. Native predators

FIGURE 69 A stick nest built by Monk Parakeets, *Myiopsitta monachus*, underneath the large nest of a Jabiru Stork, *Jabiru mycteria*, in the Pantanal, Brazil. Monk parakeets are rare among parrots in their ability to build their own nests rather than use existing cavities. Photo © James Gilardi.

on Monk Parakeets abound; they include white-eared possums and grass snakes. The biologists studying this population concluded that the stick nests are less safe than nest hollows in trees are for other parrots, an observation that seems to hold, on average, in our review of the scientific literature on wild parrots (excluding the human predator). Nest hollows, however, are simply not a reliable option in the arid scrub habitat that Monk Parakeets call home. The biologists studying the Burrowing Parrot note that the mortality of eggs and chicks in that species is particularly low, lending strong support to the predation hypothesis because Burrowing Parrots have the best of both worlds: a bona fide nest hollow in the cliff, which they themselves excavate, combined with nesting in colonies.

We might therefore postulate that safety from nest predators in the absence of trees with cavities, and perhaps other advantages, arise from their ability to nest colonially. For Monk Parakeets, there may be some social advantages, such as closer mentoring of the young, which might even stay around and help parents raise their younger siblings the next year. Biologists have observed trios building the nest chamber and caring for chicks, although no one is yet sure who that third bird is—an older offspring and sibling to the current young, or an unrelated male or female (chapter 5).

Monk Parakeets stay around the stick nest all year, maintaining and working on it. Monk Parakeets as a colony do not cooperate to build their nests, thus limiting the social advantages gained by other species of animals that live and breed together. Pairs within a compound often do the opposite of cooperating: they steal sticks from each other for their own personal chambers within the nest complex. Thus, Monk Parakeets experience a tension between acting in self-interest and the advantages of communal living. This tension is hardly unfamiliar to us and can be observed daily in human neighborhoods.

The Agapornis Lovebirds

The species of *Agapornis* provide a powerful test of these hypotheses linking nest building, colonial nesting, and the potential advantages thereof. The evolution of the species in this genus was elucidated by Eberhard using molecular methods (figure 70).

The most basal species in the lovebird clade, Grey-headed Lovebirds *A. canus,* nest in pre-existing holes in trees, as is typical of most parrots. In the next-most basal branch, Black-winged Lovebirds *A. taranta* and Red-headed Lovebirds *A. pullarius* nest in tree holes and burrow into termite nests, respectively. These species thus share the nesting habit of the majority of parrots, using existing holes and also burrowing into termite nests, a habit that, while not extremely common, is found in several parrot lineages. All of these species collect nest material, such as bits of vegetation, to line the nest cavity.

The phylogenetic tree then branches into a clade containing the Rosy-faced Lovebird *A. roseicollis* and four closely related forms all possessing white rings around their eyes. These variant populations have been named *personatus, nigrigenis, lilianae,* and *fischeri* and variously classified as species of *Agapornis* or subspecies of the nominative form, the Yellow-collared Lovebird *A. personatus* (figure 70). Females of all of these taxa build their nests entirely out of material that they collect, and the two major branches in the clade illustrate a

Lovebirds _(Agapornis)_

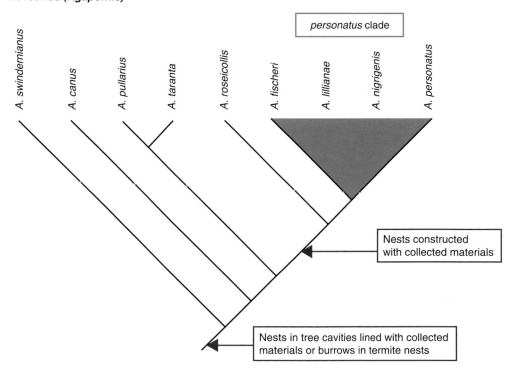

FIGURE 70 Phylogenetic tree of the lovebirds, genus _Agapornis_, illustrating the evolutionary steps from the ancestral state of using tree cavities to the derived state of constructing free-standing nests from sticks and other materials. Figure by Jessica Eberhard.

progression of nest building. The nests of _A. roseicollis_ are simple cups made of dry bits of leaves and other lightweight material, whereas the nests of the various populations of _A. personata_ are fortified by sticks and completely domed.

The overlay of these patterns of behavior on the independently derived phylogeny supports the hypothesis that nest building evolved from the habit of collecting material to line a more traditional parrot nest hole, a trait possessed by the ancestors of _Agapornis_. The nest _builders_ arose from the gene pool of ancestral nest _liners_, with the forms _A. roseicollis_ and _A. personata_ sharing a common ancestor. Perhaps the recent rapid radiation of forms engaging in the most elaborate nests is significant. Evidently, not relying on existing nest cavities was an advantage to these populations, which diversified in the arid eco-regions of Africa's Great Rift Valley. These parrots live in the dry shrublands, open-canopy woodlands, and savannas without large trees and thus without available nest cavities, much like the native environs of Monk Parakeets. Other correlations are interesting and beg more explanation. The nest-lining species are monomorphic in color (at least in the human-visible part of the spectrum) and solitary, with pairs engaging in fierce defense of their nests. The nest-constructing species are dichromatic (males and females can be told apart by coloration) and colonial.

Eggs and Chick Rearing: Parental Care and Hatching Asynchrony

Getting Started

Eventually, all nests receive their precious cargo when the pair settles down to lay eggs and raise that season's brood (figure 71). Parrots of the various species differ in the numbers of eggs that they lay in a given breeding effort, which biologists label a *clutch* for short. What determines how many eggs a female of any given species lays is a mixture of her evolutionary background and more immediate constraints with a female's lifetime. She has some control, although perhaps not conscious, over the size of her clutch. Average clutch sizes in parrots range from one (the Palm Cockatoo *Probosciger aterrimus*) or two (for example, the *Calyptorhynchus* cockatoos and many species of lorikeets) to eight eggs, or even more, in the small parakeets. Within that average for a given species, clutches vary in the number of eggs a female lays, depending on her age and her condition.

Female parrots seem to set the number of eggs they will lay that season early on in the laying process. By the time the eggs are actually in the pipeline, she does not compensate for egg loss or other contingencies by laying more or fewer eggs in a given clutch, in most species of parrots studied in the wild. Wild parrots may lay another clutch after losing all the eggs in the first clutch, but re-nesting within a single breeding season is the exception rather than the rule.

Female parrots begin to incubate soon, if not immediately, after the first egg is laid. The obvious consequence of immediate incubation is that the eggs do not all hatch at once. Rather, they hatch roughly in the order and timing of their being laid. This *hatching asynchrony* is feasible for parents that raise altricial young, because the young do not leave the nest immediately after hatching—they are far too helpless. In fact, spreading parental duties by having young not all the same age may have great advantages for the parrot parent. One idea for the adaptive benefit of hatching asynchrony is the *insurance hypothesis*, which proposes that a female lays the later-hatching eggs as a form of insurance in case some of the earlier eggs or chicks die from disease, predators, or inclement weather. A more complex alternative, discussed in detail below, is the *brood reduction hypothesis*, which proposes that the later chicks are an adaptive response of parents to the inherent unpredictability of food supplies. In this scenario, later-hatching chicks are a form of bet hedging; if conditions are good they can be provisioned adequately and survive, but if conditions are bad they may be allowed to die, thus cutting the parent's losses. Regardless of the adaptive value of hatching asynchrony, it is clear that its costs and benefits differ between parents and chicks, often putting parents and their offspring in conflict. First, let us consider the basic job of parenting, and then turn to how that might vary among first-, middle-, and late-hatched chicks.

A View from Outside the Nest Cavity

Nest attendance of wild parrots has been studied for relatively few species. We know from these species that patterns of attending eggs and chicks vary somewhat among species. In most species, only the female incubates, broods, and feeds the young chicks for the first week

FIGURE 71 A female Blue-fronted Amazon, *Amazona aestiva*, in her nest with
a newly laid egg. Photo © Igor Berkunsky.

or two after the first egg hatches, and the male's contribution is to provide food for his mate
and their young chicks during this early phase. These species include the Short-billed Black
Cockatoo *Calyptorhynchus latirostris*, Glossy Black Cockatoo *C. lathami*, Ouvea Parakeet
Eunymphicus cornutus uvaeensis, Meyer's Parrot *Poicephalus meyeri*, Crimson Rosella *Platycer-
cus elegans*, Bahama Parrot *Amazona leucocephala*, Black-billed Parrot *A. agilis*, Puerto Rican
Parrot *A. vittata*, Lilac-crowned Amazon *A. finschi*, Yellow-crowned Parrot *A. ochrocephala*,
Burrowing Parrot *Cyanoliseus patagonus*, Green-rumped Parrotlet *Forpus passerinus*, Thick-
billed Parrot *Rhynchopsitta pachyrhyncha*, the *Agapornis* lovebirds (figure 72), Eclectus Parrot
Eclectus roratus, and Vasa Parrot *Coracopsis vasa*. In other species (all cockatoos), both parents
share incubation, brooding, and feeding the very young chicks. These species include Galah

FIGURE 72 An adult Red-faced Lovebird, *Agapornis pullarius,* feeding a chick. Allofeeding of chicks by parents is a ubiquitous form of parental care in parrots. Photo © Sherry McKelvie.

or Rose-Breasted Cockatoo *Eolophus roseicapillus,* probably Long-billed Corella *Cacatua tenui-rostris,* Major Mitchell's Cockatoo *C. leadbeateri,* and Cockatiel *Nymphicus hollandicus.* In still other species, the male's attentiveness to eggs and young chicks varies with his other demands. If he does not have to forage too far, or to be vigilant against threats to the nest, he can spend more time attending the offspring with his mate early on. These species include Orange-fronted Parakeet *Eupsittula canicularis,* extra males of Eclectus Parrots and Vasa Parrots, and perhaps many other species currently thought to have female-only incubation.

Most of the remaining species of parrots have raised young in captivity, and their reproductive behaviors are well known to aviculturists. Although the general patterns of attentiveness to eggs and young are probably similar to their wild counterparts, captive parrots clearly do not have the same demands of finding scarce food and constant vigilance against predators, a luxury that might affect how parents of both sexes attend the young.

Life inside the Nest Hollow
Biologists studying wild parrots do not have it as easy as aviculturists in observing the most intimate moments of parrots' lives. A few scientists have successfully used fiber optics and tiny video cameras to observe what goes on in the privacy of the nest hollow in a few species, such as in Juan F. Masello and Petra Quillfeldt's study of Burrowing Parrots and Elizabeth

Krebs's work on Crimson Rosellas (figure 73). Most, however, have had to make do by hiding close to the nest tree, observing with binoculars, and listening carefully.

Thanks to Krebs's painstaking work on Crimson Rosellas, we now know amazing details about how wild parrots raise their young. Her findings agree with the equally detailed work on captive Budgerigars *Melopsittacus undulatus* by Judy Stamps, Anne Clark, Pat Arrowood, Barbara Kus, and their colleagues and students. What these scientists discovered are details that no one had ever imagined. Next we look into these two examples in some depth.

Crimson Rosellas

In Crimson Rosellas, the female does all the incubating of eggs and brooding and directly feeding the young until the brood is around five days old. The male, as usual, forages for her and feeds her through this period. Crimson Rosellas delay incubation more than other parrots are known to do, so that the total hatching interval is half a week, significantly shorter than the interval over which the three to eight eggs are laid (one every other day). When the oldest chicks are about five days old, the male enters the nest and joins the female in feeding the young directly. During a bout of feeding the brood of chicks, the parents regurgitate food from the crop in an average of fifteen food transfers, distributing it among the chicks. Chicks beg and position themselves for feeding during a visit by the parents, but otherwise tuck down into the nest and remain quiet. Father and mother attend the nest and chicks equally, and once brooding of the chicks is no longer necessary during the day, the two parents usually enter the nest within minutes of each other to deliver their load of food to the chicks. That is where the equality ends.

Male Crimson Rosellas are somewhat larger than females and so carry proportionately more food to the nest per load. On average, the parents delivered around 10 grams of food per visit and up to 25 percent of their body weight with the largest loads. Curiously, the average amount of food per parental visit did not depend on how old the chicks were, nor how big the brood was, but parents brought more food to the nest the more spread-out the chicks were in age. Parents brought 12–14 grams of food per visit to broods with a week's difference in age of the youngest and oldest chicks, and only 6–8 grams per visit to chicks that were only one or two days apart in age.

Male Crimson Rosellas concentrated on feeding the oldest chicks, whereas females fed chicks of all ages equally, at least when food was abundant. The sex of the chick as well as that of the parents mattered. First-hatched male chicks received more food than later-hatched male chicks and female chicks of all ages. Putting the patterns together implies that male parents tended to favor first-hatched male chicks over the others. Overall, however, Crimson Rosella parents were remarkably egalitarian to the chicks, and even the last-hatched chicks were fed as much as they needed, so chicks grew at the same rate regardless of hatch order.

This picture of family bliss (with a dose of sexism perhaps) changed when food became scarce. Krebs simulated food scarcity in her wild subjects by removing individual chicks or whole broods from the nest to deprive them of food long enough to make them hungrier than usual. She then recorded the reaction of male and female parents to the especially

FIGURE 73 A Crimson Rosella, *Platycercus elegans*. Mother and father Crimson Rosellas have different strategies for allofeeding their young. Photo © Bent Pedersen.

hungry chicks. When she caused one chick at a time to be hungrier than its brood mates, fathers were persuaded to favor hungry last-hatched chicks, the opposite of what they did in control nests. The mother, in contrast, was resistant to the demands of the hungrier chicks. Unlike the father, she did not alter her food allocation in response to one hungry chick, but rather moved around to make sure all were fed.

When the entire brood was made hungry experimentally, the parents were not able to compensate for this increased hunger, and all of the chicks lost weight relative to controls, just as would happen under a natural food shortage. Under these conditions, the mothers fed the first-hatched chick preferentially. The fathers fed the young regardless of age, perhaps because fathers respond readily to begging, to which the mother is oblivious.

Krebs concluded from her experiments that fathers tended to adjust their feeding of chicks more according to short-term changes in hunger than did mothers, who tended to see that everyone got fed while averaging over short-term changes in food abundance that might affect only one chick in a brood. The mothers in particular chose a strategy of ensuring that the last-hatched chick was fed when times were plush, but she changed to favoring the first-hatched chick in hard times. This behavior is the key ingredient in a widespread phenomenon known as *brood reduction*. Under this strategy, parents start out with an optimistic clutch, laying as many eggs as they can raise to fledging under favorable conditions, but when times get tough, they let the youngest chicks go hungry.

The chicks in her experiments had their own ideas about how things should go. Last-hatched chicks in a brood always begged more than anyone else, no matter how hungry they actually were. In the end, however, the chicks' begging did not make all that much difference in how parents of either sex allocated food, given all these complex patterns of allocation by each parent in response to changes in food abundance.

Budgerigars

Before we try to conclude what this diverse array of behavior means, we should visit the research findings on captive Budgerigars. The small size of budgies has made them convenient for behavioral studies in the laboratory because they can easily be kept in cages large enough to approximate natural conditions. The data on captive Budgerigars show us a pattern of parental care remarkably similar to that of the wild Crimson Rosellas. This concordance suggests that even parrots as domesticated as Budgerigars have not moved far from their wild ancestry. It also allows us to have confidence that controlled studies in captivity, far easier to perform, inform us reliably about parrots in the wild.

First let us review what we do know about Budgerigars in the wild. They breed in relatively large groups and appear to need to hear other Budgerigars singing to trigger physiological readiness for breeding. Scientists refer to this behavior as *socially facilitated* breeding. Because Budgerigars live in large flocks, up to tens of thousands when not breeding, getting a couple of dozen nests together for a breeding colony is not a major undertaking. Budgies move seasonally over enormous areas, giving them a reputation for nomadism. Edmund Wyndham studied Budgerigars breeding in the wild in eastern Australia and found their movements to be predictable on a large scale, if not locally. Their breeding is irregular and opportunistic, taking advantage of local abundance of food and water where Budgerigars can settle in for at least three months. In his study, Budgerigars moved into the area in mid-summer, January and February, and fledged their young by late April or early May. Local colonies could number up to twenty or twenty-five nests, with many in the same tree less than a meter apart. These tiny parrots nested in the small holes numerous in coolibah trees (*Eucalyptus microthea*), which they barely modified before moving in and setting up housekeeping.

Wild Budgerigars lay on average between four and five eggs, with some clutches as large as seven eggs. As is usual in parrots, a female Budgerigar lays an egg every one to two days. Evidently she delays incubation briefly, because on average the chicks hatch over roughly four days, still a respectable amount of hatching asynchrony considering that young fledge in four weeks. What goes on in the privacy of their wild nest holes has not been seen, but the ease of observing them in captivity has revealed those secrets to us.

As in Crimson Rosellas, the experiments by Stamps and her colleagues showed that in Budgerigars the last-hatched chicks in a clutch begged more than first- and middle-ranked chicks, and their fathers and mothers responded to that begging differently. Together, the joint feeding by both parents caused the last-hatched chicks to get plenty to eat and even catch up to the other chicks, so that all the chicks fledged within a day or two. Fathers fed chicks primarily in response to how much they begged. Females, in contrast, adjusted how much they fed chicks based on their size. Parents of both sexes were egalitarian toward their sons and daughters. However, broods with more female offspring begged at higher rates than broods with fewer females, and fathers responded by feeding these female-biased broods more. Although male and female chicks attained the same weight by fledging, broods with mostly males (which begged less, and so were fed relatively less often) took significantly longer to fledge.

At first glance, these fine nuances of feeding chicks might seem to be gratuitous sexism, but consider the evolutionary payoff. Female chicks that fledged at a younger age, because they were in female-biased broods and so were fed more and grew faster, themselves raised more fledglings in their first breeding when they reached adulthood, compared to female chicks that fledged at an older age. Male chicks showed no equivalent pattern.

Thus, on the fitness ledger sheet, fathers can put their parental investment where it does the most good, because they respond to begging as father Crimson Rosellas do. In Budgerigars, fathers end up feeding more food to daughters, on average. In turn, the scientists discovered that chicks begged harder than they needed to, trying to gain as much food as quickly as possible from their beleaguered parents. Mothers, unlike fathers, did not fall for this trick. As did mother Crimson Rosellas, female Budgerigars could fine-tune their feeding, making subtle adjustments for age and order of hatch, so that all chicks fared equally and were ready to fledge at the same time. The advantage for the mother Budgerigar of her style of feeding chicks probably accrues when food is short, which happens in the wild but not in captivity. She is not duped into feeding the chicks more than they really need, as the fathers are. For her, this savings probably makes more of a difference than it does for the father, because she is the one producing energy-rich eggs. We can only speculate as to whether she, like a mother Crimson Rosella, has the option to cut her losses in hard times and let the youngest chicks starve, allocating the scarce rations to the oldest chicks so that at least some get a good start in life. The father would not follow suit, because he responds to begging and last-hatched chicks beg more.

Based on these studies, we might then wonder, first, whether food shortages occur regularly in the wild, and second, whether wild parrots regularly use the option of (in essence) aborting the extra chicks through brood reduction when times are tough. Studies reveal that wild parrots of some species engage in brood reduction regularly in response to food scarcity. Let us first explore those cases and then ask how generally brood reduction may be practiced by wild parrots.

Brood Reduction: Hedging Bets with Babies

For a suite of reasons, not all the chicks that hatch fledge. For wild parrots, loss of offspring between hatching and fledging typically ranges between 40 percent and 80 percent of the brood, although extremes on either side occur in some circumstances. Brood reduction is not just any decrement in the size of a brood from various causes of mortality. It means specifically that the parents decide not to raise the younger chicks in a brood, to reduce the size of the brood. The idea is straightforward from an evolutionary perspective. The parents decide to cut their losses right away, before they make the considerable investment of feeding an exponentially growing chick.

In parrots, this decision apparently falls entirely to the mother. She has already decided, by the time the chicks hatch, that she is going to try to raise at least some young that season. Abandonment of the eggs before they hatch is common in parrot populations, comprising up to 50 percent, and sometimes more, of the nesting attempts in a given season. Moreover, if conditions warrant, pairs opt not to breed at all in a given season in many species of parrots. This non-breeding contingent may be more than half of the adult pairs in a population. Therefore, if the female parrot persists in her breeding effort until some chicks hatch, she is clearly attempting to raise the first-hatched chicks to fledging. The later-hatched chicks are those with which she gambles. She raises her entire optimistic brood in a good year, but she ditches the straggler chicks when she realizes she cannot do all of her offspring justice in a bad year.

In all parrot species examined to date there seems to be some degree of hatching asynchrony (figure 74). In some species, like the Crimson Rosella and Budgerigars described above, hatching asynchrony is reduced relative to what might be expected, given the laying date, by females' delaying the start of incubation until most eggs are laid. In others, hatching asynchrony is promoted by females who start incubating soon after laying the first egg, and by widely spacing their laying. It is important to note, however, that although hatching asynchrony provides a ready route to brood reduction by producing chicks of different ages, it is important to ascertain for any given species the degree to which brood reduction versus other factors are responsible for observed losses of chicks. This evidence would comprise greater mortality of later-hatched chicks compared to their older siblings as a direct result of decisions parents make to allocate resources to each of the chicks in the brood. A few recent studies of wild parrots capture the dynamics of brood reduction in wild parrots particularly well, as we next explore.

Burrowing Parrots

Masello and Quillfeldt documented brood reduction in action in Burrowing Parrots in both years of their study. As would be expected, brood reduction was most pronounced during the La Niña of 1998, when the severe drought limited the food supply of Burrowing Parrots during their breeding season.

Burrowing Parrots exhibit the hatching asynchrony that sets up the opportunity for brood reduction, and this asynchrony persists through fledging. Burrowing Parrots lay three to five eggs per clutch, with most clutches containing four eggs. Averaging over the population, the mean number of eggs in a clutch is 3.8, the mean number of hatchlings 3.4, and the mean number of fledglings 3.0. In other words, some mortality occurs, on average, in each clutch, although fledging success is higher in Burrowing Parrots than in any other wild parrot known. Still, are the lost chicks most often those that hatched last?

In the relatively good season of 1999–2000, up to 90 percent of the first-, second-, and third-hatched chicks survived, but only 80 percent of the fourth-hatched chicks and 70 percent

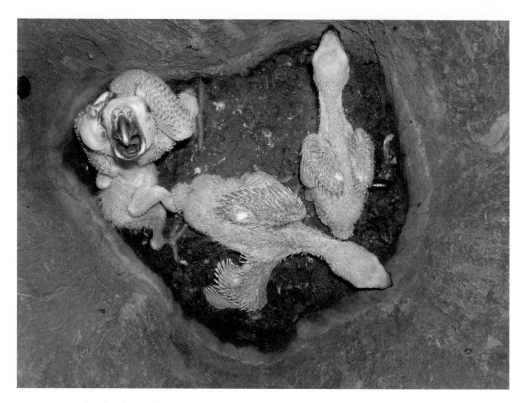

FIGURE 74 A clutch of Blue-fronted Amazons, *Amazona aestiva,* illustrating the hatching asynchrony typical of parrots. Note the range in size and degree of feather development between the youngest and smallest nestling (lower left) and the oldest and largest (bottom). Photo © Igor Berkunsky.

of the fifth-hatched chicks. In the 1998–99 La Niña year, the Burrowing Parrots laid their optimistic clutch, averaging 3.9 eggs over the population, the same as in the good season. However, only 65 percent of the hatchlings fledged that season. During the La Niña year, 80 percent of the first- and second-hatched chicks survived, 70 percent of the third- and fourth-hatched chicks, and only 35 percent of the fifth-hatched chicks. This pattern meets the requirements of evidence for brood reduction in Burrowing Parrots. The loss of the later-hatched chicks was (1) always greater than that of earlier-hatched chicks; (2) proportionately more in hard times; and (3) tied directly to food availability, or rather lack thereof, in bad years relative to good years.

Masello and Quillfeldt did not stop at tracking survival. They also recorded detailed information on the growth of the chicks. Baby parrots grow exponentially until they reach a peak weight, and then they lose some of that weight before they fledge, a phenomenon known as *mass recession* (more on this shortly). Even in the good year, the chick's pattern of growth and date of fledging also depended on the hatch order of a chick. Peak mass of first-ranked (first and second) chicks averaged 333 grams, that of middle-ranked (third and fourth) chicks 323 grams, and that of last-ranked (fifth) chicks 311 grams. This differential translated into an average fledging weight of 260 grams for first-hatched chicks and 230 grams for last-hatched chicks. Moreover, the first-hatched chicks were structurally larger, which means that the last-hatched chicks were stunted and not simply in poorer condition. The later-hatched chicks were apparently made into runts by bet-hedging parents, falling behind by up to 12 percent compared to their favored older siblings.

During the severe La Niña, Burrowing Parrot chicks attained a lower peak mass on average (264 grams vs. 246 grams, a decrease of nearly 10 percent). As predicted, the relative disadvantages to the later-hatched chicks were even greater in the La Niña year than in the normal year, in comparison to both first- and middle-ranked chicks.

These patterns corroborate the evidence for brood reduction but are silent on how the parents actually achieved this result. Did the parents discriminate against the later-hatched chicks, or did they get lost in the frenzy of feeding time, overpowered by the older chicks? It is certainly plausible that Burrowing Parrots are capable of the finesse of Crimson Rosellas, with the parents each pursuing a different feeding strategy that was modified to a brood-reduction strategy when under experimentally simulated hard times.

Short-billed Black Cockatoos

Denis Saunders studied the Short-billed Black Cockatoo for sixteen years in the southwest corner of Western Australia (box 5). Female Short-billed Black Cockatoos each lay only one or two eggs, averaging 1.8 eggs per clutch and giving this species the smallest clutch size among the parrots. Nevertheless, these cockatoos practice perhaps the most aggressive and deliberate brood reduction of any species of parrot.

Bet-hedging in Short-billed Black Cockatoos begins with the egg itself. If the female lays two eggs in a clutch, the first egg is larger than the second egg, nearly 4 grams heavier and 3 millimeters longer on average. If the female decides to lay only one egg, it is the same size exactly as the first egg in the two-egg clutches. An average egg is 5 percent of the body weight of a female Short-billed Black Cockatoo, no small investment. Perhaps the 4-gram savings is testament to her cautious optimism—she is willing to gamble on whether she can raise a second chick that season, but she is still pessimistic enough to hold something back. At Coomallo Creek (the location most resembling their historic habitat), nearly 80 percent of the 400 clutches he studied contained two eggs, so apparently most of the mother cockatoos there were willing to gamble. Yet, one- and two-egg clutches produced on average the same number of fledglings, 0.8 fledglings per nest in which at least one egg hatched. In other words, the second chick was nearly always discarded, even at Coomallo Creek, where conditions were relatively good (box 5).

Saunders discovered that the second chick hatched up to two weeks or more after the first chick and typically died within 24 hours of hatching. This long interval between the two eggs is unusual in parrots, shared primarily with some species of lorikeets that apparently have determinant two-egg clutches regardless of their size, reported anecdotally in aviculture. (Remarkably few studies have been published on wild lorikeets.) Perhaps this long interval between chicks is evidence for the insurance hypothesis, as opposed to brood reduction. Saunders discovered that the two-egg clutches had greater hatching success than the one-egg clutches. This result is not remarkable in itself. Taken with his data on fledging success in Short-billed Black Cockatoos, which does not differ between one- and two-egg clutches, the combination is also consistent with the insurance hypothesis. In other words, if the first egg or chick doesn't make it for any reason, the parents have a ready backup chick so that at least one chick can be reared that season.

Nevertheless, Saunders's impressively long-term studies of Short-billed Black Cockatoos at two very different locations in southwestern Australia (Manmanning and Coomallo Creek) suggest that when times are very good, these cockatoos will rear and fledge both chicks (box 5). That pattern suggests brood reduction, but we need to know more about the chick that did not survive before we can make that conclusion.

Other Parrots and Other Hypotheses

A study of Brown-headed Parrots *Poicephalus cryptoxanthus* in captivity by Stuart Taylor and Michael Perrin examined growth of chicks under conditions of *ad libitum* food. They found that, even under conditions of food abundance, the last-hatched chick was disadvantaged compared to its older brood-mates, in both rate of growth to fledging and asymptotic (final) weight, a result that echoes Krebs's study of Crimson Rosellas. The real test of whether brood reduction can be ruled out, however, requires a comparison of food abundance with food scarcity and the effect of each on fledging, which Taylor and Perrin did not do.

Nevertheless, Taylor and Perrin's conclusion that the insurance hypothesis best explains the patterns of asynchronous hatching has support from studies of other wild parrots. For example, Mark Myers and Christopher Vaughn followed Scarlet Macaws during their post-fledging period, a topic that we will cover shortly. For now, suffice it to say that the later-hatched chicks had the highest mortality of fledglings in that critical period just after the chicks leave the safety of the hollow. To invest the considerable resources needed to rear a chick as large as a macaw to fledging, a sizeable commitment of both energy and time, and then toss it away willy-nilly by sending a poorly prepared offspring out into the cruel world, is strong testimony against the brood-reduction hypothesis. Rather, such observations are more consistent with the insurance hypothesis or alternatives that propose that hatching asynchrony serves to make parenting less costly or more efficient. Indeed, most parrots may meet the vagaries of an unpredictable environment with a mixture of strategies, including brood reduction, depending on what factors limit breeding success in any given year. Asynchronous hatching is both a constraint and an opportunity for birds such as parrots with altricial young (box 6).

Mass Recession: Parent–Offspring Conflict?

The characteristically peaked growth curves of weights of baby parrots against time are familiar to aviculturists. When I hand-reared my first brood of baby cockatiels, I remember well the initial shock of this weight loss. Was I doing something wrong? Were the babies sick? Of course they could not grow exponentially forever. More to the point, the flaccid,

BOX 5   THE SHORT-BILLED BLACK COCKATOO: A SAD TALE OF TWO POPULATIONS

For nearly two decades, Denis Saunders and his colleagues studied two populations of the Short-billed Black Cockatoo *Calyptorhynchus latirostris,* also known as Carnaby's Cockatoo (box figure 5.1). He began his impressive work in 1969 when Australia's Division of Wildlife Research (CSIRO) asked him to determine whether these cockatoos damaged pine plantations and apple and pear orchards in the southwest corner of Western Australia, where this species is endemic. His study witnessed massive changes in land use, as fully two-thirds of the native *Eucalyptus*-dominated woodlands and interspersed dry shrublands were converted to various forms of intensive agriculture, in particular wheat cultivation. The story that Saunders's research unfolds about these impressive cockatoos unites topics that we cover in this book in chapter 2 on foraging, this chapter on life history, and chapter 7 on parrot populations and how humans affect them.

Short-billed Black Cockatoos depend on an intermixture of woodland and shrubland for critical resources: roost and nest trees, protection from predators and the extreme summer heat of this climate, and a ready source of seeds and flowers of a variety of native trees (including *Banksia, Dryanda, Grevillea,* and *Hakea*) and insect larvae they contain. The only non-native plants these cockatoos use for food are weeds in the grass genus *Erodium.* Unlike the adaptable Galahs *Eolophus roseicapillus* and Little Corellas *Cacatua sanguinea,* Short-billed Black Cockatoos do not indulge in the seeds of the agricultural crops that have come to dominate this region.

For the first decade of his study, Saunders compared the cockatoos in the more interior Manmanning area with those in the more coastal Coomallo Creek area. At

BOX FIGURE 5.1. A Short-billed Black Cockatoo, *Calyptorhynchus latirostris,* feeding on flowers in Tamala Park, Australia. Photo © Georgina Steytler.

(continued on next page)

Manmanning, Short-billed Black Cockatoos struggled in vain to continue their age-less way of life in the face of relentless conversion of native forestlands to wheat culti-vation. In contrast, Short-billed Black Cockatoos at Coomallo Creek lived in relative luxury as the main land use there, cattle grazing, left much of the woodlands intact. At the beginning of the study, the cockatoos were already not thriving at Manman-ning. This study area was seven times the size of that at Coomallo Creek but con-tained less than a quarter of the number of pairs found at the latter site. At Manman-ning, fewer than twenty pairs lived in nearly 50,000 hectares (one hectare is 100 meters on a side). At Manmanning, each pair on average fledged one offspring every three years; at Coomallo Creek, pairs fledged twice that many. At Manmanning, the young fledged at almost 80 percent of the adult's average weight, while at Coomallo Creek, the young fledged at nearly full adult weight.

Parent cockatoos at Coomallo Creek followed the typical pattern of attendance of the nest and young offspring. The female incubated the eggs and remained on the nest to brood the hatchlings, leaving for only short periods. The male's attendance to her needs permitted her to stay close to the nest, by bringing in enough food for all. When the chicks reached two weeks of age, the female stopped brooding the chicks during the day. She and the father nevertheless returned during the day to feed them at least once. In contrast, cockatoo chicks at Manmanning faced an austere existence. As soon as the first chick hatched, its mother attended it as if it were two weeks old, and often the second egg never hatched. In fact, at any time during incubation as well, she might leave the eggs and later the tiny hatchling for long periods of time as she traveled far and wide to forage for herself and her chick. When young fledged at Coomallo Creek, the family often dallied in the woodlands surrounding the nests, but at Manmanning, the family deserted the nesting area as soon as the fledgling was able to travel, and were not seen again until the next breeding season.

If the nest failed for any reason early on, the cockatoos at Coomallo Creek might make a second attempt to breed that season, but those at Manmanning never did. The statistics at Manmanning continued to remain poor for the cockatoos there relative to their kin at Coomallo Creek. Chicks at Manmanning grew more slowly on average, and growth was more variable there from chick to chick, compared to Coomallo Creek. Chicks fledged 15 percent heavier at Coomallo Creek. The proportion of nests fledging at least one (usually only one) chick at Manmanning was 35 percent, in con-trast to 65 percent of nests at Coomalo Creek. At Manmanning, pairs fledged an average of 0.4–0.5 chicks per pair per year—half the rate of cockatoos at Coomallo Creek.

When Saunders's study of nesting pairs at Manmanning ended in 1976, the hand-writing seemed to be wall for the Short-billed Black Cockatoos there. By 1990,

Manmanning was no longer home to breeding pairs of this species. What had begun as a mission to cull pest cockatoos and prevent them from harming the struggling agriculture of that area became instead one to prevent these magnificent birds from disappearing altogether from the rapidly changing Australian landscape. On January 26, 2005, Dr. Saunders was justly honored with the Order of Australia for his service to nature and conservation and especially to the black cocky.

bottom-heavy chicks with their gargantuan crops and appetites could hardly be expected to get airborne if they stayed that way until their maiden fledging voyage.

Although most parrot chicks undergo this pronounced weight-loss pattern, mass recession is the exception in altricial birds. Therefore, this phenomenon begs explanation. Nestling swallows and swifts (Apodiformes) and seabirds (Charadriiformes, family Alcidae) share this drastic weight-loss program with baby parrots. As usual, at least two alternative hypotheses come to scientists' minds. One proposes that loss of mass prior to fledging is an epiphenomenon of the natural process of development, readying what has been essentially a fetus for adult life. Mass recession could result from the entirely physiological process of converting the chick's body from a food-processing machine capable of extremely rapid growth to that of a functioning adult bird capable of flight. Yet, that hypothesis would apply to all altricial nestlings, and in most species, mass recession does not occur.

The other hypothesis proposes that mass recession results from an adaptive and calculating strategy on the chick's part. This strategy is not conscious but rather designed by natural selection to increase a chick's fitness, tempered by constraints that the parents may impose. This hypothesis states that chicks hoard extra calories, to be used to advantage as a chick matures and begins to disagree with its parents over how much it should be fed prior to fledging. We found in the previous studies that baby parrots beg for more food than they need. Adaptive strategies resulting from competition between the parents' interests and those of the chick may well be at work in parrots.

A study of Cassin's Auklets *Ptychoramphus aleuticus,* which also experience mass recession, concluded that the loss of extra weight was a cue for baby auklets that the parents were ready to kick them out of the nest. In their model, they hypothesized that irregular provisioning by parents and the chick's "departure decisions" collaborate to cause the phenomenon of mass recession. According to the model, the chicks are comparing the costs of staying in the nest (impatient parents are increasingly reluctant to bring them enough food) against the benefits (primarily the safety of the nest as opposed to taking their chances out in the cruel world). Recall that nests in holes are relatively safer than stick nests in trees and that Masello and Quillfeldt found the cliff nest holes of Burrowing Parrots to be particularly safe. The Burrowing Parrot chicks stay in the nest an average of eleven days longer than do parrots of other species the same size, and they show the most pronounced mass recession known for any parrot in the wild. These biologists concluded that their findings on Burrowing Parrots provided some support for an adaptive hypothesis of mass recession.

Yet not all parrot chicks go through mass recession. These exceptions (provided that they are exceptions, once more parrots are studied in the wild) could prove the rule, that is, support hypotheses about adaptive strategies causing mass recession. For example, Green-rumped Parrotlet chicks do not show that telltale peak of nestling weight and so do not undergo mass recession as a strategy. The overall pattern of reproduction in Green-winged Parrotlets is particularly well known—and particularly complex. Box 6 summarizes more research on these parrotlets and puts what we know about them in a broader perspective of topics that we cover later in this chapter. For now, we can surmise that Green-rumped Parrotlets either do not or cannot benefit from an adaptive strategy of mass recession, as Burrowing Parrots do. These two species might represent two ends of a continuum of strategies adopted by all parrots.

LIFE OUT IN THE WORLD

Fledging the Nest and Becoming Independent

No matter how relatively safe, nests are still not a place to spend any more time than absolutely necessary (box 6). By far most nestlings leave as soon as they can fly (or walk, as in Kakapo) and well before they are ready to make it on their own.

After the chicks leave the nest, in fact, they go through three developmentally and demographically distinct stages before they reach sexual maturity. In the first stage, the *fledgling*, chicks are newly fledged, unweaned, and clueless. They depend entirely on their parents for food and to help them learn to how to be parrots. They are also weak and uncoordinated. They must actually learn to fly and build strength and endurance to be skilled fliers. Instinct plays some role in getting chicks through this stage, but instinct alone is not enough. As in most intelligent, social species, parrots depend a great deal on learning and mentoring from their parents, just as humans do. This stage is clearly the most brutal stage of their lives, with by far the highest mortality. Up to 80 percent of fledgling birds die within a few weeks to a few months after fledging, and parrots are no exception.

Young parrots in the second stage are called *juveniles*. The youngsters are now weaned, but they are still in their first year and have much to learn. They have not yet shed their baby feathers and may be colored differently from the adults. In most species of parrots known from studies in the wild, juveniles leave their parents some time during this period, often as soon as they are fully weaned. In most species of parrots, juveniles then congregate in large flocks of their peers. While in these flocks of non-reproductive individuals, juvenile females meet up with somewhat older suitors, who will become their future mates. Males and females differ in their propensity to travel from their natal haunts, perhaps insuring that their mates will not be close relatives. These juveniles also concentrate on gaining the skills to feed on their own, the most demanding activity of their young lives. Often they weigh less than they will once they master these skills. Juveniles surely share much of what they learn with each other in these large flocks, as we inferred from studies of juvenile Keas *Nestor*

BOX 6   THE GREEN-RUMPED PARROTLET: A SMALL PARROT TELLS A BIG STORY

Much of what we know about the life history and demographics of parrots comes from the extensive research on the diminutive Green-rumped Parrotlet *Forpus passerinus* led by Steven Beissinger of the University of California, Berkeley, and his many colleagues, students, and post-doctoral researchers.

Because nearly every aspect of their wild lives has been documented for decades, the Green-rumped Parrotlet (figure 58, chapter 5) has arguably ascended to the scientific status of model organism. Study of model organisms allows scientists to seek answers to more general questions in science, ones that go beyond the peculiarities of one species. An insightful study by M. Andreína Pacheco, Steven R. Beissinger, and Carlos Bosque on the Green-rumped Parrotlet exemplifies the power of studying one organism especially well.

Their study homes in on a paradox. The tiny parrot lays an enormous clutch, which in itself is hardly mysterious. The small size of Green-rumped Parrotlets predicts a life-history strategy of greater investment in reproduction than in self: small adult size, fast growth to sexual maturity, large clutch, reproduction on an annual basis, high adult mortality, and rapid re-pairing and setting up housekeeping with a new mate (chapter 5). A large number of eggs in a single clutch means a larger degree of hatching asynchrony. Even though parrotlet chicks are able to leave the nest sooner than macaw chicks, having so many more brothers and sisters means that babies stay in the parents' nest for a relatively long time.

Herein lies the paradox. Nests, even cavity nests, are dangerous places. Baby parrots need to fledge and get out as soon as possible. Instead, Green-rumped Parrotlets grow relatively slowly for their tiny size and seem to stay in the nest longer than they need to.

Pacheco and her colleagues discovered that parent parrotlets feed their chicks nutrient-poor food relative to the nutritional optima studied by Richard Grau and Tom Roudybush at the University of California, Davis. The protein content of their diet was particularly low, at 9 percent protein by dry weight, and this finding is relevant because protein is differentially needed for growth (box 2, chapter 2). Their research affirmed that baby parrotlets took up this nutrient especially slowly, delaying their growth and subsequent fledging. Energy, in contrast, did not limit the chicks' growth, nor did accretion of key nutrients such as calcium, although phosphorus was also arguably limiting to their growth.

Their study concluded that poor diet and slow growth were causally related somehow, especially in view of the need to exit the nest quickly to avoid predators. Interestingly, the scientists did not jump to the conclusion that would be most obvious, which

(continued on next page)

is that parrotlets would prefer to grow faster and would do so, had they access to a better diet.

Instead, evidence of various kinds points the cause-and-effect arrow in the other direction. Perhaps Green-rumped Parrotlets can afford to exploit a low-nutrient resource, benefiting from reduced competition with other species for it, because slow growth itself is also beneficial. Ornithologist and ecologist Robert Ricklefs of the University of Missouri–St. Louis has turned his attention over the years to the ecology and evolution of senescence and its coupling with the evolution of life histories. His recent research has uncovered a direct statistical relation between time spent as an embryo and in postnatal growth, and longevity as an adult, in both birds and mammals. That is, for any given body size, the longer an individual takes to grow up, the longer an adult life that individual can look forward to. Longer adult life is key to reproductive success in environments that are unpredictable or low in productivity, as we have discussed in various parts of chapter 6. The longer an organism can stay in the reproductive saddle, the more chances he or she has at rearing at least some offspring that make it to reproductive life as well.

A hypothesis, then, is that hurrying up and growing quickly as an embryo might shorten the life of adult parrotlets through *intrinsic* mechanisms, causing senescence and aging prematurely, relative to the parrotlets' ability to dodge *extrinsic* sources of mortality. This slow-lane approach to life planning may be why parrotlets can subsist on low-nutrient foods—or perhaps those foods even set up the slow-growth advantages in the first place.

A study on Cockatiels *Nymphicus hollandicus* by Corinne Kozlowski and Robert Ricklefs provides some intriguing clues supporting the slow-lane life-history hypothesis even for the smallest of parrots. They examined the hormones placed into eggs by mother Cockatiels at the time of egg formation, which include testosterone, androstenedione, and corticosterone. While the amounts of testosterone and corticosterone were different between first- and last-laid eggs, the levels of androstenedione remained the same throughout the clutch and hatching order. Why put these substances in the egg yolk at all? The corticosterone may be simple product of the increasing stress on the mother with the demands of egg laying, incubation, and then feeding the older chicks. The testosterone might be there to influence the aggressiveness of begging or other behavior that could favor, or not, a chick's ability to compete with its nest mates for food, interacting somehow with the brood hierarchy created by asynchronous hatching.

The androstenedione, interestingly, plays a key role in slowing embryonic growth and delaying development. Kozlowski and Ricklefs hypothesize that the mother puts this growth retardant in the egg to lengthen the time to hatching and thus fledging in large broods, and suggest the reduction of sibling competition as ultimately being favored by natural selection. They do not state an alternative and not mutually exclusive evolutionary hypothesis: that slowing embryo growth may be part of a life-history strategy of relative longevity, all else equal. This slow-lane life-history hypothesis is in fact consistent with Ricklef's findings on correlates of longevity in birds in general. Parrots are on the slow end of the bird continuum, controlled for their particular range of body sizes. To test whether the steroid hormones placed in egg yolks are a key contributor to this strategy, we need more information on their concentrations in other species of parrots. These and other data would allow further examination of mechanisms for elongating each of these stages of parrots' lives as part of an integrated adaptive strategy of a slow life history for parrots of all sizes, from the tiny Green-rumped Parrotlet to the largest macaw

notabilis (chapter 3). This sharing is not only a great advantage of being social—it is arguably what sociality is all about. There is also some safety in numbers, because larger flocks of birds deter predators.

By its first birthday, a young parrot achieves the status of a *subadult*. In a few of the smaller species, the young birds, and in particular the females, are ready to breed within the first year, essentially skipping the subadult phase. In most species, however, age of first breeding comes later, with larger parrots maturing at later ages. This "age of first reproduction" could be from two to three years for the medium-sized parrots to four or five years for the largest. By the time they become subadults, the young birds are typically wearing their permanent adult plumage and so are indistinguishable from adults. They are becoming increasingly skilled at foraging and are in increasingly better condition. Another long and extremely important process that subadults must accomplish is strengthening and deepening a bond with a life-long mate (chapter 5). They may even begin to practice housekeeping and produce some eggs together as subadults, but these attempts to breed are only half-hearted and usually abandoned. A study of wild Eastern Rosellas *Platycercus eximius* by Meredith Smith and J. Le Gay Brereton found that subadults are physiologically capable of breeding and concluded that the delay in reproducing is entirely behavioral.

Once a parrot is seriously reproductive, it is officially termed an adult. In wild parrots, the advent of sexual maturity and reproductive activity follows pair-bonding, a process that begins in a parrot's young life while it is still immature. This Romeo-and-Juliet model is probably widespread in the relatively few species of birds that practice life-long monogamy, as parrots do (chapter 5), but it is not typical of the sequential monogamy found in many other birds.

Stories from the Wild: A Peek into the Lives of Some Wild Parrots

From here I will focus on the best-studied examples of each of the major life stages of parrots.

FIGURE 75 An inquisitive Kea, *Nestor notabilis*, perched on a mountain top in Aoraki/Mount Cook National Park, New Zealand. Photo © Andrius Pašukonis.

Crazy Keas

Before humans of European descent arrived in their native New Zealand, Keas (figure 75) apparently spent their time in the southern beech *Nothofagus cliffortioides* forests on the steep sides of the stunning mountains that grace the island chain of New Zealand. These parrots are resourceful generalists, with their own style of accommodating the radical changes in their native habits and habitats as humans move into their landscape, as we first saw in chapters 2 and 3. Their proclivity for frequenting garbage dumps makes them accessible subjects for scientists, and parrot biologists Judy Diamond and Julian Bond tell Kea tales in their delightful and informative book, *Kea: Bird of Paradox*.

Young Keas fledge at between 9 and 13 weeks and wean some time after that. While the unweaned fledglings are completely dependent on their parents for food, they pick away at food and other objects, perhaps practicing skills that they will use when they start to feed themselves. Once weaned, the juvenile Keas spend a great deal of time foraging and feeding. They are much less efficient than subadults, taking longer to eat a given type of food and including more undesirable food in their diets. The juvenile Keas also employ the resourcefulness of their kind by persuading older, unrelated Keas to feed them. They use behaviors that serve to appease the possibly aggressive older birds, such as the wing-hold display (probably borrowed from begging behaviors of the chicks).

Even the older subadult Keas are inefficient foragers, on average, presumably growing in skill progressively for the three or four years until they reach sexual maturity. At this age, most Keas are not paired, so females and males are equally responsible for feeding themselves and equally slow at it compared to the adult males who forage for themselves, their mates and chicks.

Diamond and Bond discovered that the youngsters spend a great deal of their time in play, with each other and with objects, that is to say, toys. The more complicated the toy, the more opportunities for interacting with it, and the longer the Keas will play with it. For example, Diamond and Bond report the intense interest that Keas have in rolling rocks downhill, with seemingly no other purpose than to watch them splash into the water below. They have watched Keas play tug of war with a roll of gauze, or fiddle with an endless variety of small objects. By far the favorite game of Keas is demolition, exploring an object while tearing it apart. Gangs of subadult and juvenile Keas are notorious for their destruction, dismantling any man-made object they can find in the dump, and beyond, including innocently parked cars.

The studies by Diamond and Bond on the Kea and its relatives the Kaka *Nestor meridionalis* and the Kakapo have contributed significantly to the scientific understanding of play, a quite serious and elusive topic of research in the fields of behavioral ecology and cognitive ethology. From an evolutionary perspective, play presents a paradox as well as a particular challenge to define and understand. As with all phenomena found in nature, play would not exist if this class of behaviors did not benefit an individual's evolutionary fitness. Yet, the behaviors that we recognize as play seem so frivolous and useless—a bird deliberately rolling a rock down a hill to watch it plop into a pool of water, for example. In his pivotal book, *The Genesis of Animal Play: Testing the Limits,* Gordon Burghardt meets the challenges of approaching the scientific study of play. He outlines testable hypotheses based on the comparative biology of play in all vertebrates from fishes onward, and even some invertebrates.

The upshot for the most playful of all parrots, the Kea, is that play seems to be a vitally important part of behavioral development, especially in the social vertebrates, which depend heavily on intelligence, learning, and getting along with others. Evolutionarily, play is not a frivolous side effect of youth but a deadly serious process that ensures success in survival and reproduction, the currency of natural selection. Perverting the words of George Bernard Shaw, youth is not wasted on the young. Play is of course fun, that is to say pleasurable, which is how evolution ensures that individuals will engage in an activity that enhances fitness, just as pleasure arises from the other appetites that animals have.

What intelligent young animals may be doing when they play is in part to grow neuronal pathways in the brain, in particular those connections that allow behavior to be flexible and to respond to novelty. In other words, when youngsters play, they learn how to learn. These animals live in complex, changing, and often unpredictable physical and social environments. Play allows them to build the tools they will need to look out for themselves and their offspring throughout the rest of their lives. Play not only hones specific skills but also promotes versatility. Because parrots live long lives and take a long time to mature, the years that parrots spend as juveniles and subadults are both an opportunity and investment, just as in humans.

FIGURE 76 Two Keas, *Nestor notabilis,* at play in Aoraki/Mount Cook National Park, New Zealand. Photo © Andrius Pašukonis.

Diamond and Bond recently compared Keas, Kaka, and Kakapo to find that Keas engage in more play and more complex play for a longer period of their lives than do Kaka and Kakapo, in that order. Keas engage in play (figure 76) beyond the juvenile period, and different developmental classes of Keas play with one another, in contrast to arguably most other parrots. Keas also frequently engage in object play, something not seen in the other species in their study. Object play is surely not restricted to Keas, as anyone with a pet parrot can attest, but scientific study of play in many other species of parrots has not been done.

Keas live in a particularly harsh and unpredictable environment for a parrot. They meet this challenge with a suite of traits, including a generalist diet, a large brain and intelligence, vocal learning and special dialects for the juveniles who join up to share information—and now, we find, by engaging in unusually complex play.

Cockatoos Down Under

The Australian cockatoos are particularly well known because of the diligent efforts of a generation of parrot biologists in that country, including R. J. S. Beeton, Denis Saunders, G. T. Smith, Ian Rowley, Leo Joseph, Gabriel Crowley, and their many colleagues. Thanks to their efforts, we know a great deal about cockatoos in the corella branch of the genus *Cacatua* (the Western Corella, the Long-billed Corella, and the Little Corella *C. sanguinea*) and the Galah.

FIGURE 77 A superflock of Galahs, *Eolophus roseicapilla,* in the Flinders Ranges in Australia. Such superflocks form after the breeding season as juveniles become independent of their parents and start roaming widely over the landscape. Photo © Alan Milbank.

The movements of these cockatoos across the landscape synchronize with the seasons. Each spring, when times are relatively flush, smaller flocks of a few hundred to a thousand birds settle down in a traditional area to nest and produce that year's cohort of fledglings. At first, the young fledglings and their parents do not go far from the breeding area, as the babies gingerly experience life outside of the nest. Parents assist their clumsy offspring closely, because this brief period of their lives is exceedingly dangerous, as we saw earlier. The Galah parents within an area deliberately pool their fledglings into nurseries (crèches), where they are a bit safer than if individual families were isolated and on their own.

As the young cockatoos begin to wean and become more accomplished fliers, the family groups start to join up with other families, and flocks of cockatoos enlarge and start to travel farther together. Life is still quite hazardous for the weanling juveniles. During this period, a heat wave can devastate the season's cohort of chicks. They succumb easily to dehydration because they are inefficient foragers and not in full adult condition.

If they make it to the late summer and fall, the juveniles leave their parents and blend into large flocks with subadult and adult birds as they all forage and collectively prepare for the long winter (figure 77). These flocks number in the tens of thousands of cockatoos and are highly vagrant and opportunistic. Beeton refers to them as the "roving flocks," to describe their behavior, and "superflocks," as they coalesce from the static breeding flocks that leave the nesting areas when their offspring fledge. In his engaging book *Behavioral Ecology of Galahs,* Ian Rowley estimates that these flocks are approximately half unpaired subadults and adults, a quarter established breeding pairs, and a quarter young of the year.

The next year, the cycle repeats itself for the adult cockatoos, as they return to their breeding areas, often using the same nest tree as the season before. The immature birds may well return to their general natal areas, but they remain in flocks of subadults and unpaired birds. The population fluctuates seasonally in this fashion. Subadults become reproductive adults between three and five years of age in corellas and between two and three years in Galahs. They then leave the larger roving flocks during breeding season, to return and mix again each winter in the superflocks. The corellas, as well as other cockatoos, follow an ancient pattern of breeding somewhat inland and returning to coastal areas to feed and over-winter. The cockatoos seem to be inflexible in these timeless rhythms, and as humans convert the landscape, the changes may or may not be favorable to the birds (box 5). Some cockatoos, such as the Little Corellas, Long-billed Corellas, and Galahs, are relatively accommodating and even thrive with the newfound opportunities presented by some forms of land conversion.

The chances of a parrot's surviving until the following year as the seasons wax and wane change with each life stage. The few months post-fledging hold the slimmest hope, with less than a quarter making it to the fall. Once the young corellas see their first fall, their chances of making it to their next birthday rise to about 70 percent. Hazards remain as the birds mature—they trade the stresses of inexperience for those of breeding. A mere 10 percent of male and 20 percent of female Western Long-billed Corellas may make it to breeding age. Once breeding commences, life is more hazardous for females. If they reach adulthood, the life expectancy of wild corellas is 17 years for males and 14 years for females. Galahs do not fare quite as well. Once they dodge the dangers of being juveniles, their rate of survival levels off when they are 18 months old at 0.96 per month, or a little over a fifty-fifty chance of surviving each year. At that rate, the likelihood of an 18-month-old Galah seeing its fifth birthday is slim. Life expectancy is not the same as longevity, although evolution usually sees that the two march step with one another, as we explore shortly.

The New World Parrots

Not all parrots indulge in the togetherness of the Australian cockatoos. One of the best-studied species of New World parrots is the Puerto Rican Parrot. Because it teeters on the brink of extinction, like the Kakapo, the Puerto Rican Parrot's every moment has been monitored by devoted conservation biologists and every intimate facet of its life observed. Its island existence may account for the relatively sedentary life of the Puerto Rican Parrot relative to the roving bands and superflocks of cockatoos. The very nature of these cockatoo superflocks is precisely that they are roving, for so many birds cannot support themselves if they are confined to one small area. In contrast, present-day Puerto Rican Parrots live out their lives in a few close-by valleys. It is quite possible that this sedentary behavior is a product of extremely low population size and confinement to a small patch of rainforest that is not representative of historical habitats when the species was found across the island of Puerto Rico. Nonetheless, at present their daily wanderings define their annual rhythms—they are one and the same.

For by far most of its life, a Puerto Rican Parrot lives in a small family group. A fledgling Puerto Rican Parrot stays with its parents in the general vicinity of their nest until it is fledged, as do all the other parrots that we know about. After they fledge, Puerto Rican Parrots stay with their siblings, or a single fledgling can stay with its parents, for many months. Puerto Rican parrots may gather in larger flocks briefly in late May to June, right after the young fledge. Even then, the larger flocks are composed of smaller groups of two or three birds. When the subflocks are made up of mated pairs, aggressive chasing can dominate the interflock dynamics.

Noel Snyder, James Wiley, and Cameron Kepler, in their book *The Parrots of Luquillo: Natural History and Conservation of the Puerto Rican Parrot,* remark that members of a pair rarely get farther than 3 meters from one another. A male and female mate for life and stay by each other's side around the clock until death separates them (chapter 5). Only during the few weeks of incubation and brooding hatchlings are the mates physically apart for any length of time.

We know less about free-ranging parrots on the mainland of South and Central America. Scientists are working there, too, but following elusive parrots in their natural haunts is daunting where rainforest can extend for hundreds of kilometers without paths or roads. Not surprisingly, we know relatively more about parrots that live in more arid, open habitats.

Central Bolivia, along tributaries of the Grande, Mizque, and Pilcomayo Rivers, is home to the Red-fronted Macaw *Ara rubrogenys* (figure 78). Like the Lear's Macaw *Anodorhynchus leari,* Red-fronted Macaws live in dry, open woodland and scrubland known as *chaco* in Bolivia and as *caatinga* in Brazil. And like the Monk Parakeet in the same areas, these macaws face a shortage of the large trees with suitable nest holes. For this reason, it is likely, both Red-fronted and Lear's Macaws nest in holes in the sides of exposed cliffs.

Mette Bohn Christiansen and Elin Pitter have studied Red-fronted Macaws extensively over their range. Aside from their nontraditional nest sites, Red-fronted Macaws lead the family life that we have come to expect from parrots. The parents stay with their newly fledged young and feed them until they wean. After weaning, the juveniles remain with their parents until at least the next breeding season. If a family is graced with more than one chick from that season's breeding effort, the siblings stay together and chase off other juveniles from their parents' haunts.

Juveniles remain distinguishable from adults for a year after weaning. The youngsters spend much of their time acting like young Keas, hanging out, playing, chewing on objects, and otherwise seeming to fritter away their time and efforts without the efficiency or resolve of their parents. All the while, juvenile Red-fronted Macaws remain in the same local area as their parents, often relying on their parents for various forms of help and to model what the juveniles need to be doing. Once the parents insist that the young of the year be completely

FIGURE 78 A family of Red-fronted Macaws, *Ara rubrogenys*, at a cliff in Bolivia. This species lives in an arid habitat with few large trees and so instead nests in cavities in cliffs. Photo © James Gilardi.

independent, the juveniles socialize and preen each other, strengthening the bonds between them and practicing the social skills they will use with their own future mates.

The study by Myers and Vaughn focused on the critical post-fledging period of Scarlet Macaws in Costa Rica. Scarlet Macaws take an especially long time to wean, probably due to their large size relative to other parrots. In their first 12 days after fledging, Scarlet Macaws, like Corellas, do not leave the immediate vicinity of the natal tree, staying within 1 kilometer. Perhaps this site fidelity has to do with the limited abilities of the baby macaws to travel, as well as waiting for the younger chick (if there is one) to catch up a bit to the older. The next stage, which proceeds to between 30 and 50 days of fledging, sees the family dispersing up to 12 km from the natal site and ranging in daily activities over 200–700 hectares. At 56 days, the fledgling macaws are all weaned. Accordingly, after 50 days post-fledging, the now juveniles are more adept at flying and have greater endurance. Juveniles then begin to participate in the traditional daily movements between roost and feeding areas, ranging up to 15 km on a single flight. Some 50–70 days post-fledging, the newly weaned juveniles leave their parents and join up with their peers in juvenile flocks, which will presumably coalesce with subadults and older non-reproductive macaws as they learn more skills and find their future mates.

The lives of most species of parrots in the New World (and for that matter Africa) are still cloaked in relative mystery. We have no evidence, however, that the New World parrots routinely gather in seasonally fluctuating superflocks at any time of the year, even in the New World's remaining natural areas, such as the vast unbroken expanses of rainforest in which Manú National Park is embedded.

All else equal, the average size of flock in which a parrot of a given species spends time depends on its own size. The sixteen species of parrots that live in Manú National Park range in size over three orders of magnitude, from the parakeets spanning the 25-gram Dusky-billed Parrotlet *Forpus sclateri* and the 70-gram *Pyrrhura* parakeets, to the medium-sized parrots, spanning the 110-gram Dusky-headed Parakeet *Aratinga weddellii* and the 500–800-gram *Amazona* parrots, to the largest of parrots, the *Ara* macaws, weighing over a kilogram.

A study by James Gilardi and Charles Munn in Manú found that flocks of the large *Ara* macaws and *Amazona* parrots have an overwhelming mode of exactly two. In other words, the mated pair is the fundamental and functional social unit all year round, about 70 percent or more of the time. Flocks of macaws and Amazon parrots can go as high as six or seven, but these larger flocks are one or two family groups made of the mated pair and young of the year. Less than 10 percent of their time is spent in flocks with more than three birds. The smaller parrots also spend their adult lives as mated pairs and small family groups, but in the smaller species, the pair travels alone only about 30 percent of the time. Larger groupings of small parrots are more frequent and regular, and flocks can routinely be as large as thirty birds. For example, Orange-cheeked Parakeets *Pyrilia barrabandi* and White-eyed Parakeets *Psittacara leucophthalmus* (formerly in the genus *Aratinga*) spend about 20 or 30 percent of their time in flocks of a dozen or so.

Evolution of Mortality and Senescence

Parrots are notorious for living a long time. Among animals, anyway, the long lives of parrots put them in a select crowd. Parrots are among the longest-lived of birds for their size, exceeded only by the petrels and shearwaters (Procellariiformes). Their life spans are comparable to our own, though the largest of parrots is only a tiny percentage of the body size of the typical human.

Longevity, or at least its flip side, the inevitability of death, has perplexed philosophers and scientists alike. At first glance, living to reproduce *ad infinitum* would seem to represent the highest possible fitness. As usual, things are not as simple as they first appear. Aging and death are ubiquitous among living organisms. Why should that be?

From an evolutionary or ecological perspective, there are two views of death. One view comes under the umbrella of *stochastic theories of aging*. These theories share the rather pessimistic view that every life is a minefield, and each individual life is a vain attempt to dodge all the hazards that could cause its death, such as accidents, the elements, predators, and disease. Eventually, each individual's luck runs out. The chances of living forever, given all of life's dangers, are slim indeed.

A second view is that death—or rather life span—is deliberate and carefully planned, as important in the design of the life of an individual as its birth, its pairing and the arrival of its offspring. Although an individual might not plan its demise as carefully as it chooses a mate or builds a nest, evolution stamps a plan in its DNA so that even the luckiest individual wears out and discards its spent body at the end of a long and, by the grace of the fates, productive life. Formally, natural selection works on traits conferring the flip side of death: survival and reproduction. This view of death has given rise to a suite of related hypotheses, the most relevant of which for our discussion is the *disposable soma hypothesis*. Central to this hypothesis is the trade-off in investing in one's offspring (reproduction) versus one's self (the soma, and future survival). The organism's life ends when it discards its soma (self) in favor of timely reproduction, into which it invests in full, leaving nothing to support further survival.

These evolutionary hypotheses, however, do not stand alone in permitting us to understand why trade-offs occur between investment in self versus investment in offspring. We also have to consider the exact mechanisms of aging. The business of living wears out the machinery of living things. To be alive, a cell must burn energy and build structures, processes that produce wastes. Wastes accumulate, and eventually a sufficiently old cell can no longer function. Even though new, fresh cells arise from old in all organisms (whether made of a single cell or many), mistakes can be made in transcribing DNA each and every time a cell divides. Eventually those mistakes happen, and they accumulate until the descendent cells can no longer function. Other types of mistakes can creep in through numerous vital cellular functions, too many to list here. Explaining how aging occurs goes under the label of *mechanistic theories of aging*. Parrots, with their unusually long lifespans, are a promising group for aging studies focused both on evolutionary theories and on mechanistic ones.

How long do parrots live? The record holding for parrot longevity (maximum life span) is often hard to verify, but some sources for captive parrots are reliable. Free-flying macaws at Florida's Parrot Jungle and Gardens have lived long enough to become geriatric, with creaky joints and cataracts, in their forties and fifties. Two recent compilations have examined maximum and median lifespans for parrots in captivity, with the most comprehensive of these studies conducted by Anna Young, Elizabeth Hobson, and Tim Wright from New Mexico State University and Laurie Bingaman Lackey from ISIS, the International Species Information System, which tracks animal demographic information for its member zoos. This study revealed maximum longevities routinely into the thirties, forties, and fifties for the larger species of parrots, primarily the macaws, cockatoos, and amazon parrots. The record holder of these is a Moluccan Cockatoo *Cacatua moluccensis* that lived to be 92 years old, and cockatoos in general show maximum life spans of just over 50 years. Median life spans (i.e. the most typical life span among all individuals of a species) are much lower, however, with only 30 percent of species having a median life span over 10 years. On average, males live slightly longer than females, and larger species tend to live longer than smaller ones. It is important to note that these data are taken from captivity, where husbandry and breeding practices can vary widely across institutions and over time. Thus, one must be cautious when extrapolating to wild populations.

In the wild, individuals are occasionally known to be breeding in their third and fourth decades. Some verified ages of active wild parrots include a 23-year-old Galah and a 19-year-

FIGURE 79 Richard Henry, the oldest known Kakapo (*Strigops habroptilus*), held by conservation biologist and Kakapo Rescue Team member Don Merton. Photo by Errol Nye, © Department of Conservation, New Zealand.

old White-tailed Black Cockatoo. These relatively low ages may say more about the longevity of studies following wild parrots than they do about the parrots themselves, however. Richard Henry (figure 79), the oldest living Kakapo, was already an old bird, judging by the usual signs of aging, when he was moved from his home in Fiordland to join the Kakapo Recovery Team's efforts in 1975. He was alive and actively breeding for the next 30-plus years, though he became blind toward the end. As I write this, I have just learned of Richard Henry's death in January of 2011 (more on his story in chapter 7). Surely he was much older than the 35 years he has been known to biologists, and some estimate his age at the time of his death to be at least 80. Ron Moorhouse, of the Kakapo Recovery Team, estimates the life expectancy of Kakapo to be 90 years, with an upper confidence limit of 120 years.

So, why do parrots live so long? Biologists have made a list of the routinely aged to see what distinguishes them from the rest. Size is the most universal of the so-called correlates of longevity. The larger the organism, the longer it is likely to live. The associations of size

FIGURE 80 A pair of Blue-throated Macaws, *Ara glaucogularis,* with a two-year-old chick, in the Beni region of Bolivia. Larger-bodied parrot species are typically longer-lived and have extended parental care compared to smaller species. Photo © José Antonio Díaz.

and longevity are mostly self-evident. It takes longer to get bigger, and if you live longer, you have the luxury of getting bigger. Conversely, spending from the fitness bank account, a larger individual has chosen to invest in self at the expense of producing offspring, a luxury allowed the longer an individual is likely to live. In these statements, you may notice that the arrow of cause and effect could go both ways, and indeed evolutionary theory predicts a certain amount of self-reinforcement feedback.

Within the constraints of body size, the most robust of the remaining correlates of longevity are armor, flight, island living, social life, intelligence, and herbivory. Parrots do not have armor in the sense that a turtle has a shell, but the other traits apply to parrots of all kinds. In particular, flight gives birds (and bats) the option of escaping predators and calamities with an efficacy not available to grounded mammals. Thus, flight is a bit like armor in helping animals avoid the external hazards of life in the wild. Predators are notoriously absent from islands, and many island birds such as the Kakapo discard flight, perhaps because its expensive function is no longer needed.

Likewise, social behavior and herbivory confer protection not only from predators and hazardous prey, but also, combined with communication and intelligence, give those organisms so gifted an advantage in finding scarce, unpredictable resources, as we have discovered for parrots in this and previous chapters. Although large brains are expensive to produce and require more resources (in terms of energy and protein), the payoff is large in meeting serious environmental challenges with flexible behavior (chapter 4). Parrots fit this bill well, ensuring that they will rank among the long-lived.

In fact—and paradoxically—another correlate of longevity is living where it is hard to make a go of it. At first that observation sounds contradictory, but a common successful strategy of both plants and animals in harsh environments is one of restraint and long-term investment. This and preceding chapters paint such a picture of the adaptive strategy of parrots, which we have argued make their livings exploiting resources that are far-flung and unpredictable (chapters 2 and 4). Accordingly, parrots fit the following scenario well. If an organism can get old, and big, it can ride out the perils and periods of poor fortune. It does this by delaying reproduction so it can invest in its own growth instead, later meting out a spare few offspring off and on during its long life (figure 80). This parent can take much better care of its few offspring, not only because it has few at once but also because it can accumulate wisdom and experience through the years, making it a better parent (chapter 5). It too learned from its own parents, so wisdom is culturally transmitted in the case of long-lived, intelligent, social beings, which include parrots (chapters 3 and 4). The long lives of wise parents span many breeding seasons, both good and bad. The resilient, large adult has good chances of making it through rough times. Over the years, it can be a successful parent during occasional (perhaps rare) good times, ensuring that at least some of its offspring will survive to reproduce on their own.

In this scenario, the large body size, flight, sociality, diet, and intelligence of parrots add up to lower the most critical variable in life-history evolution: *extrinsic mortality*. When the chances of an organism's living a long time increase, then a feedback loop favoring longevity is enabled. Old age itself becomes beneficial to an organism's fitness, as the ability to stay alive and get wise results in strong natural selection on traits that allow older individuals to stave off senescence and raise more offspring. Parrots share these necessary ingredients of longevity with humans. All told, parrots embody the spirit of the disposable-soma hypothesis of aging, residing on the far end of the spectrum where species do not dispose of their soma (i.e. bodies) to reproduce fast and furiously, but instead prefer to preserve their soma and reproduce sparingly and only when conditions are optimal.

One mystery remains, and that is how parrots (and other birds) avoid the various *intrinsic* hazards of living long lives, those molecular and cellular processes grouped under the mechanistic theories of aging. How do the cells in the bodies of parrots avoid oxidative damage from their high metabolic rates? How do their dividing cells compensate for shortening telomeres, and how does their DNA avoid the accumulative burden of mutations? Further investigation of the extraordinary longevity of parrots may even shed light on how longevity and long-term health can be promoted in other long-lived creatures, such as ourselves.

SEVEN

Populations of Parrots

Conservation and Invasion Biology

POPULATION ECOLOGY

It is not enough to understand the natural world; the point is to defend and preserve it.

—EDWARD ABBEY

CONTENTS

Introduction to Conservation Biology   217

Parrots in Peril   219

   How the Allee Effect Caused the Extinction of the Carolina Parakeet   219

   The Value of Biodiversity   223

   The Status of Wild Parrot Populations   223

     Box 7. The Un-Parrot and its Conservation Success Story   238

Parrots as Invasive Species   246

   Parrot Invaders: The Flip Side of Endangerment   246

   The Destructive Effects of the Trade in Wild-Caught Birds   255

   Can and Should Wild Parrots Be Harvested?   259

INTRODUCTION TO CONSERVATION BIOLOGY

My travels to the pristine rainforests deep inside the Amazon River Basin in the 1970s gave me a glimpse into what my native continent of North America must have been like before the arrival of European explorers in the fifteenth century (see preface). In another moment of insight in my career as a naturalist, I visited the old-growth hardwood forests of the Congaree River floodplain, in South Carolina, in the early 1970s, when its ownership was still in private hands. With a reverence normally reserved for cathedrals, my doctoral advisor John Terborgh, his colleague and fellow plant ecologist Egbert Leigh Jr., and I found our way into the bottomland majesty of this forest. This never-logged tract was an exceedingly rare remnant of the forest that once covered vast areas of eastern North America. In the spring of 1973, the fate of this treasure, surely as unique and valuable and irreplaceable to humankind as the Mona Lisa, hung in the balance of economic pressures. Only the next year, the Congaree River's hardwood bottomland forest was saved, as it first entered federal protection as a national landmark and then eventually became South Carolina's first national park.

FIGURE 81 A painting by wildlife artist Michael Rothman of the extinct Carolina Parakeet, *Conuropsis carolinensis,* in its former habitat in the river bottoms of the southeastern United States. Painting © ACE Coinage.

Although that story has a happy ending, the moments of wonder and heartache that I experienced that day in the Congaree have never left me. My disquiet has continued because of the silence of the forest and the absence of animals that should have been there. Among other denizens, the environs of the Congaree River bottoms should have been filled with the chatter and flash of busy parakeets browsing, exploring, playing, and tousling (figure 81). Just as the sounds of countless parrots filled my subconscious from dawn to dark in Manú National Park, so should they have in South Carolina.

This chapter expands our focus to that of whole populations of parrots. Unlike in the previous chapters, the story of parrot populations is so interwoven with that of our own species that our human presence cannot be excluded. In this chapter we review the fates of free-roaming parrot populations as they abruptly encounter ours. Our influence on wild parrots in the last millennium is a mere blink in time compared to the 50-plus-million-year history of the order and even to the shorter histories of many extant species of parrots. Yet, however short on those scales, the human effect on their populations has been profound.

In the first section of this chapter, we learn of parrots that did not survive this contact with us and we ask how many more may meet the same fate. In doing so, we explore the reasons that so many parrot populations are at risk in the modern era. In the second section, we turn the tables and consider whether there can be too many parrots. We ask why some parrots establish populations, with only a little help from us, where none existed before. We also consider whether wild parrots in their historical haunts can be so abundant that we can harvest them sustainably, and if so, whether we should. Our human role in setting the fates of wild parrots in the next century is so pervasive, I cannot help reaching the same conclusion as Edward Abbey, the latter-day American essayist and eco-warrior quoted above. In this chapter, we strive to understand wild parrots in their original environment, but simply understanding may not be enough when we hold their future in our hands.

PARROTS IN PERIL

How the Allee Effect Caused the Extinction of the Carolina Parakeet

Most of us consider parrots subtropical—an animal that would wither in the crisp frost of higher-latitude, temperate climes (chapter 1). Many of you reading this book live in these higher latitudes, and probably most are not used to seeing native parrots at feeders outside your window on a winter's day (with some exceptions, most notably those Down Under). This myth might not be so popularly held were it not for the remarkable eradication of the Carolina Parakeet *Conuropsis carolinensis,* a species that once reached far beyond the Neotropics into the northern temperate latitudes of the Americas.

This abundant little conure ranged northward in North America to the river bottomlands of New York state, and westward to the riparian forests of the great central rivers and tributaries, the Mississippi, the Missouri, and the Platte. As far as we know, these parrots did not migrate to avoid cold and snow. They were hardy, resourceful birds that graced the various temperate and subtropical forests of the mainland United States. Carolina Parakeets were not too particular about diet or habitat. They were typical parrots, feasting on a wide variety of seeds, flowers, and fruits. They apparently lived in the usual family-based flocks, nesting in tree cavities in the spring, raising one brood at most per year, and roosting together outside the breeding season. They roved here and there to take advantage of abundant food, but they seemed not to be as nomadic as the Cockatiels *Nymphicus hollandicus* and Budgerigars *Melopsittacus undulatus* of Australia. Rather, they frequented the same haunts but with seasonal local migrations following whatever resources they might need for their activities at that time of year, much as the Short-billed Black Cockatoos *Calyptorhynchus latirostris* and many other species of parrots do (chapter 6).

Biologists debate exactly how numerous these North American parrots were. Journals of explorers, pioneers, settlers, and sightseers alike spoke frequently of encounters with flocks of Carolina Parakeets, from the Europeans' first introduction to them as early as 1580 until their disappearance at the turn of the last century. Daniel McKinley, their primary historian and a trained biologist, reviewed these many accounts with skepticism, for no one took data or recorded specific numbers, not even the early scientists such as John James Audubon. In a series of papers published over twenty-five years, McKinley chronicled as best he could the Carolina Parakeet's historical existence and tried to distill out as much quantitatively reliable information as the written accounts would allow. The conclusion is that Carolina Parakeets were certainly abundant enough to be found over wide areas, and were common at the boundaries of their known range. In other words, they were not rare visitors outside of strongholds in the southernmost coastal-plain forests, at least not at first encounter with European immigrants and travelers. What happened to them?

Scientists debate to this day what could have brought about the demise of a common species that seemed well adapted to this biome,[1] one similar to those where other parrots (native and naturalized) continue to thrive to this day. Most extinctions of parrots have occurred on islands, as we shall soon explore. The Carolina Parakeet remains arguably the only widespread, generalist, continent-dwelling species of parrot that has met with extinction.

The last captive Carolina Parakeet died in 1918. Although his death is popularly considered to be the extinction of this species, Carolina Parakeets may well have survived in the wild for several more decades. Noel F. R. Synder and Rod Chandler uncovered a remarkable history of the final demise of North America's native parrot, as Synder reports with Keith Russell in *The Birds of North America* and in his thorough treatise on this species, *The Carolina Parakeet: Glimpses of a Vanished Bird.* The very last stronghold of the Carolina Parakeet may well have been the river bottomlands of North and South Carolina (the very place that struck me as so empty in 1973), where sightings of little green parrots lasted into the 1940s.

The Extinction Vortex

To be sure, the reasons for the extinction of the Carolina Parakeet are forever hidden in the mists of time. Despite the difficulty of re-creating events for which contemporary records are lacking, Snyder's premise is that we as conservation biologists need to keep trying to understand, and I concur. His chapter entitled "Post Mortem of a Conservation Failure" is offered in the hopes that a lesson can be learned from this terrible loss and that other species may be spared the same fate. I will not reiterate Snyder's excellent analysis, nor the analysis of Robert Askins in *Restoring North America's Birds: Lessons from Landscape Ecology.* Instead I bring their findings together here with a bit of a new twist.

It is not difficult to enumerate the pressures on the Carolina Parakeet in North America after the arrival of Europeans. The most obvious of these depredations on the Carolina Parakeet was the relentless shooting for fun, food, retaliation (for raiding crops), scientific museum specimens, and adornment, in approximately that order. Without doubt, some Carolina Parakeets were also removed from the wild by entrepreneurs and local residents

---

1. *Biome* is a term referring to a collection of vegetation types and ecosystems typical of certain climatic zones. Thus, the temperate-forest biome occurs at similar latitudes on different continents and comprises ecosystems dominated by deciduous trees.

for sale as pets, especially destined for Europe but also for northern cities in the United States, and as eggs for both private and scientific egg collections. More insidious perhaps was land conversion, most notably clearing of the eastern deciduous forests. Perhaps especially harmful to parakeet populations was the logging of southern bottomland old-growth forests (like that of the Congaree River floodplain), containing large rotting trees with suitable nest cavities, and of canebrakes, the dense thickets of *Arundinaria gigantea,* which may have provided Carolina Parakeets with bonanzas of seeds from this large American bamboo. The parakeet's range was notably coincident with the distribution of both types of vegetation.

Lesser factors, but still adding up, included competition for those increasingly rare nesthole cavities in large trees by exotic introduced European honeybees (not only by the bees themselves but in the aftermath of honey hunters' destroying the trees to obtain the spoils). Finally, Snyder brings to our attention the possibility of contagion of avian diseases from domestic fowl, against which the parakeet had no evolved resistance.

Askins presents a compelling argument that the Carolina Parakeet was more of a specialist in the eastern deciduous forest than were most North American bird species. In fact, their population stronghold was in the southern coastal and river bottomland hardwood forests, linking them with two other mainland bird species lost from North America after the arrival of Europeans: the Ivory-billed Woodpecker *Campephilus principalis* and Bachman's Warbler *Vermivora bachmanii.* The eastern deciduous forest was relatively slowly cleared until its nadir in the 1870s, at about 50 percent of its former landscape coverage. Since then, the eastern deciduous forest has rebounded in its extent, albeit as still relatively young second growth. Not so the southern bottomland old-growth forests. This ecosystem was reduced by nearly 80 percent of its former landscape coverage and has not regained any significant ground since the last great tracts were clearer prior to World War II.

Carolina Parakeet populations were in an alarmingly steep decline in the late 1800s. Given the modern landscape-level analyses of Askins and his colleagues, we can now suspect that the decline was not so mysterious. By applying various ecological theories (principally the theory of island biogeography), conservation biologists predicted that with a reduction to 50-percent coverage 16 percent of bird species native to the eastern deciduous forest would become extinct—this is indistinguishably close to the actual percentage, 18 percent. As Askins points out, hitting the predicted number of extinctions can satisfy us that habitat loss probably precipitated the extinctions. This fit to prediction, however, does not account for which of the several hundred bird species were included in that 18 percent. Thus, we need once more to consider the particular biology of the Carolina Parakeet to understand why this species was extirpated and not another.

Both Snyder and McKinley, and others, have reviewed all of the above known impacts on the Carolina Parakeet to sleuth out the cause of the final decline of this species. In Snyder's evaluation of the last credible sightings of Carolina Parakeets in the wild in the 1920s and 1930s, he reports that family groups with newly fledged youngsters were a feature of the local flocks. Because breeding individuals and offspring were observed until the populations winked out, and by then the historical pressures on their populations were no more, their final demise has greatly perplexed the many biologists who have pondered their extinction.

I argue here that it need not. The loss of individuals from all of these causes was additive, not compensatory. By *additive* I mean that all the different causes of mortality piled on top of each other to do more damage in concert than any did alone. In the alternative scenario, *compensatory* mortality would mean that one cause of mortality was simply killing birds that probably would have died from another source of mortality anyhow, so adding the two forms together would not make overall mortality much worse. In the case of the Carolina Parakeets, those that died from being shot could be added to those leaving the population to be sold as pets. If they were not captured or killed, the remaining parakeets had a tough time finding nest cavities to raise their babies, and so fewer babies were produced. The cultural transmission of information about where to find food, good nest sites, and safety from predators had been lost with the relentless removal of older, experienced individuals from the population (chapters 3–6). No longer were flocks big enough to provide protection from natural predators. The Carolina Parakeet had slipped below the population-level extinction threshold, as predicted by pioneering ecologist W. C. Allee, in which the growth rate of a species actually becomes negative, and the species was caught in the grip of the so-called *extinction vortex*.

What density was the extinction threshold for the Carolina Parakeet? And when was that threshold density crossed for the average population of this species over its wide range? Reports suggest that populations on the periphery of the range had entered the vortex by the early 1800s. By the mid-to-late 1800s the entire species was in its grip. To save the Carolina Parakeet from oblivion would have been extremely difficult indeed, even if someone had decided to try, and even with modern methods of conservation.

In other words, I argue that why the Carolina Parakeet became extinct is no mystery. The full answer is not simply a shift at the end to some elusive cause of extinction, delivering the final coup, nor some undiscovered, unaccounted-for scenario that was the real reason the bird went extinct. I argue that everyone who has advanced a credible hypothesis is correct. The Carolina Parakeet is gone precisely because of *all* of these proposed causes of their demise. The species was hammered relentlessly and mercilessly from all sides. All of the killing, capturing, converted habitat, introduced competitors, and diseases added up, pushing the populations down the slippery slope and over the precipice of the extinction vortex.

In this sense, the end of the Carolina Parakeet began many years before the last, lonely wild parakeet left our physical world. Even by twenty-first-century standards, the Carolina Parakeet would have been difficult to save had a concerted effort begun as late as the 1890s. The Kakapo *Strigops habroptilus*, the Puerto Rican Parrot *Amazona vittata*, Spix's Macaw *Cyanopsitta spixii*, the Thick-billed Parrot *Rhynchopsitta pachyrhyncha*, and more, all give testimony to how difficult is the task of saving a species once the population's size hovers around the extinction threshold. All the king's horses and all the king's men, representing all the modern technology, modern sensibilities, and modern resources, struggle today to save these still-extant endangered species; their fates are hardly secure.

Surely we must all agree with Noel Snyder that the loss of the Carolina Parakeet was a massive conservation failure. I cannot improve on Snyder's poignant observation: "The loss of this spectacular species, perhaps the most brilliant of all our original wildlife species, remains, and will presumably always remain, a profound aesthetic, biological, and social tragedy. That no real effort to preserve the parakeet was ever made represents a perpetual reminder of the dark side of our own species' history."

The Value of Biodiversity

Why does the extinction of a species somehow mean more than simply the collective deaths of many individuals? Only humans, it would seem, can reflect on the species to which they belong and worry about the continuation of humankind, with its unique civilization and culture. None of the last lonely individuals of the Carolina Parakeet, the Passenger Pigeon, the Ivory-billed Woodpecker, or Spix's Macaw (figure 82), nor those of countless more terminated lineages, bemoaned the passing of their kind. At most they bemoaned their own loneliness or grieved the loss of a mate, in those species, such as the parrots, which pair monogamously for life (chapter 5).

What can parrots tell us about why some species become extinct while others continue to thrive or even expand, as we humans transform the biosphere? What can we learn from the parrots, so that our species will not repeat its past mistakes, letting the Carolina Parakeet, Spix's Macaw, and others slip away from this world? Next we explore patterns in populations of parrots, seeking lessons for how our species can coexist with wild parrots, and indeed with all the other living inhabitants of the only biosphere known to us.

The Status of Wild Parrot Populations

Some 371 (or so) recognized species of parrots have been described since Linnaeus first began assigning names to living organisms (chapter 1). Nineteen of these are designated as extinct by the World Conservation Union (IUCN), which annually monitors the Earth's biodiversity in its Red List of Threatened Species. Spix's Macaw and another species, the Glaucous Macaw *Anodorhynchus glaucus,* are not yet classed as extinct in the wild by the IUCN because a remote possibility exists that they could persist in nature in areas not yet searched. Both species, however, are viewed as extinct in the wild by the majority of conservation biologists. Extensive searching has failed to discover any individuals of either species in what little remains of the habitat suitable for them.

Since parrots have been interacting with humans, then, 21 (6 percent) of the recognized species have become extinct as a direct result of human activity (table 4). Another 122 species (35 percent) are currently at some level of risk, leaving 228 (61 percent) not immediately threatened by human activities. Can we discover any patterns in parrot populations that explain why these species are at risk, or not?

FIGURE 82 The last wild Spix's Macaw, *Cyanopsitta spixii*, in the caraiba woodlands of eastern Brazil. This species now only exists as a small captive population. Photo © Luiz Claudio Marigo.

The Numbers Give Up Their Secrets

We could explore this question by comparing three groupings of species: (1) those presumed or known to be extinct in the wild; (2) those currently at some degree of risk; and (3) those not immediately threatened.[2]

2. The IUCN recognizes seven categories of status for any given species, based on the likelihood of species-wide extinction and the speed of the remaining population's trajectory toward extinction. In order of decreasing risk, these are (grouped by the three categories used in the text): (1) *extinct* (EX) and *extinct in the wild* (EW); (2) *critically endangered* (CR), *endangered* (EN), *vulnerable* (VU), and *near threatened* (NT); (3) *least concern* (LC) and all remaining species known not to be at risk.

TABLE 4　A list of species of parrots described from verified specimens or credible historic descriptions that are now extinct in the wild

Human impact was the direct cause of these extinctions.

Causes of extinction: H, hunting; T, traded as pets or as eggs: L, habitat loss; C, competition with introduced species; P, predation by introduced predators and disease.

Genus	Species	Common name	Extinction by or soon after:[1]	Cause of extinction
Island species				
Vini	*sinotoi*	none	1300	H
Vini	*vidivici*	none	1300	H
Nestor	*productus*	Norfolk Island Kaka	1851	H,T
Eclectus	*infectus*	none	1000[2]	H
Cyanoramphus	*uleitanus*	Raiatea Parakeet	1773	H,L,P
Cyanoramphus	*zealandicus*	Black-fronted Parakeet	1842	H,L,P
Mascarinus	*mascarinus*	Mascarene Parrot	1834	H,T
Psittacula	*exul*	Newton's Parakeet	1875	H,L
Psittacula	*wardi*	Seychelles Parakeet	1883	H,L
Lophopsittacus	*bensoni*	Mauritius Grey Parrot	1764	H
Lophopsittacus	*mauritianus*	Broad-billed Parrot	1680[2]	H
Necropsittacus	*rodericanus*	Rodrigues Parrot	1763[2]	H
Ara	*atwoodi*	Dominican Green-and-yellow Macaw	1791[3]	H,T
Ara	*erythrocephala*	Jamaican Green-and-yellow Macaw	1847[3]	H
Ara	*gossei*	Jamaican Red Macaw	1765[3]	H
Ara	*guadeloupensis*	Lesser Antillian Macaw	1760[3]	H,T
Ara	*tricolor*	Cuban Macaw	1885	H,T
Aratinga	*labati*	Guadeloupe Parakeet	1750[3]	H
Amazona	*martinicana*	Martinique Parrot	1779[3]	H
Amazona	*violacea*	Guadeloupe Parrot	1779	H
Mainland species				
Psephotus	*pulcherrimus*	Paradise Parrot	1927	H,T,L
Anodorhynchus	*glaucus*	Glaucous Macaw	1961	L,H
Cyanopsitta	*spixii*	Little Blue Macaw	2000	L,T
Conuropsis	*carolinensis*	Carolina Parakeet	1938	H,T,C,P

[1.] Last confirmed record in the wild *or* last known individual died or removed from the wild.

[2.] Known from fossil or subfossil bones and, in some species, travelers' accounts. For species known only from bones, dating is based on stratigraphy and can only roughly estimate last known record.

[3.] Known not from physical specimens but recorded in sufficiently detailed and credible descriptions to warrant listing as extinct taxa in *Threatened Birds of the World* (Stattersfield and Capper 2000). Fuller (2001) lists some of these as hypothetical, including them in a list of ten other hypothetical taxa not listed in this table. Steadman (2006b) estimates that the number of actual extinctions in the Pacific island area (Near and Remote Oceana) is double that known. Therefore, this list is in all likelihood incomplete.

The causes of extinction for each of the parrot species that have met this fate already are the familiar litany that we reviewed with the Carolina Parakeet (table 4). Most of these now-extinct species of parrots were hunted to low densities for food and for trade as pets. For others, loss of vital habitat for some reason or other caused the species to lose ground, through loss of food supply or nesting sites or familiar places for the birds to gather or avoid predators. For a few, species other than humans competed with the parrots for nest sites or food or preyed on the adults, nestlings, or eggs. But most of these inroads on density of parrot populations affect nearly all known species of parrots, so, again, why did the 6 percent lose the battle with humans so quickly, and why are another third of all known species of parrots imminently at risk?

We can sleuth out some answers to these questions using the quantitative approaches of population ecology. Jamie Gilardi and I analyzed data for each of the 371 species of parrots listed by Sibley and Monroe (see the preface), which include recognized species that became extinct following the wave of European expansion that began in the sixteenth century.[3] We tested for patterns that might explain why any given species of parrot might have become extinct or be at risk by the twenty-first century. Potential causes of a population's vulnerability (groups 1 and 2) or lack thereof (group 3) that we considered included the following characteristics of each species: average body size, size of that species' geographic range, and whether a species occurs primarily on a water-bound island (as defined by geographers) or mainland.

Statistical analyses yielded some straightforward answers that I will summarize here. Figure 83 presents an overview of all 371 species of parrots plotted on a single graph, arranged according to the size of each species' geographic range on the *x* axis (from small to large) and the average body size characteristic of that species on the *y* axis (from small to large). Both axes are expressed as the logarithm (to base 10), which means that the differences along these axes are in orders of magnitude and not simple linear scaling. The degree to which each species is at risk is denoted by color: closed red circles for extinct in the wild; open yellow circles for at risk (critical, endangered, vulnerable, and near threatened); and blue diamonds for least concern. Another pattern apparent enough in this scattergram is that the yellow and especially the red points tend to be concentrated toward the upper-left-hand quadrant representing smaller range sizes and larger body sizes.

Overall, we found the following patterns. On mainland areas, larger species tend to have larger geographic ranges than do smaller species. We might expect such a trend because larger individuals require larger home ranges to provide for themselves, and somehow this need translates to the entire population's being spread over a wider area. However, the opposite trend holds for island-dwelling species; body size and range size are inversely related on

3. We did not include extinctions of species prior to the sixteenth century in table 4, even if these extinctions were rather recent and presumably caused by human activities, such as the island-dwelling species of *Vini* and *Eclectus* that were probably driven extinct by the expansion of the Polynesians and other peoples across Oceania.

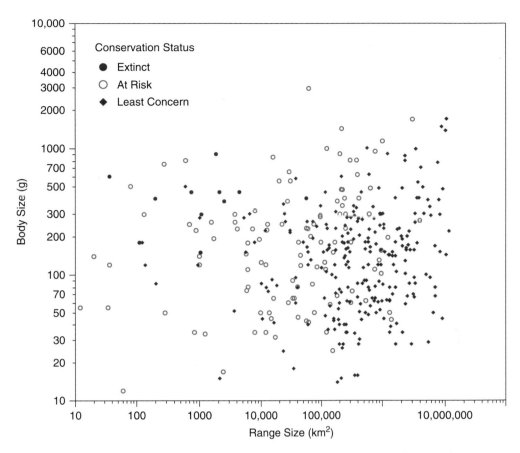

FIGURE 83 Body size of parrot species, in grams, plotted against range size, in km². Both axes are plotted on logarithmic scales. Extinct species are plotted as closed red circles; species at risk of extinction, open yellow circles; and species of least conservation concern, blue diamonds. Figure by Tim Wright from unpublished data compiled by Cathy Toft and James Gilardi.

islands. In other words, the larger species of parrots that live on islands occupy smaller islands, on average, than do smaller island-living species of parrots. This nonintuitive result is consistent with the tendency for island-dwelling animals to evolve to larger body sizes relative to their mainland-dwelling counterparts (recall that the Dodo *Raphus cucullatus* was just a giant, flightless pigeon found on a small island).

These relationships then put the other patterns in better context. Parrots of any size occurring over smaller areas are more at risk for extinction than those with larger geographic ranges, as summarized in figure 84. Here we see that, on average, extinct species and those at risk for extinction have much smaller ranges than species of least conservation concern. Note that these are the historical (pre-Columbian) geographic ranges, not present-day, so they are not confounded by the decrease in range size often seen as a species shrinks in numbers.

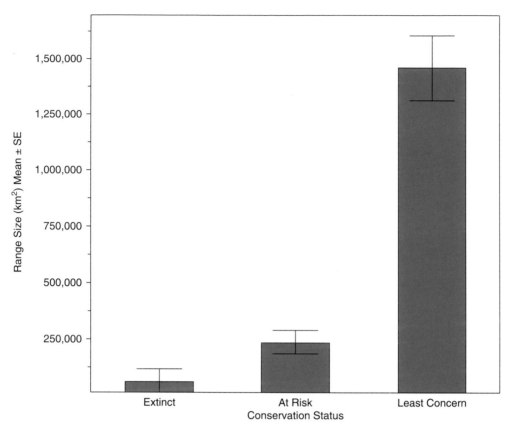

FIGURE 84 The range size of parrot species classified as extinct, at risk of extinction, or of least conservation concern. Extinct and at-risk species have much smaller range sizes, on average, than those of least concern. Figure by Tim Wright from unpublished data compiled by Cathy Toft and James Gilardi.

A corollary of that pattern is that more species living on islands have become extinct than have those living on mainland areas (figure 85), probably because geographic ranges on islands are necessarily restricted in size and, further, populations are cut off from immigration. Of the species of parrots, 153 (41 percent) occur only on islands. Of those, 17 (11 percent) have become extinct. In contrast, only 4 (2 percent) of the 218 species of parrots occurring on continental mainlands have gone extinct in the wild. Thus, island living entails a high risk of extinction for parrots. In addition, figure 85 shows that disproportionately more island species remain at risk in the twenty-first century compared to mainland species.

Turning now to the effect of body size, no small species of parrot has become extinct in post-Columbian times, even though more than one-third of parrot species weigh under 100 grams. The mean body size of parrots that have become extinct is just over 400 grams, and the mean size of those at risk is around 300 grams (figure 86). Thus, large parrots are particularly at risk for extinction, even though they tend to have larger geographic ranges, on continents at least. What is going on here?

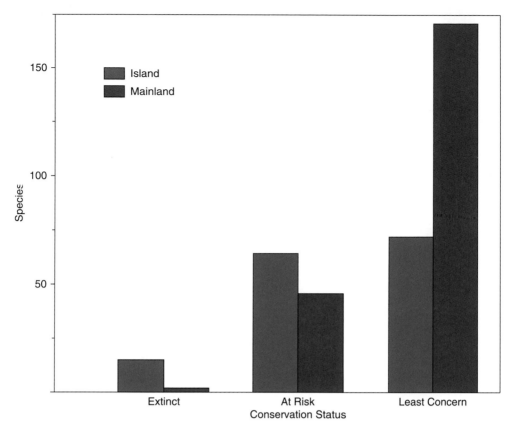

FIGURE 85 A comparison of the number of species classified as extinct, at risk of extinction, or of least conservation concern that are found on islands (blue bars) versus mainland areas (red bars). Much higher proportions of extinct and at-risk species are found on islands than of species of least concern. Figure by Tim Wright from unpublished data compiled by Cathy Toft and James Gilardi.

This seeming paradox has its resolution in the fact that many island species evolve to very large body sizes. It is those large parrots living on islands that are far more at risk for extinction than any other category. You can see this in the red circles dominating the upper-left-hand side of figure 83. In contrast, those blue points in the upper-right quadrant of figure 83 represent the large parrots living on continents. Many of these may be at less risk than small and medium-sized parrots because of their larger geographic ranges. To understand more about these broad statistical patterns, let us next examine specific examples of parrot populations on continents and islands in turn.

Parrots on Continents

Of the four extinct species of parrots occurring on mainlands, the medium-sized Carolina Parakeet (box 1, in chapter 1) had an uncharacteristically large range for an extinct species, on the order of 900,000–1,000,000 km². Thus, the Carolina Parakeet is an outlier with

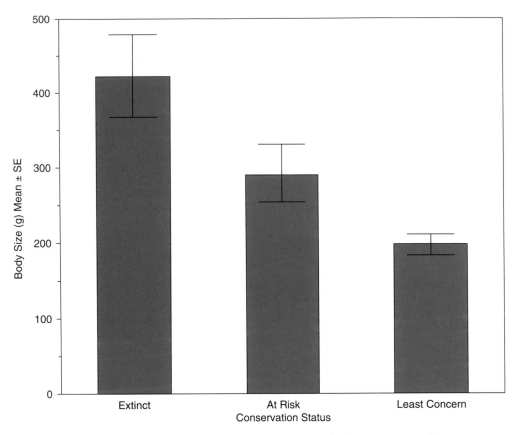

FIGURE 86 Body size of parrot species classified as extinct, at risk of extinction, or of least conservation concern. Extinct and at-risk species have much larger body sizes, on average, than those of least concern. Figure by Tim Wright from unpublished data compiled by Cathy Toft and James Gilardi.

respect to the extinction patterns described above, and its extinction is yet more paradoxical than just being one of the expected 18 percent of species of birds to become extinct with a 50-percent reduction of temperate forest in North America.

The other three mainland extinctions, Spix's Macaw, the Paradise Parrot *Psephotus pulcherrimus,* and the large Glaucous Macaw, occurred over ranges that were restricted to perhaps unusually specialized habitats, even if they were not bounded by water as are bona fide islands. In other words, these other three mainland extinctions might be considered as having occurred within the bounds of *habitat islands.* A habitat island gets its name because when the matrix surrounding a specific habitat is hostile to the population within that habitat, the matrix can act as a barrier to dispersal and isolate that population as surely as a an ocean does for a real island. These habitat-matrix islands (or *patches*) have critical consequences for the small and isolated populations in them.

Spix's Macaw (figure 82) was restricted to a specific type of vegetation that was itself limited in extent, namely the gallery forests in the arid *caatinga* of northern Brazil. Spix's Macaw's journey with humans is chronicled by Tony Juniper in his gripping book *Spix's Macaw: The Race to Save the World's Rarest Bird*. Sadly, that race was lost, representing another massive conservation failure. Spix's Macaw was apparently always confined to the particular open-gallery woodlands of large caraiba trees, *Tabebuia caraiba,* surrounded by *caatinga* scrublands and limited to the floodplains of the wandering São Francisco River system that empties eastward into the Atlantic Ocean, bypassing the mighty Amazon River's drainage. The climate in this area of Brazil is tropical but dry, producing an open-savanna type of forest, where plants can come by a little more water for their growth by hugging the floodplains of rivers. This area of Brazil is both isolated and not very favorable for human enterprise, so it remains lightly populated. Nevertheless, deforestation marches along there, like almost everywhere, fueled by the lumber industry that targets the large caraiba. Unfortunately, the caraiba also supports Spix's Macaw, providing safe roosts, nest hollows, and food. Thus, the relatively small and delimited range (on the order of 400,000 km^2) got even smaller as logging increased. This trend made the relatively large (400-gram) parrot more vulnerable to extinction, and no match for the ravenous high-end collectors rich enough to afford such a status symbol.

Likewise, the Glaucous Macaw was relatively restricted to another dry subtropical eco-region, the Uruguayan savanna, where the chatay *Butia yatay* and other colonial palm trees thrive along the banks of major river systems. This eco-region turned out to be good for raising cattle, which in principle could coexist with large macaws, but with overgrazing, clearing, and unnatural fire, the chatay itself became threatened in this area. The Glaucous Macaw was therefore not only a habitat specialist but a food specialist. Although at 200,000 km^2 its range was much larger than that of the average island population of parrots, its distribution was spotty within this range and limited to patches of the chatay. With an effectively small range size and large body size (900 grams), the Glaucous Macaw was definitely vulnerable. Pressure from humans hunting the Glaucous Macaw for food was more than a species with these characteristics could withstand. The sister species of the Glaucous Macaw, the Lear's Macaw *Anodorhynchus leari* (figure 87), is itself endangered. Its restriction to scrubland (*caatinga*) habitat and its historical range size (200,000 km^2) mirror those of the Glaucous Macaw, and not surprisingly the Lear's Macaw persists by a precarious toehold.

We know less about the Paradise Parrot, which disappeared from the radar in Australia in the late 1920s. Those accounts that do survive indicate that this parrot, too, was restricted to a specific habitat type—interestingly open woodland savanna, an eco-region much like that of the native haunts of the Spix's and Glaucous Macaws. Its historic range was relatively small, at 50,000 km^2, for a population on a continent (in the bottom 10 percent of continental species' range sizes), putting the species' population at risk despite its relatively small size (just under 200 g). Again, why the Paradise Parrot, and not the other twenty or so other parrots with equal or smaller ranges? The answer may be simple. The entire eco-region that was home to the Paradise Parrot is itself endangered. The savanna type labeled the *brigalow*

FIGURE 87 A pair of Lear's Macaws, *Anodorhynchus leari*, perched in a licuri palm in Brazil. This endangered species is restricted to the dry *caatinga* scrubland of northeastern Brazil. Photo © Sam Williams.

belt, after its overstory dominant, the brigalow *Acacia harpophylla*, contains a variety of unique vegetation types. This eco-region is sandwiched between the moister tropical climes on the coast and the menacing dryness of the vast Australian interior. The eco-region was massively cleared to less than 7 percent of its original cover of brigalow for agriculture and livestock grazing. The area was also intentionally "improved" for those purposes with the introduction of non-native forage plants. As a result, numerous native Australian plant species and vegetation communities within the region are threatened, endangered, or fully extinct. The extinction of the Paradise Parrot was collateral damage in this no-holds-barred land conversion. Apparently it nested in the large termite mounds that would have been frequent in unmanaged scrublands and savannas but which could not persist with clearing and grazing. This "most beautiful" (*pulcherrimus*) little parrot was also hunted and traded, but the biggest threat to its population was loss of feeding and nesting areas with the invasion of exotic species. The Paradise Parrot slipped away, along with an inestimable number of species of plants and animals unique to the brigalow eco-region.

The other continental parrots critically at risk but still extant in the wild comprise five species with small ranges for continental species, combined with large bodies in the case of two of them. Four of the five critically endangered species are Neotropical. Two are large macaws: the Lear's Macaw (over 900 grams), already mentioned above, and the Blue-throated Macaw *Ara glaucogularis* (800 grams; figure 88), another macaw with an ecology similar to that of the Glaucous Macaw. Both species are dependent on a specific species of palm. For the Blue-throated Macaw, both macaw and palm occurred in naturally flooded savanna that has lately been drained and cleared for cattle grazing. The Blue-throated Macaw's historic range size of around 400,000 km² is now much reduced and at best under the average for continental parrots. The third is the Yellow-eared Parrot *Ognorhynchus icterotis* (less than 300 grams), a monotypic form that has not yet been included in any molecular phylogenetic analysis, so that its affinities are unknown. Like the macaws, it is also highly dependent on palm trees, in this case wax palms in the genus *Ceroloxyn*, and it occurs at specific mid-elevations in the Andes, in elfin forests where the palm occurs, over a historic range of 90,000 km².

The fourth of the critically endangered continental species is the Indigo-winged Parrot *Hapalopsittaca fuertesi* (125 grams), which had been known only from the type specimen collected in Columbia and presumed extinct until its recent rediscovery in two protected locations. The Indigo-winged Parrot shares a restricted distribution in mid-to-high-elevation cloud forest in the Andes with its two congeners, also at risk, the Rusty-faced Parrot *H. amazonina* (endangered) and the Red-faced Parrot *H. pyrrhops* (vulnerable). These habitat islands are produced by the variable elevations of peaks of great mountain chains. Plants respond to climate, which in turn varies in related ways with both latitude and altitude. Thus, mountain peaks constitute "islands" of certain vegetation types surrounded by "seas" of lower-elevation habitats of radically different vegetation types. Populations of animals that do not disperse readily can become genetically and reproductively isolated from other populations on these mountaintop islands. If the habitat type is distinct enough and isolated enough, these patches of land can mimic all the ecological properties of islands surrounded by water.

FIGURE 88 A pair of Blue-throated Macaws, *Ara glaucogularis,* at a nest in Beni, Bolivia. This critically-endangered species is threatened by loss of its dry forest habitat, which has been cleared for cattle pasture, and by capture for the pet trade. Photo © Darío Podestá.

The range of the Indigo-winged Parrot is on the order of 10,000 km^2, which places it solidly within the distribution of range sizes of parrots restricted to water-defined islands. Its at-risk congeners also occur over smaller ranges than does the one species in this genus that is not known to be at risk, the Black-winged parrot *H. melanotis,* with a historic range of 112,000 km^2. All species in this genus are restricted to middle and high elevations in the Andes, and none is common.

The fifth critically endangered species of parrot occurring on continents is Australia's Night Parrot *Pezoporus occidentalis.*[4] The Night Parrot shares close phylogenetic affinities with its congener, the Ground Parrot *Pezoporus wallicus* (figure 26), but not the superficially similar but unrelated Kakapo (chapter 1). All three species are ground dwelling and nocturnal (odd for parrots). Apparently living on the ground is dangerous, so all of these species evolved subtle camouflage-colored feathers and nocturnal habits. The ground-living habit might therefore be related to why Night Parrots are highly vulnerable, having greater susceptibility to predators, especially the exotic placental mammals that colonized Australia along with European immigrants. With all three species, their nocturnal habits make them equally as hidden from the attention of predators as from that of curious humans, making their population status difficult to determine. The Night Parrot remains more of a mystery bird now than the Kakapo. The early settlers of the Australian outback ran into this bird frequently enough in the 1870s and 1880s, but it seemed suddenly to disappear from human view after that. The last verified encounter in the wild took place in the 1920s, until a sad Night Parrot road kill was discovered in 1990 in an area where they were rumored to exist. The well-preserved, months-old corpse, obligingly cleaned up like a museum specimen by helpful ants, was unequivocal proof that the species was still extant. This encouragement gave sufficient energy for modern-day ornithologists to redouble their efforts and keep looking. One of the best recent sightings of live Night Parrots was that of April 12, 2005, by Brendon Metcalf and Rob Davis, who were consultants preparing an environmental impact report for a strip-mining project in a remote area of north Western Australia. The thrill of seeing living members of a species once declared extinct surely can only be described as sublime.

Oddly enough, biologists cannot ascertain whether Night Parrots are exceedingly rare, or more abundant but exceedingly difficult to find; this is reminiscent of the controversies over whether the Ivory-billed Woodpecker has actually been rediscovered. In the case of each, their native habitat is remote and inhospitable, as well as threatened in and of itself. Night Parrots live in the vast Australian interior substeppes, which are dominated by bunch grasses and salt-tolerant shrubs, where populations of all flora and fauna occur at low densities over wide areas. The Night Parrot is apparently nomadic, tracking flushes of seed production in this otherwise stingy landscape. The Night Parrots are additionally vulnerable

4. The Australians are responsible for another critically endangered species, the Orange-bellied Parrot *Neophema chrysogaster.* We lump this species in with the island inhabitants because its sole breeding grounds are in Tasmania, to which the parrots migrate over water (an act shared with few other parrots), returning to the southern coasts of continental Australia to over-winter.

because they must seek out local water holes in this arid land, where predators also congregate. The inherently low density and precarious existence in a fragile landscape surely predispose a population of grounded parrots to be at risk in the face of human encroachments. It seems wise therefore to list this species as critically endangered.

Other populations were not included in our study because they were subsumed under a name for a widely distributed metapopulation. One such example is the Cape Parrot *Poicephalus robustus*, which is a food specialist and arguably restricted to habitat islands, making it especially vulnerable to exploitation by humans for the trade.

Parrots on Islands

Of the described species of parrots extant when humans arrived on the scene, 41 percent were endemic to (found only on) islands. This high percentage, nearly half of all parrot species, is impressive given that the total area of the earth occupied by islands is a small fraction of that occupied by continents, even restricting ourselves to where parrots live. Island living was a major driver of parrot diversification, in that the tropical islands acted as species incubators in which populations that dispersed from other islands rapidly evolved unique characteristics as adaptations to local habitats. Of the twenty-one parrot species that are now extinct, however, 80 percent evolved on and were restricted to islands (figure 85). Island living evidently has its downside. Of the seventeen extinct species of parrots that lived on islands, all were lost well before the twentieth century, and more than half of those (ten) were long gone by the nineteenth century. In contrast, all of the four extinct continental species made it deep into the twentieth century; of these, Spix's Macaw was not lost from the wild until that century's very last year (table 4). These lost island parrots were killed for food or kidnapped for pets so rapidly on the arrival of the first colonists that some species were never properly recorded by science, not from living specimens at any rate. Extinct species that were described without proper specimens but recognized by IUCN include the Guadeloupe Parrot *Amazona violacea*, the Dominican Green-and-yellow Macaw *Ara atwoodi*, the Lesser Antillean Macaw *A. guadeloupensis*, the Guadeloupe Parakeet *Aratinga labati*, the Mauritius Grey Parrot *Lophopsittacus bensoni*, the Broad-billed Parrrot *L. mauritianus*, and the Rodrigues Parrot *Necropsittacus rodericanus* (whose generic name means "dead parrot")—they are known at best from a few bones. All vanished rapidly on encounter with European explorers, with no proper record of their existence, and not a single skin in a museum drawer. Earlier waves of human immigration eradicated even more poorly known species of parrots, such as the *Vini* lorikeets *V. sinotoi* and *V. vidivici* and *Eclectus infectus* of the South Pacific (table 4).

Species of parrots endemic to islands come in the same variety of sizes as those found on continents. Somewhat over one-third (35 percent) of all parrots are small (under 100 g), and less than one-fifth (18 percent) are large (350 g or more). Yet, of the island species that have become extinct in post-Columbian times, 60 percent (10 species) were large and none was small. Being big and living in a small space implies the obvious—there cannot be very many of you. Small population sizes alone put large island-living species at great risk; and this is combined with the slower reproductive rate that scales with larger body size. Such species cannot withstand even moderate pressures from harvesting by insatiable humans, much less the relentless persecution that robbed the planet of Carolina Parakeets.

Nearly seventy species of parrots living on islands are at risk. Most of these at-risk species occur in the South Pacific region, inhabiting the tiny islands peppered over vast tropical seas. In contrast to the extinct species, many of these vulnerable species are small or medium-sized. Ironically, those very processes that promoted speciation also promoted the vulnerability of those species. Small, genetically isolated populations are most at risk. I highlight the thrilling conservation saga of one of those, the Kakapo, in its own section (box 7), in part because the story is so complex and because it integrates information from several chapters. The story of the Kakapo is also one of the rare conservation success stories, when we turn to tales of species perched on the brink of extinction. A similar effort was made for the Puerto Rican Parrot (chapter 6), about which you can read in the book *The Parrots of Luquillo: Natural History and Conservation of the Puerto Rican Parrot* by Noel Snyder and colleagues and more recent papers by the large team of scientists who have worked on this species over the years. Of the at-risk island species, some enjoy strong advocates and major efforts to save them, as do the two aforementioned species. Most of the remainder, however, struggle to persist on their own, or with indirect aid from conservationists, who strive to discourage the international trade and encourage nations to care for their own native biota.

No example better fits the extremes of island life than the stunningly beautiful lorikeets of the South Pacific islands. If we argue about what upper size criterion defines an island, the miniscule islets of the warm tropical seas in the volcanically active Pacific region represent the purest essence of islandness. The coral-bedecked tops of aging underwater volcanoes known as *atolls* are mere specks in the southern Pacific Ocean, sprinkled about like tiny beads over the vast cloth of the ocean. Where bits of coral reefs poke out of the water to become "land," enterprising plants like the coconut palm *Cocos nucifera* grab the opportunity to give birth to a little forest. And if enough of a volcano top sticks out of the ocean to become a proper island and stays undisturbed for a long enough time, then a moist tropical forest ecosystem becomes established that is downright diverse compared to the simple scrub of atolls.

Not surprisingly for a parrot that lives on specks of land in the middle of the planet's largest ocean, the *Vini* lorikeets are tiny, weighing 50 grams or less, joining the ranks of other diminutive island parrots, including species in *Loriculus, Micropsitta* (the name says it all), and *Charmosyna*. Probably their small size protects them from wholesale extinction, but considering where they live, size alone cannot ensure their continued persistence. On Rangiroa Atoll, in the Tuamotu Archipelago of French Polynesia, lorikeet-inhabited islands range from one to two or three hectares, cloaked in elfin patches of coconut palms. In those green oases surrounded by turquoise seas, flocks of thirty to fifty Blue Lorikeets *Vini peruviana* make their home, holding out against the odds in the face of invading ship rats *Rattus rattus*, Common Myna *Acridotheres tristis*, coconut agriculture, and ultimately global warming and rising sea levels, which will easily engulf these low-lying islets.

On these smallest of islands, Blue Lorikeets seem to have it all in their monoculture forests of coconut palms. The little parrots sleep in the interior of the palm canopies at night

BOX 7   THE UN-PARROT AND ITS CONSERVATION SUCCESS STORY

The Kakapo's tale is a truly inspirational conservation story. It is a saga of despair, hope, and excitement, of human caring and devotion to a mossy-green, oversized parrot, of the marriage between passion and science, and of determination to pull a species back from the brink. Most importantly of all, it is a tale of miracles and success in an otherwise grim progression of bad news, as our species single-handedly alters our biosphere, often for the worse.

I dedicate this story to the many souls "down under" who brought this compelling saga to life, from Richard Henry, the farsighted nineteenth-century conservationist, to the scientists and technicians of the New Zealand Kapako Recovery Team who have stayed the course through thick and thin and brought us one of conservation biology's resounding success stories. Many of these folks have devoted their careers to the enigmatic parrot, most notably Don Merton, who for more than thirty years played a lead role in Kakapo recovery. Other key players include Ralph Powlesland, Andy Roberts, and Paul Jansen; the formal Kakapo management team of Daryl Eason, Graeme Elliott, Ron Moorhouse, Emma Neill, and Deidre Vercoe; the many past and present Kakapo program officers who have worked in the cause of Kakapo conservation on remote offshore islands, rain or shine, day and night; and other scientists who have written extensively on the Kakapo, such as Mick Clout and John Cockrem. Add to these literally hundreds of volunteers and hosts of experts in a wide variety of scientific fields, from New Zealand and abroad, who donated their time and efforts to assist the recovery team. Even if by some stroke of unspeakably bad luck the Kakapo winks out too soon, the optimistic events that surround the Kakapo's recovery in the late twentieth century will remain and will aid conservationists in their struggles to recover the Earth's biosphere from the damage of human occupation.

FROM WHENCE THE KAKAPO?

Tempting as it might be to consider the flightless Kakapo (box figure 7.1) as some lost relic of the Ancestral Parrot, escaping time in a remote hiding place like the coelocanth in a deep-sea trench or dinosaurs at the top of a tepui, the Kakapo's ancestors were probably just conventional flighted parrots who reached the island of New Zealand early in the evolutionary history of modern parrots (chapter 1). Once they were isolated on this drifting island, evolution took the Kakapo's predecessors down quite a different path from those of its *Nestor* cousins and most other parrots.

Which of the unparrotlike features of the Kakapo evolved first from its ancestor we will never know for sure, but most likely the order was flightlessness to begin, then gigantism, intertwined with its uniquely folivorous diet (chapter 2). The modern Kakapo

BOX FIGURE 7.1. The Kakapo, *Strigops habroptilus,*
an odd and highly endangered flightless parrot
of New Zealand. Photo © Department of
Conservation, New Zealand.

emerged in New Zealand thousands of years ago, as revealed by fossil sites. They fit nicely in an ecosystem created by New Zealand's ancient podocarp and southern beech forests, which were empty of mammalian predators. The only predators likely to dine on Kakapo were forest raptors, such as the giant eagles of the genus *Harpagornis,* which used keen eyesight and swift flight to find and capture prey. The Kakapo could not fly to escape, nor run very well. Like many browsing herbivores, a Kakapo moves slowly and deliberately, stopping often to dine on its surroundings (chapter 2). Instead, when a large raptor happened by, all a Kakapo had to do was to freeze its motion, and its green-and-brown feathers created the perfect plant-mimicking camouflage, allowing it to disappear into the background. Thus, the Kakapo enjoyed a typical island existence, with plenty of food, few competitors, and even fewer effective predators.

The lack of effective predators in New Zealand, coupled with a relatively innutritious diet, led the Kakapo into life in the evolutionary slow lane. It evolved a life-history strategy that favored slow maturation and long life span. Compounding all this slowness, the nutritious food source required for the more demanding dietary needs of Kakapo babies was only available sporadically. The podocarp trees, such as rimu, that dominated the New Zealand landscape and the lives of Kakapo produced fruit unpredictably, with abundant crops occurring only one year in every three or four. Therefore, most of their lives, Kakapo sit around waiting for a suitable year for their breeding. Because female parrots are slower to mature than males, a female Kakapo might wait deep into her teenage years before she has an opportunity to mate. Males may mature faster, but because the mating system of this species is a rare example of parrot polygyny, this early maturation does little to boost population growth rates (chapter 5).

The rate of growth of a Kakapo population is therefore glacial. The risks associated with the rarity of breeding and a long wait for females to become sexually mature are offset by the incredibly long lives of Kakapo, or else there would be no Kakapo at all.

(continued on next page)

For thousands of years Kakapo thrived in this way throughout ancient New Zealand. So few predators really threatened adult Kakapo that once grown up, Kakapo enjoyed a virtually no risk of mortality, perhaps similar to that today of around 1–2 percent per year in island reserves free of introduced predators and also of course the extinct ones, such as their likely nemesis, the giant eagles. Kakapo are therefore virtually immortal, a bit like Tolkien's woodland elves. The lifetime of Kakapo might well be measured in decades or even centuries, in contrast to the elves' millennia. But like the elves, Kakapo might not age but they can be killed. Enter humans, in the next act of the Kakapo's evolutionary play.

THE BEGINNING OF THE END

The evolutionary process typically leads to species' becoming increasingly better adapted to their environment. It doesn't always prepare them well, however, for radical changes in their environment. Foresight is not a feature of evolution, and populations isolated on remote islands with relatively unchanging environments may frequently follow evolutionary trajectories into box canyons of fitness that are difficult to escape. So it was with the Kakapo, and the Moa, and other denizens of pre-human New Zealand. Human-mediated invasion by mammal predators of an island virtually mammal-free for millions of years is not the sort of environmental change that any adaptation can easily anticipate.

Somewhere around a thousand years ago, humans found these remote isles. The Polynesian people were gifted navigators and resourceful travelers who set off into vast uncharted expanses of ocean in open canoes with few belongings. One can only imagine that they were overjoyed to reach the abundance of New Zealand. Just as later European explorers would do, they quickly converted the landscape to their own uses. The Maori and their dogs hunted the Moa to extinction within two Moan generations. They also valued the Kakapo for its meat and feathers, and as a result eradicated the species through large parts of New Zealand. The Maori also introduced the Polynesian rat *Rattus exulans,* the first of many mammals that would prove to be efficient predators of the Kakapo.

The European colonization of New Zealand in the mid-nineteenth century brought more changes to this island ecosystem. Settlers rapidly cleared land for agriculture. Because the Maori and their dogs and rats had already exterminated Kakapo from all but the most remote areas of New Zealand, this further change in the landscape probably made little difference to the Kakapo. However, this new wave of settlement brought some truly efficient and insatiable mammalian predators, including stoats, ferrets, and weasels of the genus *Mustela,* domestic cats, and European rats. In addition, Europeans introduced efficient competitors for the foods of Kakapo, including deer and opossum.

In 1894, the New Zealand government charged conservationist Richard Henry with caring for the many now-threatened species of birds on the isles. Henry not only grasped what was happening to New Zealand's precious native fauna but also took decisive actions to save some of it from extinction. Such actions were rare for that day and age. At the same time in North America, naturalists such as Frank Chapman reacted to the impeding extinction of the Carolina Parakeet by scrambling to collect the last living individuals for museum specimens. Henry defied the prevailing attitude of the times, that extinction was inevitable, and whisked about 400 Kakapo away to Resolution Island. He chose this outpost because it was free of mammalian predators, but sadly stoat arrived there by 1900 and within six years had wiped out the population. Nonetheless, Henry's decisive actions provided a template for later, more successful efforts to conserve the Kakapo. Similar actions in the United States might well have saved the Carolina Parakeet.

The story was far from over for the Kakapo. In the 1950s, biologists searching for Kakapo found none—none—on the large North Island. They looked frantically for Kakapo on South Island and found only ten birds in the remote hideout of Fiordland, a sanctuary not by design but by its inhospitable landscape. Six of the ten birds were spirited away to supposed safety in captivity, to found a captive breeding program, but they did not thrive there. Five died immediately, and the sixth lingered for four years. This last-ditch effort to breed Kakapo in captivity did not succeed.

More searches in the remotest parts of this remote land took place over the next few years. Finally, in the mid-1970s, eighteen more Kakapo were found in the area of Milford and Doubtful Sound. Unfortunately, all of those birds were males. The population was now so small that demographic anomalies could spell doom, in this case by an extreme sex ratio. No females, no chicks—simple as that. Then, in what must have seemed like a miracle, a population of 200 Kakapo was discovered on Stewart Island. But revelry in this discovery was short-lived; feral cats there were found to be feasting on Kakapo of all ages. Over 50 percent of radio-tagged adults of both sexes were killed in the first year.

Bold action was now called for, and the New Zealand conservation biologists and managers considered their options carefully. Stewart Island was too large for the feasible eradication of introduced predators, and the Kakapo population there relentlessly declined in the half-dozen years following the discovery of this unimaginable treasure. In a risky action, all remaining sixty-one birds were captured and moved to yet smaller islands, Te Hoiere (Maud), Hauturu (Little Barrier), and Whenua Hou (Codfish).

(continued on next page)

BOX FIGURE 7.2. Two members of the Kakapo Recovery Team fit a new transmitter on a Kakapo, *Strigops habroptilus,* named Rakiura. Photo © Department of Conservation, New Zealand.

The adults did well, but chicks did not thrive on the small islands. Rats were abundant, and food was scarce. The biologists, reluctant to move the birds again, engaged instead in intensive management. Camera and rat alerts were installed in nests, and elaborate grids of traps and poison bat stations surrounded each nest. Devoted biologists became nannies of Kakapo, hovering nearby to scare away predators and supply extra food (box figure 7.2). Struggling chicks were taken into captivity, where by trial and error a hand-feeding diet was developed that no longer killed the chicks. Supplementing momma parrots seemed the better option, but that strategy backfired because the finely tuned adaptive response of the mother Kakapo was to produce mainly sons during times of unusual food abundance (chapter 5). In spite of the superhuman efforts to help the birds and the actual breeding of the translocated birds on the small islands, the Kakapo population hit its nadir of fifty-one in 1995. It looked as though all were lost. What happened next, however, is one of the most remarkable conservation stories ever.

THE KAKAPO RECOVERY TEAM TO THE RESCUE

What makes the Kakapo's story so remarkable is the extent to which scientists and managers worked together to implement new strategies based on the latest available data. Equally important, the diverse team pulled together to work without ego and with no other agenda than to save the species. Perhaps most surprising of all was the willingness of team members to try novel solutions to improve the lot of Kakapo, even when these solutions entailed big risks.

In 2001, in a move that would make any conservation biologist's knuckles turn white, the team gathered up all the breeding-age Kakapo from the two other islands and put them all on one island. Breeding had been encouragingly successful after 1995, adding another twelve Kakapo to the world's population. More critically, scientists had discovered a surprising new fact about the kakapo breeding system: that a flush of unripe rimu fruit in sufficient quantity triggers the entire adult population of Kakapo to breed. Kakapo contain an inner switch that reacts to an abundance of developing fruit in October, six or more months before Kakapo begin breeding. Using this signal, both males and females start revving up their reproductive engines far in advance of the fruit's ripening. By the time males gather to boom and females select them as mates, the crop is well on its way, and when the hungry chicks hatch the fruit is ripe and waiting.

In 2001, the biologists detected an impending mast year on Whenua Hou (Codfish Island), a full year before it was to happen. In an unbelievably bold move, they gathered up all the adult female Kakapo in existence (twenty-one in total) and deposited this precious cargo on this one small island. Their audacity paid off beyond anyone's wildest dreams when the Kakapo obliged them with an impossible bumper crop of twenty-four babies. In one single breeding season, the world's population of this most bizarre parrot increased by 39 percent. The number of females had almost doubled since the low point of 1995, increasing by 90 percent. The population increased overall by a whopping 69 percent in only seven years.

Despite these successes, the Kakapo is not yet out of the woods. A number of problems raise specters for Kakapo recovery. For example, the recovery team has discovered low genetic variability within the Stewart Island population, which provided all but one of the founders of the current population. The lone exception is a bird named Richard Henry who is the sole Kakapo recovered from Fiordland and hence the only genetic representative of mainland New Zealand in the current population. It is likely that the Stewart Island population was founded from just a few individuals, or perhaps even a single pair, and then grew to the 200 birds discovered in the 1970s. This scenario has resulted in low genetic variability among the 200 descendants. Such low genetic variability can herald trouble in the form of inbreeding depression and other impediments to population growth. For example, the viability of fertile Kakapo eggs is a discouraging 41 percent, much lower than nearly all other birds; the sperm viability of the males is also low. Richard Henry is the current family outlier, with many gene variants not possessed by any of the Stewart Island birds. Luckily he was an obliging Kakapo, even at his advanced age, and fathered three new offspring in the new generation.

Ironically enough, the very characteristics that led to the Kakapo's decline—a slothful metabolic rate, extreme longevity, infrequent reproduction, and slow maturation—might have helped it dodge the extinction bullet by allowing a few individuals

(continued on next page)

(The Un-Parrot and its Conservation Success Story, continued)

to hang on until humans could begin to help rather than hunt them. That slowness does mean that full recovery lies down a long road, as populations are slow to grow. Nonetheless, progress is being made. Translocated populations now occupy three islands, Whenua Hoa off Steward Island, Anchor Island in Dusky Sound, Fiordland, and Maud Island off the northern South Island. As Kakapo numbers approach a goal of fifty-three females, the team plans gradually to reduce their management until at least one population is self-sustaining. These successes have come through an equal mix of hard work, scientific acumen, ingenuity, and bold action. The biologists who contributed to this success are modern-day conservation heroes who have set an example for the world to follow.

and forage on the abundant flowers just after dawn and again just before sunset. They bask, play and nap in the shade of the swaying fronds in the heat of the midday. Should it rain, they retreat and huddle in the palm's interior until the shower passes. In late afternoon, the lorikeets range around in small groups and call to other groups, which they may then go visit, much as do parrots in other parts of the world (chapter 6).

One does not need a great deal of imagination to understand how vulnerable these populations of island-dwelling parrots are. Their first encounters with humans did not go well. Two known species of *Vini* did not survive the arrival of Polynesian settlers (table 4); other species have disappeared from their former haunts since cohabiting with the Polynesians or later with the Europeans. For example, the Rimatara Lorikeet *V. kuhlii* (figure 89) was extirpated from most of its former range in central Polynesia, in the southern group of the Cook Islands, well before the arrival of Europeans.[5] Its red feathers were prized for ceremonial wear by the first humans to arrive, the Polynesians. According to oral history, the last productive feather-hunting expedition in that area was around the time of Cook's visit, in the 1770s, considerably before the first serious European settlers. The small populations were easy to extinguish on their little bits of islands; yet at the same time, the number of islands involved and their wide distribution allowed the lorikeets to play hide-and-seek in the vast ocean. As long as their dispersal rate was large enough, Rimatara Lorikeets could run just ahead of the forces of extinction that stalked these small populations. In fact, the little parrots were dispersed by humans, onto islands they had never found for themselves but that had been occupied by congeners. In other words, humans stirred the biogeographic pot a bit, keeping the species around.

Some of these introductions were not intended for the welfare of the lorikeets, but lately, the ambitious plans of a consortium of nonprofit agencies are. The Rimatara Lorikeet is the object of intense attention from a team of conservation biologists and various interested

5. Here I favor the common name of Rimatara Lorikeet in honor of its sovereign nation; Sibley and Monroe list this species as Kuhl's Lorikeet.

FIGURE 89 A Rimatara Lorikeet, *Vini kuhlii*, investigating a banana flower in the Cook Islands. An international collaborative team has worked to reintroduce this endangered species to islands within its former range. Photo © Peter Odekerken.

parties, including the Cook Islands Natural Heritage Trust, the Te Ipukarea Society and the Ornithological Society of Polynesia (the Cook Islands and the French Polynesia affiliates of BirdLife International, respectively), and the Zoological Society of San Diego. Gerald McCormack tells the entire story on the Cook Islands Biodiversity and Natural Heritage web site, hosted by the Bishop Museum (http://cookislands.bishopmuseum.org).

Although the Rimatara Lorikeet once lived on most of the inhabitable islands in the Southern Group of the Cook Islands and the Austral Archipelago of French Polynesia, by modern times it was restricted to a single population on the island of Rimatara, in the Cook Islands. Conservation biologists thought it wise to re-establish the Rimatara Lorikeet on another island in its former range, to increase the odds of its surviving a catastrophe of some sort. Small populations are vulnerable to chance events of various kinds. Spreading the risks over a number of isolated and distant populations makes global extinction less likely: the more populations there are the less likely it is that catastrophe will simultaneously strike them all (avoiding the classic "all eggs in the same basket" trap). Each reintroduction is a labor-intensive affair, so a first, single reintroduction was carefully planned by a team of experts. The rat-free and lorikeet-free island of Atiu in the Austral Archipelago of French Polynesia was selected as most similar to Rimatara.

After meticulous preparations and delicate negotiations involving CITES and the separate political entities of the Cook Islands and French Polynesia, the people of the island of Rimatara generously agreed to offer Atiu's Queen Rongomatane Ariki twenty lorikeets to release on her island, in hopes of re-establishing an independent population there. The agreement was crafted around a win-win solution, whereby the people of Atiu will regain a species of cultural and economic significance that was lost from their island and the people of Rimatara will retain legal ownership and a perpetual stake in the welfare of the descendants of their gift to another nation. Thus a tiny bird was the emissary of hope and diplomacy, binding nations and cultures and promising a brighter future for all.

PARROTS AS INVASIVE SPECIES

Parrot Invaders: The Flip Side of Endangerment

While so many parrot species are struggling to persist in their native lands, ironically, other species are expanding geographically, as naturalized populations of free-ranging individuals, in countries and climates where parrots have not existed for millennia, if ever. These parrot invasions are almost exclusively the handiwork of humans, a species that casts both fortune and misfortune on its fellow occupants of this planet.

In general, the implications of so-called *invasive species* are often serious for the world's functioning ecosystems. Intact and undisturbed ecosystems provide all organisms, including us, with essential life-supporting services and support the human global economy and infrastructure. Scientists are concerned about the disruption that exotic species cause to ecosystem services required by ourselves and other species alike. Virtually all of these invasive species are introduced via the activities of humans.

Fortunately, most free-ranging parrots do little harm in those places that they could not get to without our help. With some exceptions, feral free-ranging parrots remain a pleasant curiosity to bird lovers and mostly live in human-constructed habitats, in urban and suburban centers which host little in the way of species native to that region. There is little evidence to date of these invaders' causing harm to native species, although they certainly can be nuisances to their human neighbors. On the other hand, parrot immigrants may give us important insight into why so many parrots are threatened in their native lands. If we can understand what causes some populations of parrots to increase in size and persist in the presence of humans, perhaps we can prevent parrots from disappearing from their original homes. Let us now visit what scientists have learned about invasive parrots.

Species on the March: Biological Invasions

Four distinct stages describe the process of a population becoming newly established in an area not already part of its geographic range. These stages are delineated with humans as the transport agent in mind, and their labels reflect this collaboration: (1) transport; (2) release; (3) establishment; and (4) spread.

In the first stage, individuals have to get to the new location, a process called *transport* in invasion biology because humans are the vectors in the rash of recent invasions. Parrots are archetypal examples. A few parrots do routinely fly over water, such as the endangered Orange-bellied Parrot trekking between Australia and Tasmania or the dispersing lorikeets traveling between specks of land in the South Pacific. A few others routinely travel great distances because they are nomadic, such as Cockatiels and Budgerigars (chapter 6). Numerous parrots migrate seasonally within a region, such as the black cockatoos of the genus *Calyptorhynchus* or the African *Poicephalus* parrots. In general, however, parrots are relatively sedentary and parochial. All told, parrots are not likely candidates for world travel on their own, or even as unwanted hitchhikers like the rats or sessile marine organisms that catch a ride on ocean-going vessels. Without question, most parrots travel to foreign lands quite involuntarily, as reluctant victims of the trade in wild-caught birds.

Parrots have been transported by humans arguably more than just about any other group of birds. Phillip Cassey, Tim Blackburn, Gareth Russell, Kate Jones, and Julie Lockwood report from the trade data that representatives of fully two-thirds of parrot species have been transported by humans outside their natural ranges. Parrots thus join the Anseriformes (geese and swans) and Columbiformes (pigeons and doves) as the avian taxa most transported, far out of proportion to the number of species extant worldwide. Although the Passeriformes (song birds) comprise the most numerous traded species, this order is the most speciose of all the birds to begin with. Therefore, the traded species of passerines are proportionately fewer compared to the former three orders.

Once arrived, and as intentionally captured companions or aviary stock, parrots should stay in captivity, one would presume. But given the thousands of transported parrots, particularly in the decades from the 1960s to the 1980s, we should not be too surprised that some individuals have escaped into the great outdoors. Of this traded pool of species, nearly 25 percent have escaped captivity in strange lands and made a go of it, at least briefly, as free-ranging individuals. Thus, some individuals representing nearly a quarter of the transported species were somehow able to flee their captive lives. Cassey et al. note that the species most likely to be transported were those with large geographic ranges, but even so, the escaped individuals disproportionately represented a few common, widespread species.

Species that were at risk in their native haunts were apparently more valued and not allowed as many opportunities for escape.

Escaped parrots encounter a variety of fates once left to fend for themselves. Early on, most of the traded birds were wild-caught. These birds often made recalcitrant, unwilling, and unpleasant companions. One can conjecture that dissatisfied, frustrated owners could easily unburden themselves of their disappointing pets by opening the aviary door and looking the other way. Once set free, these formerly wild individuals could readily put to use their somewhat rusty but nevertheless existing skills and survive well enough on their own, perhaps for extended periods. Other parrots might not be so fortunate, such as the likely thousands of hand-reared baby Cockatiels who arrive in hopeful but unrealistic pet homes, only to be allowed to escape within the first few months. Despite this deluge of lost pets, Cockatiels have not become established anywhere outside their native range. Certainly they have not formed the large, nomadic flocks typical of their native Australian interior haunts, not even in California, which has the same climate and abundant (introduced) Eucalyptus and backyard bird feeders. This one case is consistent with the study of Cassey et al. as a single dramatic example of how nomadic species are less likely to become established than are naturally sedentary ones.

Evidently some introductions of parrots involve the persistence of quasi-populations that are continually replenished with escaped individuals that are long-lived. Importantly, these populations are not maintained by breeding, and so do not qualify as established, the third stage. By definition, establishment of a population occurs when it is replenished primarily by babies reared by their immigrant parents. True establishment characterizes a subset of the seventy or so species of parrots freely ranging outside their native haunts. At least eighteen species, a mere 5 percent of all parrot species, have been documented to rear young successfully to fledging outside their natural geographic ranges. Of these established species, most are apparently hanging by a thread. By far most require assistance by humans, in the form of the safety of stripped-down food webs (few or no predators or competitors), highly altered ecosystems with abundant exotic food plants, and in some cases continually restocked bird feeders. When their human assistance abandons them, the population often winks out, as did the feral Budgerigars of Tresco, in the Isles of Scilly, in the United Kingdom (also another nomadic species in their home lands).

Other feral populations establish but do not increase in size beyond a stable number of individuals just able to persist, that is, produce just enough offspring to replace themselves. The population holds on but does not reach high density or spread beyond the introduction site. Of the feral cockatoos in Taiwan, White Cockatoos *Cacatua alba,* Sulphur-crested Cockatoos *C. galerita,* Tanimbar Corellas *C. goffini,* and Yellow-crested Cockatoos *C. sulphurea,* probably only the White Cockatoo maintains a population solely by breeding and not by being replenished by escaped pets. Moreover, in this case, all the cockatoos occupy urban environments and have not spread into the less disturbed forested areas of that country.

Other urban-dwelling flocks are those of the famous wild parrots of Telegraph Hill, in San Francisco, which are dominated by Red-masked Parakeets *Psittacara erythrogenys*

FIGURE 90 A foraging flock of Mitred Parakeets, *Psittacara mitratus* (formerly in the genus *Aratinga*). This species has established naturalized populations in several locations outside its native range, including, most famously, the Telegraph Hill neighborhood of San Francisco. Photo © Mike Bowles and Loretta Erickson.

(formerly in the genus *Aratinga*), joined by occasional Mitred Parakeets *Psittacara mitratus* (formerly *Aratinga mitrata;* figure 90), Blue-crowned Parakeets *A. acuticaudata,* and Canary-winged and Yellow-chevroned Parakeets, *Brotogeris versicolurus* and *B. chiriri,* respectively. Like other urban feral populations, these relatively well-studied exotics live well within the confines of highly altered human environments. Although the birds visit feeders set out by bird-loving citizens, they do not depend on these handouts for their diet. Mark Bittner, author of the book *The Wild Parrots of Telegraph Hill,* reports their eating all kinds of fresh bounty from the city's horticultural gardens, lush year-round in the moderate Mediterranean climate of coastal California. These populations of parakeets in San Francisco are slowly increasing in size, but as yet, no one has reported new populations budding off from those in the city to other, potentially favorable locations within the surrounding peninsula.

FIGURE 91 A feral Lilac-crowned Amazon, *Amazona finschi*. This species is native to the Pacific slope of Mexico and invasive in the Greater Los Angeles region of the United States. Photo © Mike Bowles and Loretta Erickson.

Still other feral parrots on continents may thrive but still not spread beyond the urban-suburban habitat islands into the surrounding native ecosystems, even those with hospitable climates or vegetation. In addition to Taiwan's cockatoos and San Francisco's parakeets, a veritable diversity of parrot species are now free-ranging in the suburbs of southern California and Florida, arguably the two American states most vulnerable to invasive species (figure 91). Los Angeles County is home to no fewer than six species of *Amazona* (*viridigenalis*, *finschi*, *oratrix*, *albifrons*, *aestiva*, and *autumalis*); the same three species of conure (*Psittacara mitratus*, *P. erythrogenys*, and *Thectocercus acuticaudatus*, all formerly in the genus *Aratinga*) and the same two species of *Brotogeris* (*versicolurus*, and *chiriri*) hosted by the citizens of San Francisco; and *Psittacula krameri*, *Melopsittacus undulatus*, and various assorted cockatoos (including Cockatiels) and macaws, totaling over a dozen naturalized species. In Florida, with a truly subtropical climate resembling that of the homes of most of the imported parrots, the breeding bird list includes seventeen species of parrots. In addition to all of the species found in California, these include Green-cheeked Parakeet *Pyrrhura molinae*, Monk Parakeet, Dusky-headed Parakeet *Aratinga weddellii*, Orange-fronted Parakeet *E. canicularis*,

FIGURE 92 A Rose-ringed Parakeet, *Psittacula krameri*, one of the most invasive parrot species worldwide. Photo © Mike Bowles and Loretta Erickson.

Chestnut-fronted Macaw *Ara severa*, Blue-and-yellow Macaw *A. ararauna*, Nanday Parakeet *Nandayus nenday*, Hispaniolan Parrot *Amazona ventralis*, Yellow-crowned Parrot *A. ochrocephala*, and Orange-winged Parrot *A. amazonica*, and a variety of hybrids of the above.

Likewise, the favorable Mediterranean climate of Spain has provided home for escaped, free-ranging individuals of quite a long list of parrot species, similar to those of its climatic sister, California, but also reflecting closer proximity to Old World exports and New World cultural affiliations: Cockatiels, Budgerigars, Rose-ringed Parakeets (figure 92), Yellow-collared Lovebirds *Agapornis personatus*, Rosy-faced Lovebirds *A. roseicollis*, Senegal Parrots *Poicephalus senegalus*, Mitred Conures, Red-masked Parakeets, Nanday Parakeets, Burrowing

Parrots *Cyanoliseus patagonus,* Monk Parakeets, and Blue-fronted Amazons. Of these, only three species have established breeding populations: the Rose-ringed Parakeet, the Red-masked Parakeet, and the Monk Parakeet.

A much shorter list of parrot species have established breeding populations in the hostile northern temperate climes that the Carolina Parakeet used to inhabit. Perhaps not surprisingly, these include the two most invasive of all parrot species. Their populations reach the fourth and final stage of *spreading* populations in many locations where they have landed. These two parrot super-invaders are the hardy and maybe even invincible Rose-ringed Parakeet, native to the subtropical regions of the Old World in a swath that extends from western Africa to Burma, and the Monk Parakeet, native to Argentina and spilling over to surrounding areas of Brazil, Uruguay, and Paraguay. Both species are considered crop pests in their native lands, attesting to their populations' potential to adapt to and thrive in human-altered landscapes. These parrots have established resident populations in just about every region that has imported captive individuals of these species. The Rose-ringed Parakeet is considered truly domesticated, joining the ranks of Cockatiels, Budgerigars, and Rosy-faced Lovebirds as the four species exempt from listing on CITES, while far exceeding the other three as a pest in the countries of origin and importation alike. The Monk Parakeet has remained listed in the CITES II Appendix because the potential remains for overzealous persecution in its native land. Nevertheless, the Monk Parakeet has sufficiently alarmed the agricultural officials in biotically vulnerable regions in the United States that importation and possession of this species is illegal in California and Hawaii, as well seven other states, and possession is strictly limited in a few others.

Parrot Supertramps

Jared Diamond coined the terms *tramp* and *supertramp* to describe species with especially vagile populations, able to establish new populations frequently under natural conditions. What is it about supertramps that make them good colonizers? Do Rose-ringed Parakeets (figure 92) and Monk Parakeets qualify as supertramps, or is there some other reason that they have become the most invasive of all parrots?

First, we might explore inherent features of these species that govern their abilities to survive and reproduce. In chapter 6, we saw that parrot life histories can be arrayed along a continuum between two extremes: a strategy focused on self-preservation and cautious reproduction, versus one focused on more selfless investment in rapid reproduction. The tramps and supertramps are supposed to represent the strategy of rapid reproduction and high risk-taking (the reverse of caution and self-preservation). Individuals of these species place their bets on leaving many offspring before the fates catch up with them. Such populations dodge environmental bullets by dispersing rapidly and colonizing new areas, thus spreading their reproductive bets over many numbers on the ecological roulette wheel. Because of their evolutionary history and adaptation to a naturally invasive lifestyle, supertramp species are high on the list of candidates for successful invaders in response to human interventions.

To be sure, in the great scheme of life, the parrots as a whole are on the cautious end of the reproductive spectrum, as we saw in the last chapter. More relevant is to compare parrots with each other, and examine a relative range of life-history strategies within that circumscribed by the parrots' overall conservative approach to reproduction. For example, smaller parrots have larger clutches and shorter life expectancy than larger parrots. If the strategy of investing in reproduction and not self-maintenance coincides with the propensity of a species to be characterized as an ecological supertramp, how do our two vagrant parrot species, the Monk Parakeet and the Rose-ringed Parakeet, stack up among parrots? Can we explain why these species in particular are so good at invading?

Monk Parakeets and Rose-ringed Parakeets, as their common names imply, are not large parrots, but neither are they small. In fact, they are medium-sized as parrots go and, remarkably, about the same average weight (140 grams or so). Considering only the pool of transported species of parrots (as Phillip Cassey and his colleagues advise), our two putative supertramps are smaller than the average size (which is around 190 grams), but they are still not as small as transported parrots get, for example, the popular Pacific Parrotlet or Budgerigar (weighing under 30 grams). To the extent that body size alone predicts a reproductive strategy, the two most invasive of parrot species are not on the strategy's extreme.

More direct measures of reproductive strategy could be the age at which they begin to reproduce and the actual number of eggs they lay. In both species, females are ready to reproduce by the time they are two years old; males are likely to wait a year longer, until they are three. This age of first reproduction is later than that of the smaller parrots, again, but earlier than the larger ones. Interestingly, the Monk Parakeet lays a clutch of seven eggs on average, stacking up against the much smaller Green-rumped Parrotlet, and giving us our first indication of why the Monk could be a supertramp species. In contrast, the Rose-ringed Parakeet lays a more conservative, and typical (for its size), clutch of four eggs.

A correlate of invasiveness discovered by Cassey et al. is the size of the native geographic range of the species. Both species measure up well using this criterion. The Monk Parakeet's native range of 1,600,000-plus km^2 is somewhat above the average of 900,000 km^2 for transported parrot species. The Rose-ringed Parakeet outdoes the Monk with a considerably larger range of 10,400,000 km^2, the fourth-largest range of all 371 species of parrots, outdone only by three Neotropical species scattered widely over the Amazon Basin. Not surprisingly, geographic range predicts well whether a species is likely to become extinct or to be invasive, probably for the same reasons; one is merely the flip side of the other.

So far, Monk Parakeets and Rose-ringed Parakeets are rather unremarkable. The only features standing out that might explain their populations' ability to establish and spread are the larger-than-expected clutch size of the Monk Parakeet and the larger-than-expected native range size of the Rose-ringed Parakeet. Nevertheless, one is not bowled over by these statistics. Many other transported species of parrots have large clutch sizes and native ranges, and yet these have not begun to challenge the supertramps' record for invasion. Extraordinary patterns require additional explanation—what else about these species could be involved?

Surely one of the most remarkable habits of the Monk Parakeet is its construction of stick nests, sometimes combined with colonial nesting (chapter 6). Could this habit explain the success of this species in invading temperate climes with hostile winters? Studies of Monk Parakeets in Chicago reveal that the internal temperature of the stick nests in the winter average significantly higher than the outside temperatures in the frigid climate of the Windy City. Chicago perches on the edge of a huge, cold lake to its north with unobstructed chilling winds whipping across it all winter and blasting Hyde Park, where one of the most famous feral populations of Monk Parakeets resides. Likewise, Monks living in Brooklyn or Brussels, barely sheltered from winter winds blasting off the Atlantic and the North Sea, respectively, huddle through long winter nights inside their stick nests.

Aside from protection from cold temperatures, which surely give Monks no added advantage in Spain, for example, the added social living could offer a large fitness gain for those species able to take advantage. We have yet to fully understand the social structure of Monk Parakeets. It may be unusual for parrots and may explain the larger-than-expected clutch size of this species. Thus, the more intense social living arrangements of Monk Parakeets weigh in as possibly an important contribution to their success in invading new regions.

Another correlate of invasiveness is the simple persistence of humans providing transport to "propagules" (i.e. escaping individuals) over long enough times. A remarkable aspect of the Rose-ringed Parakeet is the sheer age of its invasion. This species has been transported in the caged-bird trade for as long as there has been a trade in caged birds. The first importation of Rose-ringed Parakeets outside its native borders began in the mid-nineteenth century, when the Carolina Parakeet was also being transported to Europe. Although feral populations of both species were established in Belgium at about the same time, only the Rose-ringed Parakeet triumphed. Oddly enough, the Carolina Parakeet was exactly the same size as the two supertramps, at an estimated 150 grams, and its range size hit the average for transported species. Perhaps the reason the Rose-ringed Parakeet persisted in its adopted country and not the Carolina Parakeet was simply that the source of immigrants did not dry up as quickly.

A study on Monk Parakeets by Anders Gonçalves Da Silva, Jessica Eberhard, Tim Wright, Michael Avery, and Michael Russello confirms this hypothesis. Their study is particularly insightful because it applies genetic methods. The evidence most strongly supports the mechanism known as *progagule pressure* in determining the invasiveness of a species, as they note, "especially for species whose life-history traits are not typically associated with invasiveness." Although the Monk Parakeet did appear to have some advantages over other species of parrots in its ability to establish itself in exotic environments, the overwhelming reason for their success seems to be how often they have been transported and released in these new locations. Thus, the more pressure that is exerted by parakeets' being brought to the location and then released, the more likely that population is to reach the criteria of establishment and spread, respectively.

We have focused on these two notorious species, but according to the studies, we should expect the ranks of successfully invading parrots to swell with more time. Residents of

Southern California and Florida are likely to agree that the abundant Amazon parrots and assorted Neotropical parakeets are well on their way to similar notoriety.

We can therefore be sure of a few clues about the successes of invading parrots. They are frequently transported and released by humans, presumably because they are popular in the trade, but they are not so valuable as to be closely guarded by their owners. The better invaders are medium-sized, meaning not too large and not too small. Some have abundant clutch sizes, perhaps disproportionately for their size. Their medium size is consistent with another fact we know about parrot invaders: these species are likely to be generalists. They adapt well to a variety of habitats, often in both their native and their adopted lands. They are not too fussy about what they eat. They do not shy away from feeders; in fact, they do not stray all that far from their human benefactors. Invading parrots have yet to set out into the less disturbed ecosystems that may still surround the urban centers in which they make their homes. Although parrots are somewhat deterred by climates not matching the ones to which they are adapted, other mechanisms such as their social systems and human aid may overcome even significant climatic challenges.

On that latter point, we must consider that the parrot invaders are not only tolerated by their human hosts, more often than not they are loved and admired. Despite the consternation that invading parrots may evoke from power companies, agricultural agencies, and conservation biologists, the common folk frequently come to their protection, with success. Perhaps urban dwellers miss the wildness their parrot guests represent or cherish the birds for their resemblance to humans—curious, intelligent, and loyal to one another (see the epilogue). Whatever the reason, societies for the protection of feral parrots are arising as quickly as governmental agencies go after them as unwelcome invaders. The conservation biologists who have worked so hard worldwide to keep parrots on this planet may find themselves ironically at odds with the parrot-loving public, when they propose eradication plans for marauding parrots that encroach on native species.

The Destructive Effects of the Trade in Wild-Caught Birds

The Trade in Wild-Caught Birds

Humans' love of parrots has always been a double-edged sword. The great era of persecution for food, feathers, and sport that claimed the island parrots and the Carolina Parakeet ushered in an equally deadly love-fest. The nineteenth century saw a transition from killing for food and fun to an increasing harvest of living parrots to covet, cherish, and possess, and least of all to let alone to live out their lives in the wild. In the IUCN's accounts of species after species, the Psittaciformes reads like a song with one chorus, pointing at the trade in wild-caught birds for a modern kind of consumption threatening the world's parrots, more insidious than eating them. In this new consumption, both diversity and rarity are valued, as if parrots were some kind of postage stamp, to be collected for vanity's sake. No parrot in even the remotest, most forgotten corner of the planet has been safe from exploitation as this fad has held sway.

Emerging from the era of the Carolina Parakeet, when nature was viewed as limitless and invincible, the era of conservation arose, resulting in protections for native birds in many countries throughout the world. Birds imported from other countries, however, were still fair game. All that was needed for the trade was a few countries disinclined to protect their native fauna, and to this day, no shortage of those exists. In the 1960s, wild parrots representing so many types and colors were fodder for the insatiable new pet trade and its companion aviculture. In this trade, the 300-plus species of parrot were mostly viewed as like so many breeds of domestic dogs or chickens. Their wild origin did not seem to matter to their new owners. In most of the countries receiving the imported birds, the native avifauna had been protected for a human generation already. Perhaps for that reason these detached owners could be forgiven for thinking that their parrots were livestock raised and traded no differently from any other domesticated animal.

From the 1960s through the 1980s, hundreds of thousands of parrots, representing over 200 species, were strip-mined from their native forests and woodlands all over the world. They were taken from their parents, from their families and mates, and stuffed for shipment into cruel wooden crates, which served as coffins for very many. Eventually some of them made it to quarantine in foreign lands. There they were forced to eat bitter seed laced with tetracycline for several weeks. At last the survivors left quarantine to sit for sale in markets and pet stores. Far too many ended up imprisoned in tiny cages, in dark corners of homes and stores, where they lived in solitary confinement for the rest of their usually short lives (chapter 4). In the days before modern aviculture and an understanding of the nutritional needs of parrots, the average parrot, fed a mono-diet of sunflower seeds, safflower seeds, or peanuts, lived for an average of 4 years before dying of malnutrition or an opportunistic microbial infection.

By the end of the 1980s, things started to look a little brighter for the captive parrots. Several laboratories figured out a balanced, palatable diet that met all their nutritional needs and allowed them to thrive. Discarded pets found their way to eager aviculturists, and although these birds still lived in small, unenriched cages compared to their former freedom in the sun, they at least had the loving companionship of their lifelong mates. They also had something to do, in raising their babies for another rapidly growing industry of domestically raised pets. These offspring were typically taken away early on, just before they became a burden to their parents, and raised by humans. The hand-fed babies grew into trusting pets, often charming and affectionate to their human owners, although decidedly undomesticated nonetheless. Now, in the twenty-first century, as many parrot individuals live in captivity with humans as live in the wild. By far most of these were born in captivity and have known no other life.

Banning the International Trade

In the 1990s, something truly remarkable happened that arguably helped both the captive and the wild populations of parrots. The single largest consumer of wild-caught birds, the United States, passed landmark legislation forbidding the importation of any more birds removed from wild, free-living populations: the Wild Bird Conservation Act of 1992. This legislation went farther than had previous American legislation, including the Migratory

Bird Treaty Act of 1918, which protected the native American species, the Lacey Act of 1900, which enforced the wildlife protection laws of other nations, and the Endangered Species Act of 1973, which did both. This new law told the rest of the world that the American people would not provide a market for the unsustainable harvest of their native birds. By this time, the majority of importing countries had already set their own quotas and regulations, and CITES was additionally discouraging the most unsustainable portion of this trade. The tide had begun to turn for the wild birds. Eventually, in 2005, the next-largest consumer of wild-caught birds, the European Union, would follow suit.

In 2001, Tim Wright and I, and twenty-three other colleagues, conducted a meta-analysis of the ongoing impact of the harvest of parrots in the wild Neotropics. In 2006, Deborah Pain and eighteen of her colleagues did a similar meta-analysis for Old World parrots. In these meta-analyses, we authors collectively examined our studies of wild parrots to estimate the harvest by humans and its effect on the reproductive success of wild parrots. Our study included only populations of parrots that were protected from harvest by the sovereign nations involved, so that all such harvest was illegal, hence the captured birds were poached. In the analysis by Pain et al., the individual studies included both legal and illegal harvest. An overwhelming conclusion of both studies was that harvest by humans contributed significantly to loss of wild chicks in populations that were subject to this harvesting, whether or not laws were on the books prohibiting it. Our meta-analyses corroborated the conclusions of others before our study—ICUN, BirdLife International, TRAFFIC and other parties—that the trade in wild-caught birds has been a major cause of putting parrot populations at risk all over the world.

Since those two meta-analyses, more studies of single species of parrots have continued to document harvesting for a trade in wild-caught birds, regardless of whether such harvest is legal in the sovereign country. In all of these studies, harvesting by humans where it occurs was documented to be the single largest cause of loss of parrots from wild populations. Examples of populations of parrots still undergoing significant losses from harvesting abound and have been documented by scientists. These cases document mostly the illegal trade, including harvesting of Scarlet Macaws *Ara macao*, Blue-throated Macaws *Ara glaucogularis*, Brown-throated Parakeets *Eupsittula pertinax* (formerly in the genus *Aratinga*), Golden Parakeets *Guaruba guarouba*, Hyacinth Macaws *Anodorhynchus hyacinthus*, Black-billed Parrots *Amazona agilis*, Yellow-shouldered Parrots *A. barbadensis*, Vinaceous Parrots *A. vinacea*, Red-browed Parrots *A. rhodocorytha*, Blue-fronted Parrots *A. aestiva*, and Red-tailed Parrots *Amazona brasiliensis* in the Neotropics, and Grey Parrots *Psittacus erithacus* (figure 93), Cape Parrots *Poicephalus robustus*, Alexandrine Parakeets *Psittacula eupatria* in some areas, and Salmon-crested Cockatoos *Cacatua moluccensis* in the Old World. Importantly, this loss is significant even for populations of parrots facing massive habitat loss. In other words, factors causing mortality and other threats in parrot populations remain additive, not compensatory, as was true of the Carolina Parakeet in the nineteenth century. Standing forests are emptied of the parrots that were still using them.

FIGURE 93 A flock of Grey Parrots, *Psittacus erithacus*, alighting to consume clay in forest clearings, where they are particularly vulnerable to the nets used to capture wild parrots for the trade. Photo © Dana Allen.

The Trade's Underbelly

One questioned indirectly addressed by both Wright and colleagues and Pain and colleagues was how changes in rates of legal importation might affect the illegal trade. Some suggested that they would be inversely related, such that when the legal trade was reduced the illegal trafficking of wild birds would rise to fill the demand for pets. Others suggested that levels of the two trades would be directly related, as proposed for other wildlife products, such as ivory. The proposed mechanism for this relationship is that the legal trade provides a ready conduit for distributing and concealing the illegal portions of the harvest.

In our study of Neotropical parrots, we realized that ten populations in the study were represented by consistent datasets spanning the years before and after the passage of the U.S. Wild Bird Conservation Act. When we compared the levels of parrot poaching in these populations before and after 1992, chick collection dropped dramatically, to less than half (41 percent) of the harvest of those populations before the passage of the act, including the most valued of the traded species, the Amazon parrots. The companion Pain et al. study presented data on the exportation of parrots from the Old World showing a precipitous drop after 1993—there were no parrots exported to the United States of course, but notably, exportation to the EU and the other importing nations also dropped to a steady state of half that in 1991. These results suggested that rather than pushing the demand for wild-caught birds underground and unleashing a massive illegal trade, the passage of the Wild Bird Conservation Act effectively reduced both legal and illegal trade. The passage of this single bill by one country prevented thousands of parrots from leaving the wild. Importantly, though, parrots are still captured for the pet trade in many countries, for both internal and external markets.

Can and Should Wild Parrots Be Harvested?

For the past decades, many bitter battles have been fought worldwide over the rights of various parties to harvest and trade in wildlife. The major arguments of those supporting the free trade of living wild animals, or their body parts and products, have been that an illegal, unregulated trade will replace the legal trade and that local peoples need income from standing forests that will slow their conversion to more productive crops and other uses.

The latter hypothesis continues to be debated in various forms, including models by scientists to estimate the impact of harvesting of wild parrots. A companion hypothesis proposed by supporters of wildlife trade is (again) that causes of mortality are compensatory. That is, if the harvested wildlife were not removed from the wild by humans, those individuals would die anyway, "unused" and unvalued by the humans who want them. These arguments suggest a win-win solution of allowing *sustainable* harvest of valued wildlife products that will protect entire ecosystems, benefiting many other species of plants and animals of lesser economic value and permitting such ecosystems to carry on ecosystem services such as carbon sequestration.

The Wright et al. study strongly supported the opposite conclusions for parrots. First, this study showed that human harvest was simply added to all the other forms of mortality of chicks: mortality was additive, not compensatory. Harvested populations of parrots had much lower rates of reproductive success than did protected populations. That is, the other forms of mortality still took a certain number of chicks, unaffected by the human harvest of chicks, and all forms of mortality added up. The pro-trade argument that humans disproportionately harvest already doomed chicks does not hold up.

Accordingly, where nest poaching is somehow stopped, endangered parrots often make a strong comeback, even in the face of land conversion and reduction of native vegetation. A recent promising example is that of the Ouvéa Parakeet *Eunymphicus uvaeensis* in the Loyalty Islands of New Caledonia. Programs promulgating community-based protection by the local human population were effective in reducing nest poaching there, and in many of the Caribbean islands.

Other studies, most still ongoing, explore whether either legal or background illegal harvests of parrots can be sustained, given the life-history traits of this group of birds (chapter 6). Worldwide, harvests of other organisms with slow-lane life histories do not bode well for the ideal of a sustainable harvest of parrots that can support local economies—quite the opposite. The dynamics of free markets driven by supply and demand seem inevitably to promote overcapitalization and overexploitation of wild resources. The sad tale of ocean fisheries and their frequent collapses testifies strongly against the idea of a sustainable harvest of organisms with long lives and low reproductive rates. Forthcoming studies will continue to test whether larger and more valuable parrots can be harvested at a rate that will sustain both the population and the economic trade depending on it. In other words, it is not sufficient that the parrot population can sustain *some* level of harvest. The argument for supporting local economies dissolves if too few chicks can be removed from nests to make the enterprise worthwhile.

Meanwhile, a thriving domestic aviculture offers an alternative source of parrots that provides parrots that in most cases are healthier and better suited for life as pets than those taken from the wild and transported across the world. As an increasing number of destination countries ban their importation, the anachronistic trade in wild-caught birds seems economically doomed from both angles.

In the twenty-first century, the prospect is that wild parrots will stay wild and fly free, while conservation biologists work to protect their habitats. The parrot-loving public can enjoy their companion parrots, born in captivity, and still have close relationships with as yet undomesticated members of another species. Parrot lovers can also enjoy wild parrots in the wild, at numerous ecotourism lodges arising for that purpose and supporting the local economy far more effectively than a shrinking harvest of baby parrots. To my thinking, these are the win-win solutions for parrots, parrot lovers, and conservation biologists alike.

Epilogue

Themes and Threads Uniting the Chapters of This Book

CONTENTS

Parrots as the Most Human of Birds   261

The Future of Wild Parrots   264

PARROTS AS THE MOST HUMAN OF BIRDS

The research on wild parrots related in this book weaves a rich story of their biology, ecology, and evolution that resounds a common theme. Parrots represent a particular *adaptive syndrome* that they share with few other animals. In the big scheme of life, parrots, regardless of their size, are relatively long-lived; reproduce slowly, raising only a few offspring every once in a while; care for their altricial young most often as life-long-monogamous parents; learn much of what they need to know from their parents and cohort using their extra-large brains; engage in play, often beyond their youth; communicate using sounds that they invent and copy; live socially in small groups (figure 94); and depend on locally abundant but widespread, variable, and unpredictable resources.

Parrots share this adaptive syndrome with some primates (monkeys, apes, and humans), cetaceans (whales and dolphins), elephants, and corvids (crows and ravens), to greater or lesser degrees. The conclusions I reached after preparing the material presented in the previous chapters have occurred also to scholars with primary expertise in the other taxa. For example, an argument for adaptive similarities among primates, parrots, and cetaceans is presented by my colleague Katharine Milton, an anthropologist at the University of California, Berkeley, in a chapter of an edited volume entitled *On the Move: How and Why Animals Travel in Groups*. Other scholars have likewise done comparative analyses of various pairs of these taxa, sounding the same themes that we have explored for parrots.

The match seems especially good between parrots and one particular species—our own. We humans converge on this suite of adaptive traits with parrots by a mixture of descent

FIGURE 94 A small flock of Blue-headed Parrots, *Pionus menstruus*, socializes in southern Pará, Brazil. Photo © Carlos Yamashita.

from a common ancestor (homology) and convergent evolution by natural selection (analogy). Recall that natural selection is the process by which the traits of organisms are molded in response to a given environment, so that these organisms may function as best they can.

Thus one can say that parrots are the most human of birds, backed by strong support from science. Wait, you might say. Is not such a statement blatantly anthropomorphic? The label of anthropomorphism is thrown around quite a bit these days in many quarters. In this epilogue, it is fitting to put anthropomorphism in an appropriate context, as the bulk of scientific studies on parrots point to these evolutionary parallels with humans.

Anthropomorphism may be defined as the error of incorrectly attributing human traits to other species. The key word here is *incorrectly*. An opposite and I argue equally significant error is that of anthropocentricism. Anthropocentricism may be defined as the error of incorrectly attributing human-like traits only to humans. Which error is worse? One could ask whether one of these errors is more costly or damaging to scientific enterprise than the other. Historically, the error of anthropomorphism has gotten the most attention and attracts most of the energy of scientific criticism. Given that scientists wish not to err at all, can we ask whether it is acceptable to commit major errors of anthropocentricism to avoid even slight errors of anthropomorphism?

The best way to simplify the arguments, in my opinion, is to recognize anthropomorphism and anthropocentricism as type I and type II errors, respectively. Scientists pay a great deal of attention to the kinds of errors that they could make, and this is one way to organize error. Type I error is claiming that something is when it is not. Type II error is claiming that something is not when it is. Traditionally, scientists strive harder to avoid type I error than type II error. Type II error can be passed off simply as uncertainty, as in "We do not yet know the answer but we will eventually." This kind of type II error is fairly benign and conservative, if not a tad boring. I refer to this type of error as *weak type II error*.

A potentially pernicious type II error occurs when a scientist claims that something is not, period, as opposed to simply not yet known. This kind of type II error is on equal footing with type I error. A claim of absence is presented as fact. A claim of absence as fact requires the same rigorous standard of proof as any claim subject to type I error, that is, a claim of presence as fact. I therefore call this type of error *strong type II error*.

A growing number of authors (cited in the notes to this epilogue) think that the answer to the above question is a resounding no: it is not acceptable to make errors of anthropocentricism to avoid anthropomorphism. To avoid anthropomorphism at all costs can lead to the strong type II error of anthropocentricism. Many scholars now take the position that the past focus on avoiding anthropomorphism creates an unacceptable barrier to study, thought, and discussion that harms progress in many fields. We are all the poorer in our understanding of both human and nonhuman animals for this past bias. Any statistician can tell you that a trade-off exists between avoiding type I and type II errors in all areas of scientific study. A claim cannot be realistically made with high probabilities of avoiding both types of error simultaneously. In other words, we lose so much knowledge that we could have gained had we used a reasonable criterion for both errors, such as the 95-percent confidence that is conventional in the sciences.

Turning back to parrots, we are therefore not being anthropomorphic when we recognize that parrots share certain traits with humans. Although a few of these traits, such as basic brain structure (chapter 4), follow from our common vertebrate heritage, most have arisen by convergent evolution, as discussed throughout this book. Natural selection has molded parrots and people in response to the environments in which our ancestors lived. We share large brains and dependence on learning because we are social animals, adapted to solving problems in groups of individuals rather than striking out on our own. Parrots and humans (and other primates, and cetaceans, corvids, and elephants) are social and intelligent because the resources on which they depend are locally abundant but scarce and unpredictable on larger spatial and temporal scales. Harvesting such resources efficiently is aided by a good memory, flexible social interactions, and perhaps the sharing of information about resources. Corollaries to sociality and dependence on learning are long lives and infrequent reproduction. In this way, a few offspring can be carefully attended so that they acquire sufficient skills and knowledge before they become independent.

Parrots also happen to share our inventive use of vocalizations, presumably as a more flexible way to communicate in this social environment. After thinking long and hard about why parrots evolved vocal learning, I have arrived at the hypothesis that humans first evolved vocal learning for similar reasons. I doubt that humans evolved vocal learning in anticipation of building great civilizations and libraries—evolution does not work that way. Rather, I believe that humans evolved vocal learning to enhance their use of song, and therefore some aspects of music, for the same reasons that parrots added vocal learning to enhance their communications. Because of the many evolutionary parallels between humans and parrots, I hypothesize that human songs and music were first used to woo mates and establish group identity and cohesion. Copied sounds are used primarily for these purposes in other animals with vocal learning, and human songs still fulfill these functions to this day. Recent studies establish that parrots are able to synchronize with a beat, that is, *entrain* their movement to music, as reported in several articles cited in the notes. In these publications, the authors support the hypothesis that musical ability and vocal learning are intimately tied in both parrots and humans.[1]

Thus, vocal learning may well have evolved and had the same original functions in humans as we know it to have in parrots. Once vocal learning was in place in humans, it could then be co-opted for flexible communication among group members to solve problems. This same process I hypothesize to be in play in parrots, albeit to a vastly simplified degree compared to this trait in humans. Parrots of some species in particular seem capable of using copied sounds as labels and connecting these labels syntactically, as best exemplified in Grey Parrots.

A lesson I therefore hope to share from the threads woven though this book is that one group of birds, the parrots, shares much in common with our own species. This conclusion arises from a solid body of rigorous scientific study. It is not anthropomorphism to so conclude but rather poses an insightful instance of evolutionary convergence. To say that humans are the most parrot-like of mammals would be equally correct.

We humans love parrots and are fascinated by them because they are like us, and we are like them. My final hope is that this love and fascination can be transformed into providing better lives for them both in captivity and in the wild.

THE FUTURE OF WILD PARROTS

The human population of the planet continues to grow unabated, despite slight decreases in the population growth rate of some developed countries. Some of the largest rates of increase in human subpopulations remain those on the interfaces with the rare few undisturbed ecosystems. At the same time, the highest per capita consumption of the planet's resources occurs in the developed world, where people live in highly disturbed regions but still reap the bounty of the rest of the planet. Will there be any space, or any resources, for anything wild? Will any species remain not counted among our domesticated plants and animals or

---

1. Only animals with vocal learning— specifically fourteen species of parrots, plus humans and elephants, so far—are known to possess the ability to entrain with a beat.

undomesticated species of use to us and intensely managed? Will our efforts to preserve populations of truly wild parrots in their native environments even matter in another human generation?

Environmental scientists, including conservation biologists, do not know the answers to these questions. The immediate future is in better focus than the inevitable but complex consequences of human-induced climate change and usurpation of the planet's ecosystems. The world's wild parrots today still face the dual threat of habitat destruction and harvest for the trade in wild-caught birds. If we take a pessimistic view, that *in situ* conservation is so hopeless that we are best advised to continue strip-mining wild parrots from their remaining habitats no matter how undisturbed, then we have at best a self-fulfilling prophecy. At worst, caraiba trees still grow on the banks of the São Francisco River in eastern Brazil—they have not all been felled, nor will they be any time soon—but the forest now is empty of Spix's Macaws.

A more prudent approach would be for us to strive to keep wild parrots in the wild, because putting captive-bred parrots back to replace what was lost from wild places has not worked very often. Wild parrots, with their long lives, dependence on learning, and possibly even cultural transmission of knowledge from current to future generations, cannot easily be replaced by captive-bred counterparts. It is best, then, to keep wild parrots flying free.

NOTES

PREFACE

Some accounts of the pioneering naturalists to explore North America include Allen (1814); Audubon (1869); and Fremont (1845).

Although subsequent work on the evolutionary relationships of birds continues to refine our understanding of parrot taxonomy and that of birds in general, the studies by Sibley and his colleagues Ahlquist and Monroe remain landmarks in the field (Sibley and Ahlquist 1990; Monroe and Sibley 1993). A more recent taxonomy for parrots can be found in del Hoyo, Elliot, and Sargatal (1997). Much general information about parrots can be found in the encyclopedic volumes by Forshaw (1989, 2006, 2010), the species guide by Juniper and Parr (1998), and recent works by Cameron (2007, 2012).

CHAPTER 1

Studies examining higher-order relationships among avian orders and where the parrots fit in the avian family tree include Andreina Pacheco et al. (2011); Forshaw (1989); Gibb et al. (2007); Hackett et al. (2008); Harrison et al. (2004); Mayr (2008a); McCormack et al. (2013); Poe and Chubb (2004); Pratt et al. (2009); Sibley and Ahlquist (1990); Slack et al. (2007); Suh et al. (2011); van Tuinen and Hedges (2004); van Tuinen, Stidham, and Hadley (2006); van Tuinen et al. (2003); and Wang, Braun, and Kimball (2012).

Phylogenetic studies focused on the evolutionary history and taxonomic relationships of the parrots include Adams et al. (1984); Astuti et al. (2006); Brereton (1963a); Brown and Toft (1999); Chambers and Boon (2005); Chambers et al. (2001); Christidis, Shaw, and Schodde (1991); Christidis et al. (1991); de Kloet and de Kloet (2005); Eberhard (1998b); Eberhard and Bermingham (2005); Eberhard, Wright, and Bermingham (2001); Groombridge et al. (2004); Joseph and Wilke (2006); Joseph et al. (2011, 2012); Kirchman, Schirtzinger, and Wright (2012); Kundu et al. (2012); Leeton et al. (1994); Mayr (2010); Mayr and Clarke (2003); Miyaki et al. (1998); Remsen et al. (2013); Ribas, Joseph, and Miyaki (2006); Ribas and Miyaki (2003); Ribas, Miyaki, and Cracraft (2009); Ribas et al. (2005, 2007); Russello and Amato (2004); Schirtzinger et al. (2012); Schodde et al. (2013); Schweizer, Guentert, and Hertwig (2013); Schweizer, Seehausen, and Hertwig (2011); Schweizer et al. (2010); Smith (1975); Tavares, Yamashita, and Miyaki (2004); Tavares et al. (2006); Tokita (2003); Tokita, Kiyoshi, and Armstrong (2007); White et al. (2011); and Wright et al. (2008).

The timing of the diversification of birds, and of parrots in particular, has been examined by Bottke, Vokrouhlicky, and Nesvorny (2007); Brochu, Sumrall, and Theodor (2004); Brown et al. (2008); Clarke et al. (2005); Cooper and Penny (1997); Cracraft (2001); Cracraft and Prum (1988); Dyke (2001, 2003); Dyke, Nudds, and Benton (2007); Ericson, Anderson, and Mayr (2007); Ericson et al. (2006); Feduccia (1999, 2003a, 2003b); Schulte et al. (2010); and Zhou (2004).

Information about the parrot fossil record can be found in Boles (1993), (1998); Dyke and Cooper (2000); Dyke and Mayr (1999); Lieberman (2003); Mayr (2002, 2008a, 2009, 2014); Mayr and Daniels (1998); Mayr and Gohlich (2004); Mlikovsky (1998); Mourer-Chauviré (1992); Stidham (1999, 2009); and Waterhouse (2008).

Information on Budgerigar ecology and evolutionary relationships is found in Christidis et al. (1991); Mayr (2008b); and Wyndham (1983, 1980a, 1980b), while analyses of body-shape evolution in parrots include Miyaki et al. (1998) and Schweizer, Hertwig, and Seehausen (2014).

CHAPTER 2

General information on diet types, plant defenses, and granivory can be found in Abbas et al. (2006); Atanasov (2007); Banko et al. (2002); Bosque and Pacheco (2000); Brereton, Mallick, and Kennedy (2004); Brice, Dahl, and Grau (1989); Bryant (2006); Burton (1974); Cameron (2005); Cameron and Cunningham (2006); Cannon (1979, 1984a); Caviedes-Vidal et al. (2007); Cipollini (2000); Eriksson (2008); Freeland and Saladin (1989); Hrabar and Perrin (2002); Izhaki (1993, 2002); Izhaki and Safriel (1989, 1990); Jumars (2000); Karasov and Martinez del Rio (2007); Koutsos, Matson, and Klasing (2001); Langer (1988); Levey and Del Rio (2001); Marsh, Wallis, Andrew, and Foley (2006); Marsh, Wallis, McLean, et al. (2006); McNab (2003); McNab and Ellis (2006); McNab and Salisbury (1995); McWilliams, Afik, and Secor (1997); Milton (1980); Mishra (2007); Moegenburg and Levey (2003); Montagna and Torres (2008); Morowitz and Smith (2007); Parrado-Rosselli, Cavelier, and van Dulmen (2002); Schaefer, Schmidt, and Bairlein (2003); Schaefer, Schmidt, and Winkler (2003); Singer, Bernays, and Carriere (2002); Terborgh (1992); Terborgh et al. (1990); Traveset, Rodriguez-Perez, and Pias (2008); Visalberghi et al. (2008); and Witmer and Van Soest (1998).

For information on typical parrot diets in Africa, Asia, and the Americas, including geophagy, see Berg (2007); Bollen and van Elsacker (2004); Bollen, Van Elsacker, and Ganzhorn (2004); Bonadie and Bacon (2000); Bonilla-Ruz, Reyes-Macedo, and Garcia (2007); Borsari and Ottoni (2005); Boyes and Perrin (2009b, 2010a); Brightsmith et al. (2010); Brightsmith and Muñoz-Najar (2004); Brightsmith, Taylor, and Phillips (2008); Bucher, Bertin, and Santamaria (1987); Cant (1979); Carciofi (2008); Coates-Estrada, Estrada, and Meritt (1993); Cochrane (2003); Cotton (2001); da Silva (2005, 2007); Desenne (1995); Enkerlin-Hoeflich and Hogan (1997); Francisco, de Oliveira Lunardi, and Galetti (2002); Francisco et al. (2008); Freeland (1973); Fule, Villanueva-Diaz, and Ramos-Gomez (2005); Galetti (1993, 1997); Galetti and Rodrigues (1992); Gilardi et al. (1999); Gilardi and Munn (1998); Gilardi and Toft (2012); Greenberg (1981); Haugaasen (2008); Heatherbell (1992); Higgins (1979); Janzen (1981, 1983); Jordano (1983, 1987, 1989); Khaleghizadeh (2004); Lee et al. (2010); Martuscelli (1994); Masello et al. (2006); Matuzak, Bezy, and Brightsmith (2008); May (1996); McDiarmid, Ricklefs, and Foster (1977); Mee et al. (2005); Ndithia and Perrin (2006); Ndithia, Perrin, and Waltert (2007); Paranhos, de Araujo, and Marcondes-Machado (2007); Perrin (2005); Pizo, Simão, and Galetti (1995); Powell et al. (2009); Ragusa-Netto (2007, 2008); Renton (2001, 2006); Rodriguez-Ferraro (2007); Roth (1984); Sazima (1989, 2008); Schubart, Aguirre, and Sick (1965); Selman, Perrin, and Hunter (2002); Silvius (1995); Simão, dos Santos, and Pizo (1997); Symes and Perrin (2003b); Taylor and Perrin (2006b); Toyne and Flanagan (1997); Trivedi, Cornejo, and Watkinson (2004); Valdes-Pena et al. (2008); Vaughan, Nemeth, and Marineros (2006); Vicentini and Fischer (1999); Wallace 1869; Warburton and Perrin

(2005b); Wendelken and Martin (1987); Wermundsen (1997); Wetmore (1935); Wirminghaus, Downs, Perrin, and Symes (2001a); Wirminghaus, Downs, Symes, and Perrin (2001, 2002); Yamashita (1987, 1997); Yamashita and Machado de Barros (1997); Yamashita and Valle (1993).

Diets of typical granivorous Australian parrots can be found in Allen (1950); Barker and Vestjens (1980); Brown (1984); Bryant (1994); Cameron (2007); Cannon (1981, 1983); Cleland (1969); Filardi and Tewksbury (2005); Fleming, Gilmour, and Thompson (2002); Garnett and Crowley (1995, 1997); Green and Swift (1965); Higgins (1999); Jones (1987); Joseph (1982a); Joseph, Emison, and Bren (1991); Leslie (2005); Long (1984, 1985); Long and Mawson (1994); Lowry and Lill (2007); Magrath and Lill (1983, 1985); Maron and Lill (2004); McFarland (1991d); McInnes and Carne (1978); Nixon (1994); Pratt and Stiles (1985); Robinet, Bretagnolle, and Clout (2003); Robinson (1965); Romer (2000); Saunders (1980); Scott and Black (1981); Simpson (1972); Smith and Moore (1991); Symes and Marsden (2007); Temby and Emison (1986); and Wyndham (1980b).

Studies of parrots with unusual diets, including dietary specialists, nectarivores, folivores, and omnivores, include Baker, Baker, and Hodges (1998); Beggs (1988); Beggs and Mankelow (2002); Beggs and Wilson (1987); Bell (1966, 1968); Bellingham (1987); Best (1984); Brereton, Mallick, and Kennedy (2004); Brice, Dahl, and Grau (1989); Bryant (2006); Butler (2006); Cameron (2005); Cameron and Cunningham (2006); Cannon (1979, 1984a); Chapman and Paton (2005, 2006, 2007); Churchill and Christensen (1970); Clarke (1967); Clout (1989); Cottom, Merton, and Hendricks (2006); Crowley and Garnett (2001); Davis (1997); Del Rio et al. (2001); Diamond and Bond (1991); Díaz and Kitzberger (2006, 2012); Díaz and Peris (2011); Elliott, Dilks, and O'Donnell (1996); Fleming et al. (2008); Ford, Paton, and Forde (1979); Franke, Jackson, and Nicolson (1998); Frankel and Avram (2001); Gajdon, Fijn, and Huber (2006); Garrod (1872); Gartrell (2000); Gartrell and Jones (2001); Gartrell et al. (2000); Greene (1998, 1999); Gueneau et al. (2006); Hasebe and Franklin (2004); Hingston, Gartrell, and Pinchbeck (2004); Hingston, Potts, and McQuillan (2004); Hofmann (1989); Homberger (1981, 2002, 2003); Hopper (1980); Hopper and Burbidge (1979); Houston et al. (2007); Joseph (1982b); Karasov and Cork (1994, 1996); Kearvell, Young, and Grant (2002); Kennedy and Overs (2001); Kirk, Powlesland, and Cork (1993); Lanning and Shiflett (1983); Livezey (1992); Lotz and Schondube (2006); Mack and Wright (1998); Magrath (1994); Mbatha, Downs, and Penning (2002); McDonald (2003); McWhorter and Lopez-Calleja (2000); McWhorter, Powers, and del Rio (2003); Medway (2005); Moorhouse (1997); Napier et al. (2008); Nicolson and Fleming (2003); O'Donnell (1993); O'Donnell and Dilks (1986,1989, 1994); Pacheco et al. (2004, 2008); Pepper (1997); Pepper, Male, and Roberts (2000); Pryor (2003); Pryor, Levey, and Dierenfeld (2001); Richardson and Wooller (1990); Schnell, Weske, and Hellack (1974); Shepherd, Ditgen, and Sanguinetti (2008); Smith and Lill (2008); Snyder, Enkerlin-Hoeflich, and Cruz-Nieto (1999); Snyder, Koenig, and Johnson (1995); Stock and Wild (2005); Trewick (1996); Tsahar et al. (2006); Tsahar et al. (2005); Waterhouse (1997); Westfahl (2008); Wilson, Grant, and Parker (2006); and Wolf et al. (2007).

Further information on anatomy and phylogenetic relationships of species discussed in this chapter can be found in Einoder and Richardson (2006); French and Smith (2005); Güntert and Ziswiler (1972); Joseph et al. (2011); Leeton et al. (1994); and Murphy et al. (2011).

CHAPTER 3

Studies of vision and visual signals in parrots include Arnold et al. (2002); Bennett (2006); Bennett and Thery (2007); Boles (1991a, 1991b); Bowmaker et al. (1997); Brush (1990); Dyck (1971a, 1971b, 1977); Finger, Burkhardt, and Dyck (1992); Goldsmith and Butler (2003, 2005); Hausmann et al. (2003); Homberger and Brush (1986); Hudon and Brush (1992); Krukenberg (1882); Masello et al. (2004); McGraw (2005); McGraw and Nogare (2005); McGraw et al. (2004);

Pearn, Bennett, and Cuthill (2001, 2002, 2003); Pini et al. (2004); Prum, Andersson, and Torres (2003); Prum et al. (1999); Santos, Elward, and Lumeij (2006); Santos et al. (2007); Stradi, Pini, and Celentano (2001); Toral (2008); Völker (1936, 1937, 1942); Wilkie et al. (2000); and Zampiga, Hoi, and Pilastro (2004).

Information on the hearing abilities of parrots, and the production and perception of their vocal signals, can be found in Ali, Farabaugh, and Dooling (1993); Amagai et al. (1999); Beckers, Nelson, and Suthers (2004); Brittan-Powell and Dooling (2004); Brown, Dooling, and O'Grady (1988); Dent and Dooling (2004); Dent, Larsen, and Dooling (1997); Dent et al. (1997); Dooling (1986); Dooling, Brown, et al. (1987); Dooling, Gephart, et al. (1987); Dooling, Park, et al. (1987); Dooling and Saunders (1975); Dooling and Searcy (1981, 1985); Dooling et al. (1990, 1992, 2001, 2002, 2006); Eda-Fujiwara and Okumura (1992); Farabaugh, Dent, and Dooling (1998); Farabaugh and Dooling (1996); Larsen, Dooling, and Michelsen (2006); Lavenex (1999); Leek, Dent, and Dooling (2000); Lohr, Wright, and Dooling (2003); Manabe, Sadr, and Dooling (1998); Okanoya and Dooling (1987, 1990a, 1990b, 1991); Park and Dooling (1985); Plummer and Striedter (2000); Weisman et al. (2004); Wright et al. (2003); and Yamazaki, Ohi, and Satoh (2000).

Descriptions of the vocal repertoires of parrots and similar animals have been provided by Bradbury (2003); Brereton and Pidgeon (1966); Brittan-Powell, Dooling, and Farabaugh (1997); Fernandez-Juricic and Martella (2000); Fernandez-Juricic, Martella, and Alvarez (1998); Hardy (1963); Jurisevic (2003); Jurisevic and Sanderson (1994); Martella and Bucher (1990); Nair et al. (2009); Nicolas, Fraigneau, and Aubin (2004); Pidgeon (1981); Power (1966); Rendall and Owren (2002); Symes and Perrin (2004a); Taylor and Perrin (2005); Toyne, Flanagan, and Jeffcote (1995); Van Horik, Bell, and Burns (2007); Venuto et al. (2001); and Wirminghaus, Downs, Symes, and Perrin (2000).

Studies of vocal learning and its role in the communication systems of parrots include Baker (2000); Balsby and Bradbury (2009); Bond and Diamond (2005); Bradbury and Balsby (2006); Bradbury, Cortopassi, and Clemmons (2001); Brockway (1962, 1974); Chan and Mudie (2004); Cortopassi and Bradbury (2006); Couzin (2006); Cruickshank, Gautier, and Chappuis (1993); Dahlin and Wright (2009); Deecke, Ford, and Spong (2000); Eda-Fujiwara, Watanabe, and Kimura (2002); Farabaugh, Linzenbold, and Dooling (1994); Foote et al. (2006); Guerra et al. (2008); Hile, Plummer, and Striedter (2000); Hile and Striedter (2000); Hile et al. (2005); Jahelkova, Horacek, and Bartonicka (2008); Kanesada et al. (2005); Kleeman and Gilardi (2005); Kondo, Izawa, and Watanabe (2010); Lengagne et al. (2000); Manabe (1997, 2008); Manabe and Dooling (1997); Manabe, Staddon, and Cleaveland (1997); Masin, Massa, and Bottoni (2004); McElreath, Boyd, and Richerson (2003); McElreath et al. (2005); Messing (2008); Moravec, Striedter, and Burley (2006, 2010); Pepperberg (1994b); Ramos-Fernandez (2005); Rowley (1980b); Salinas-Melgoza and Wright (2012); Sanvito, Galimberti, and Miller (2007); Saunders (1983); Scarl and Bradbury (2009); Searby and Jouventin (2004); Searby, Jouventin, and Aubin (2004); Sewall (2009); Soltis, Leong, and Savage (2005); Striedter et al. (2003); Trillmich (1976a); Tyack (2008); Vehrencamp et al. (2003); Volodina et al. (2006); Wanker (2002); Wanker and Fischer (2001); Wanker, Sugama, and Prinage (2005); Wanker et al. (1998); Weib et al. (2007); Wright (1996, 1997); Wright and Dahlin (2007); Wright, Dahlin, and Salinas-Melgoza (2008); Wright, Rodriguez, and Fleischer (2005); and Wright and Wilkinson (2001).

Roosts as information centers are discussed in Heinrich (1989).

CHAPTER 4

The evolution and ecology of general cognitive abilities in parrots and other animals have been studied by Allen and Bekoff (1995); Bekoff, Allen, and Burghardt (2002); Burish, Kueh, and Wang

(2004); Butler et al. (2005); Corfield (2008); Emery (2006); Evans (2002); Fouts, Jensvold, and Fouts (2002); Griffin (2001); Heyes and Saggerson (2002); Iwaniuk, Clayton, and Wylie (2006); Iwaniuk, Dean, and Nelson (2004), (2005); Iwaniuk and Hurd (2005); Iwaniuk, Hurd, and Wylie (2006), (2007); Iwaniuk and Nelson (2002, 2003); Iwaniuk, Nelson, and O'Leary (2001); Iwaniuk et al. (2004); Lefebvre et al. (1998); Lorenz (1952); Moore (2004); Schuck-Paim, Alonso, and Ottoni (2008); Schuck-Paim, Borsari, and Ottoni (2009); Tinbergen (2005); van Woerden, Isler, and van Schaik (2009); Wilson (2002).

The selective pressures leading to the evolution of vocal learning in parrots are discussed by Hile et al. (2005); Jarvis (2006); Lachlan and Slater (1999); Manabe, Kawashima, and Staddon (1995); Rendall and Owren (2002); Robbins (2000); and Sanvito, Galimberti, and Miller (2007)—while the neural mechanisms underlying this ability have been examined in Brauth, Liang, and Hall (2006); Brauth, Liang, and Roberts (2001); Brauth, Liang, et al. (2003); Brauth, Tang, et al. (2003); Brauth et al. (1987, 2002); Brittan-Powell et al. (1997); Downing, Okanoya, and Dooling (1988); Durand, Liang, and Brauth (1998); Durand et al. (1997); Eda-Fujiwara et al. (2003); Feenders et al. (2008); Haesler et al. (2004); Heaton and Brauth (2000a, 2000a); Jarvis and Mello (2000); Jarvis et al. (2002, 2005); Konishi (2006); Lavenex (2000); Mello (2002, 2004); Mello, Velho, and Pinaud (2004); Plummer and Striedter (2000, 2002); Reiner, Perkel, Bruce, et al. (2004); Reiner, Perkel, Mello, and Jarvis (2004); Roberts, Brauth, and Hall (2001); Roberts, Hall, and Brauth (2000, 2002); Roberts et al. (2001); Scharff and Haesler (2005); Scharff and White (2004); Striedter (1994); Striedter and Charvet (2008); Wada et al. (2004); Webb and Zhang (2005); Webster et al. (1990); White et al. (2006); Wild, Reinke, and Farabaugh (1997); Yamazaki, Ohi, and Satoh (2000); and Zhang, Webb, and Podlaha (2002).

The cognitive abilities of captive African Greys are examined in Giret, Monbureau, et al. (2009), Giret, Peron, et al. (2009), and Giret et al. (2010), and in the comprehensive body of research by Irene Pepperberg and colleagues: Patterson and Pepperberg (1994, 1998); Pepperberg (1981, 1983, 1984, 1985, 1986, 1987a, 1987b, 1988a, 1988b, 1988c, 1988d, 1990a, 1990b, 1991, 1992, 1993a, 1993b, 1994a, 1994b, 1997, 1998, 1999, 2001a, 2001b, 2002a, 2002b, 2002c, 2002d, 2004a, 2004b, 2005, 2006a, 2006b, 2006c, 2006d, 2007a, 2007b); Pepperberg, Brese, and Harris (1991); Pepperberg and Brezinsky (1991); Pepperberg and Funk (1990); Pepperberg, Gardiner, and Luttrell (1999); Pepperberg and Gordon (2005); Pepperberg and Kozak (1986); Pepperberg and Lynn (2000); Pepperberg and McLaughlin (1996); Pepperberg, Naughton, and Banta (1998); Pepperberg and Neapolitan (1988); Pepperberg and Shive (2001); Pepperberg and Wilcox (2000); Pepperberg, Willner, and Gravitz (1997); Pepperberg et al. (1995, 1998, 2000); and Warren, Patterson, and Pepperberg (1996).

For more information on how parrot cognitive abilities impact their captive welfare, see Bradshaw et al. (2005); Garner (2005); Garner, Meehan, and Mench (2003); Garner et al. (2006); Langen et al. (2011); Lewis et al. (2007); Mason (2010); Mason et al. (2007); Meehan, Garner, and Mench (2004); and Swaisgood (2007).

CHAPTER 5

General references on mating systems include Akcay and Roughgarden (2007); Iyer and Roughgarden (2008); and Lebas (2006).

Aspects of the mating systems of typical parrots, including monogamy, divorce, mate choice and the occurrence of extra-pair copulations, have been studied by Arrowood (1988, 1991); Arrowood and Saunders (1991); Baltz and Clark (1996, 1997); Beggs and Wilson (1991); Beissinger (2008); Bond, Wilson, and Diamond (1991); Brockway (1964); Bucher et al. (1991); Dilger (1962); Eberhard (1998a); Ford (1980); Garnett, Pedler, and Crowley (1999); Gnam (1991a); Griggio, Zanollo, and Hoi (2010); Hardy (1963); Hile and Striedter (2000); Hile et al. (2005); Koenig

(2001); Masello and Quillfeldt (2003a); Masello et al. (2002); Massa, Galanti, and Bottoni (1996); McFarland (1991a, 1991c); Meredith, Gilmore, and Isles (1984); Moore (1994); Moorhouse et al. (1999); Moravec, Striedter, and Burley (2006, 2010); Oren and Novaes (1986); Power (1967); Renton (2004); Rodríguez Castillo and Eberhard (2006); Rogers and McCulloch (1981); Rowley (1990); Rowley and Chapman (1991); Santos, Elward, and Lumeij (2006); Saunders (1982); Scarl (2009); Serpell (1981, 1982, 1989); Skeate (1984); Smith (1991); Smith and Rowley (1995); Snyder, Wiley, and Kepler (1987); Spoon, Millam, and Owings (2004, 2006, 2007); Taylor and Parkin (2009); Taylor and Perrin (2004); Trillmich (1976b); Waltman and Beissinger (1992); Zampiga, Hoi, and Pilastro (2004); and Zhang et al. (2010).

The unusual mating systems of such exceptional species as the Kakapo, the Eclectus Parrot, and the Vasa Parrot are described by Clout, Elliot, and Robertson (2002); Clout and Merton (1998); Ekstrom et al. (2007); Elliot, Merton, and Jansen (2001); Heinsohn (2007, 2008a, 2008b); Heinsohn and Legge (2003); Heinsohn, Legge, and Barry (1997); Heinsohn, Legge, and Endler (2005); Heinsohn, Murphy, and Legge (2003); Merton, Morris, and Atkinson (1984); Miller et al. (2003); Powlesland and Lloyd (1994); Powlesland et al. (1992); Robertson et al. (2006); Seibels and McCullough (1976); Tella (2001); Theuerkauf et al. (2009); Trewick (1997); and Wilkinson and Birkhead (1995).

CHAPTER 6

General information on life-history evolution can be found in Akcay and Roughgarden (2009); Barrickman et al. (2008); Beissinger and Stoleson (1997); Boege (2009); Bonsall (2006); Burghardt (2005); Gavrilov and Gavrilova (2004); Ghalambor and Martin (2001); Gonzalez-Lagos, Sol, and Reader (2010); Holmes, Flückiger, and Austad (2001); Isler and van Schaik (2006, 2008, 2009); Kirkwood (2002, 2008, 2011); Martin (1993, 2002); Neill (2010); Partridge (2010); Ricklefs (2000, 2006, 2010a, 2010b, 2010c); Summers and Crespi (2010); Terman and Brunk (2006); van Woerden, Isler, and van Schaik (2009); Veran and Beissinger (2009); Wasser and Sherman (2010); Weinert and Timiras (2003); and Wiersma et al. (2007).

Information on nest site selection and construction in parrots comes from Berkunsky and Reboreda (2009); Berovides Alvarez (1986); Bianchi (2009); Bonebrake and Beissinger (2010); Brightsmith (2000, 2005a, 2005b); Burger and Gochfeld (2005); Cameron (2006); Carrara et al. (2007); Conway (1965); Cornelius et al. (2008); de las Pozas and Gonzalez (1984b); Dilger (1960, 1962); Eberhard (1998a); Emison, White, and Caldow (1995); Gibbs, Hunter, and Melvin (1993); Gnam (1991b, 1991c); Goldingay and Stevens (2009); Guittar, Dear, and Vaughan (2009); Harrison (1971); Heinsohn, Murphy, and Legge (2003); Khan, Beg, and Khan (2004); Koenig, Wunderle, and Enkerlin-Hoeflich (2007); Lanning (1991); Lanning and Shiflett (1983); Legge, Heinsohn, and Garnett (2004); Marsden and Jones (1997); Marsden et al. (2000); Mawson and Long (1994); Meyers (1996); Monterrubio, Enkerlin-Hoeflich, and Hamilton (2002); Monterrubio-Rico and Enkerlin-Hoeflich (2004); Monterrubio-Rico et al. (2009); Murphy and Legge (2007); Ndithia, Perrin, and Waltert (2007); Nores (2009); Ortega-Rodriguez and Monterrubio-Rico (2008); Ortiz-Catedral and Brunton (2009); Ortiz-Maciel, Hori-Ochoa, and Enkerlin-Hoeflich (2010); Pell and Tidemann (1997); Powell et al. (1999); Reed and Tidemann (1994); Renton and Brightsmith (2009); Renton and Salinas-Melgoza (1999); Sanchez-Martinez and Renton (2009); Santharam (2004); Sanz (2008); Saunders (1979, 1990); Saunders, Smith, and Rowley (1982); Snyder and Taapken (1978); Symes et al. (2004); Thompson and Karanja (1989); Walker, Cahill, and Marsden (2005); Warburton and Perrin (2005c); Weaver (1987); White et al. (2005); and Wiley (1985a).

Studies of egg laying, nest attendance, brood care, nutrition, and brood reduction in parrots include Andreina Pacheco, Beissinger, and Bosque (2010); Beissinger and Stoleson (1991);

Beissinger and Waltman (1991); Brockway (1968); Bucher (1983); Bucher, Bertin, and Santamaria (1987); Budden and Beissinger (2004, 2005, 2009); Christiansen and Pitter (1994); de las Pozas and Gonzalez (1984a); Garnett, Pedler, and Crowley (1999); Garnetzke-Stollmann and Franck (1988); Gnam and Rockwell (1991); Guedes (1994, 1995); Hardy (1963); Koenig (2001); Kozlowski and Ricklefs (2010); Krebs (1998, 1999, 2001); Krebs, Cunningham, and Donnelly (1999); Krebs and Magrath (2000); Masello and Quillfeldt (2002, 2003b); Masello et al. (2009); Morbey et al. (1999); Murphy, Legge, and Heinsohn (2003); Ortiz-Catedral and Brunton (2008); Philip, Oommen, and Baiju (2010); Plischke et al. (2010); Power (1967); Reillo, Durand, and McGovern (1999); Renton (2002b); Rivera-Ortiz et al. (2008); Robinet and Salas (1999); Rodríguez Castillo and Eberhard (2006); Roudybush and Grau (1986, 1991); Sanz and Rodriguez-Ferraro (2006); Saunders (1977a, 1977b, 1982, 1986); Saunders and Smith (1981); Saunders, Smith, and Campbell (1984); Siegel, Weathers, and Beissinger (1999a, 1999b); Smith (1991); Smith and Brereton (1976); Smith and Rowley (1995); Stamps (1990); Stamps et al. (1985, 1987, 1989); Stoleson and Beissinger (1997, 1999, 2001); Symes and Perrin (2004b); Taylor and Parkin (2008); Taylor and Perrin (2006a, 2008a); Thompson (1990); Tovar-Martinez (2009a, 2009b); Waltman and Beissinger (1992); Wilson, Field, and Wilson (1995); Wilson, Wilson, and Field (1997); Wirminghaus, Downs, Perrin, and Symes (2001b); and Wyndham (1980a).

Insight into movements and behavior during the fledgling, juvenile and non-breeding-adult stages of parrot lives are provided by Amuno, Massa, and Dranzoa (2007); Beeton (1985); Berg and Angel (2006); Bonadie and Bacon (2000); Bond and Diamond (1992); Bonilla-Ruz, Reyes-Macedo, and Garcia (2007); Boussekey, Saint-Pie, and Morvan (1991); Boyes and Perrin (2009a, 2010b); Brereton (1963b); Buckland, Rowley, and Williams (1983); Cannon (1984b); Caparroz, Miyaki, and Baker (2009); Chapman, Chapman, and LeFebvre (1989); de Moura, Vielliard, and da Silva (2010); Diamond and Bond (1991, 1999, 2003, 2004); Diamond et al. (2006); Drechsler, Brugman, and Menkhorst (1998); Emison and Nicholls (1992); Evans, Ashley, and Marsden (2005); Gajdon, Fijn, and Huber (2006); Garnetzke-Stollmann and Franck (1991); Gilardi and Munn (1998); Goncalves da Silva et al. (2010); Hardy (1965); Hesse and Duffield (2000); Huber and Gajdon (2006); Leite et al. (2008); Lindsey, Arendt, and Kalina (1994); Lindsey et al. (1991); Mac Nally and Horrocks (2000); Magrath and Lill (1985); Marini et al. (2010); Marsden and Pilgrim (2003); Masello et al. (2006); McFarland (1991d); Munn (1988); Munn, Thomsen, and Yamashita (1990); Myers and Vaughan (2004); Perez and Eguiarte (1989); Pizo, Simão, and Galetti (1995, 1997); Renton (2002a, 2004); Robinet, Bretagnolle, and Clout (2003); Rodríguez-Estrella, Mata, and Rivera (1992); Rowley (1980a, 1983, 1990); Salinas-Melgoza and Renton (2007); Sandercock and Beissinger (2002); Sandercock et al. (2000); Saunders and Dawson (2009); Skead (1964, 1971); Skeate (1985); Smith and Moore (1992); Snyder, Wiley, and Kepler (1987); South and Pruett-Jones (2000); Stahala (2008); Stamps et al. (1990); Stock and Wild (2005); Strahl et al. (1991); Symes and Marsden (2007); Symes and Perrin (2003a); Taylor and Perrin (2004, 2008b); Wanker (2002); Wanker, Bernate, and Franck (1996); Warburton and Perrin (2005a); Westcott and Cockburn (1988); Wilson (1993); Wirminghaus, Downs, Perrin, and Symes (2001a); Wyndham (1983); Yamashita (1987); and Yamashita and Machado de Barros (1997).

Parrot life spans have been documented by Brouwer (2000); Clubb and Clubb (1992); Munshi-South and Wilkinson (2006, 2010); and Young et al. (2012).

CHAPTER 7

Information on general avian conservation can be found in Allee (1949); Armsworth and Roughgarden (2001); Askins (2002); Blackburn, Cassey, and Duncan (2004); Blackburn, Cassey, et al. (2004); Blackburn, Jones, et al. (2004); Blackburn et al. (2005); Bulte and van Kooten (1999); Butler (1989, 2006); Cemmick and Veitch (1987); Clegg and Owens (2002); Clout (2006); Clout and Craig (1995); Clout, Elliot, and Robertson (2002); Clout and Merton (1998); Cockrem (2002, 2006); Cockrem and Rounce (1995); Cokinos (2000); Cottom, Merton, and Hendricks (2006); Courchamp, Clutton-Brock, and Grenfell (1999); Curnutt and Pimm (2001); Elphick, Roberts, and Reed (2010); Fitzpatrick et al. (2006); Franklin and Steadman (1991); Fuller (2001); Halliday (1980); Ishii (1999); Jacobs and Walker (1999); Jennings, Reynolds, and Mills (1998); Jones (2008); Leopold (1966); Mayr and Diamond (2001); Orensanz et al. (2004); Quammen (1996); Roberts, Elphick, and Reed (2010); Roughgarden and Smith (1996); Shirley and Kark (2009); Stattersfield and Capper (2000); Steadman (2006b); Steadman and Kirch (1998); and Vogel et al. (2009).

Studies of conservation issues in parrots include Abbas et al. (2006); Amuno, Massa, and Dranzoa (2007); Attrill (1981); Baker and Whelan (1994); Bamford and Mahony (2005); Barre et al. (2006, 2010); Beissinger and Snyder (1992); Best et al. (1995); Boles, Longmore, and Thompson (1991, 1994); Brightsmith et al. (2005); Bucher (1992); Cameron (2009); Carrara et al. (2008); Christian et al. (1996); Cockle et al. (2007); Collar (1997, 2000); de las Pozas and Gonzalez (1984b); Drechsler (1998); Drechsler, Brugman, and Menkhorst (1998); Engebretson (2006); Engeman et al. (2006); Fairfax and Fensham (2000); Faria et al. (2008); Fensham, McCosker, and Cox (1998); Fisher (1986); Forshaw (1970, 1981, 1989, 2006); Forshaw, Fullagar, and Harris (1976); Frynta et al. (2010); Garnett et al. (1993); Gerischer and Walthier (2003); Gnam (1990, 1991a); Gnam and Rockwell (1991); Goodland (1987); Graves and Restrepo (1989); Guedes (2004); Hesse and Duffield (2000); Hoppes and Gray (2010); Johnson, Tomas, and Guedes (1997); Jordan and Munn (1993); Juniper (1991, 2002); Juniper and Parr (1998); Juniper and Yamashita (1990); Kinnaird et al. (2000, 2003); Krabbe (2000); Kuehler et al. (1997); Lee, Finn, and Calver (2010); Leech, Gormley, and Seddon (2008); Leite et al. (2008); Loyn et al. (1986); Mac Nally and Horrocks (2000); Manning, Lindenmayer, and Barry (2004); Marini et al. (2010); Maron et al. (2010); Marsden, Pilgrim, and Wilkinson (2001); Mawson and Long (1996); McCormack (1997, 2006); McCormack and Künzlé (1996); McFarland (1991b, 1991d); Menkhorst and Isles (1981); Monterrubio-Rico et al. (2009); Moorhouse et al. (2003); Moyle, Vardon, and Noske (1997); Munn (1992); Neuhauser (2003); Nunes and Galetti (2007); Olson and Lopez (2008); Ortiz-Catedral and Brunton (2009); Ortiz-Maciel, Hori-Ochoa, and Enkerlin-Hoeflich (2010); Pomarede (1988); Ridgely (1989); Rios-Munoz and Navarro-Siguenza (2009); Rivera et al. (2010); Robinson and Paull (2009); Rodriguez, Rojas-Suarez, and Sharpe (2004); Rowley (1997); Russello et al. (2010); Salaman (1999); Salaman et al. (2002); Sandercock et al. (2000); Sanz and Grajal (1998); Sanz et al. (2003); Saunders (1982, 1990); Saunders and Heinsohn (2008); Saunders and Ingram (1987); Selman, Hunter, and Perrin (2000); Sick and Teixeira (1980); Smales et al. (2000); Smiet (1985); Snyder et al. (1994, 2000); Stahala (2008); Steadman (1993, 2006a); Steadman and Kirch (1990); Steadman and Zarriello (1987); Tschudin et al. (2010); Watling (1995); Wiedenfeld, Molina, and Lezama (1999); Wilcove (1996); Wiley (1981, 1985a, 1985b, 1991); Wilson (1993); Wilson et al. (1994); Wirminghaus, Downs, Symes, and Perrin (1999, 2000); Yamashita (1992); Yamashita and Valle (1993); and Ziembicki (2005).

The biology of the highly endangered Kakapo has been studied by Best (1984); Bryant (2006); Eason and Moorhouse (2006); Eason et al. (2006); Elliott (2006); Elliott, Merton, and Jansen (2001); Elliott et al. (2006); Farrimond, Clout, and Elliott (2006); Farrimond and Elliott (2006); Gibson (2007); Grzelewski (2002); Harper and Joice (2006); Harper et al. (2006); Hutching (1997); Imboden, Jones, and Aktinson (1995); Jansen (2006); Johnson (1976); Karl and Best (1982); Kirk, Powlesland, and Cork (1993); Lee, Wood, and Rogers (2010); Livezey (1992); Lloyd and Powlesland (1994); McNab and Salisbury (1995); Merton and Empson (1989); Merton, Morris, and Atkinson (1984); Miller et al. (2003); Minot, Robertson, and Miller (2000); Moorhouse and Powlesland (1991); Powlesland and Lloyd (1994); Powlesland, Lloyd, and Grant (1986); Powlesland, Merton, and Cockrem (2006); Powlesland et al. (1992, 1995); Raubenheimer and Simpson (2006); Robertson (2006); Robertson et al. (2000, 2006); Sutherland (2002); Tella (2001); Towns and

Williams (1993); Trewick (1996, 1997); Triggs, Powlesland, and Daugherty (1989); Walsh, Wilson, and Elliott (2006); Williams (1956); Williams and Merton (2006); Wilmshurst et al. (2004); Wilson, Grant, and Parker (2006); and Wood (2006).

Causes of extinction of the Carolina Parakeet and other aspects of its biology are examined in Allen and Sprunt (1936); Brantley and Platt (2001); Kirchman, Schirtzinger, and Wright (2012); McKinley (1978, 1980a, 1980b, 1980c); Nowotny (1898); Prestwich (1966); Snyder (2004); and Snyder and Russell (2002).

Parrot invasions and their consequences are discussed in Bittner (2004); Blackburn and Cassey (2004); Brochier, Vangeluwe, and van den Berg (2010); Bucher et al. (1991); Bull (1971); Burger and Gochfeld (2009); Butler (2005); Caruso and Scelsi (1994); Cassey, Blackburn, Duncan, and Chown (2005); Cassey, Blackburn, Duncan, and Gaston (2005); Cassey, Blackburn, Jones, et al. (2004), Cassey, Blackburn, Russell, et al. (2004); Chiron, Shirley, and Kark (2010); Clavell et al. (1991); Clergeau, Vergnes, and Delanoue (2009); Dangoisse (2009); De Schaetzen and Jacob (1985); Dumser (1987); Gonçalves da Silva et al. (2010); Gochfeld (1973); Harrison (1971); Hurley (2001); Hyman and Pruett-Jones (1995); Iriarte, Lobos, and Jaksic (2005); Khan, Beg, and Khan (2004); Kumschick and Nentwig (2010); Lin and Lee (2006); Long (1981); Major and Parsons (2010); Martin (1989); Martin and Bucher (1993); Munoz and Real (2006); Murgui and Valentin (2003); Neidermyer and Hickey (1977); Oliverieri and Pearson (1992); Peris and Aramburu (1995); Pithon and Dytham (1999a, 1999b, 2002); Pranty (2009); Pranty, Feinstein, and Lee (2010); Ramzan and Toor (1972, 1973); Runde and Pitt (2007a, 2007b); Shieh et al. (2006); Shields, Grubb, and Telis (1974); Shirley and Kark (2006); Shwartz, Shirley, and Kark (2008); Shwartz et al. (2009); Simon (1990); Simpson and Ruiz (1974); Sodhi and Eguchi (2004); Sol et al. (1997); South and Pruett-Jones (2000); Spano and Truffi (1986); Strubbe and Matthysen (2009a, 2009b, 2009c); Strubbe, Matthysen, and Graham (2010); Toor and Ramzan (1974); Van Bael and Pruett-Jones (1996); Waring (1996); Weathers and Caccamise (1978); Weiserbs (2010); and Wright et al. (2010).

The conservation impact of the trade in wild parrots, and the complex issues surrounding sustainable harvesting, have been examined by Arrowood (1981); Beissinger and Bucher (1992); Bowles et al. (1992); Bruning (1985); Cahill, Walker, and Marsden (2006); Cooney (2005); Cooney and Jepson (2006); Desenne and Strahl (1991); Dold (1992); Duffy (1990); Fischer (2004); Gilardi (2006); González (2003); Imboden (1992); Iñigo-Elias and Ramos (1991); James (1990); Jarry (2003); Juste (1996); Knights and Currey (1990); Lambert (1993); Pain et al. (2006); Pillay et al. (2010); Roet, Mack, and Duplaix (1981); and Wright et al. (2001).

EPILOGUE

Discussions of the role of music in language evolution, and of the more general question of what makes us human, are in Kako (2001); Miller (2000); Mithen (1996); Okanoya (2007); and Szamado and Szathmary (2006).

Important comparative insights into the cognitive abilities of other brainy animals like primates, cetaceans, elephants, and corvids are provided by Barkow (2001); Barrett-Lennard et al. (2001); Bates, Poole, and Byrne (2008); Bradshaw and Casey (2007); Bradshaw and Finlay (2005); Byrne and Bates (2010); Emery (2006); Garnett and Lindenmayer (2011); Heyes (2003); Lynn and Pepperberg (2001); Marino et al. (2007); Milton (2000); Mithen (1996); Moore (2004); Pepperberg (2002c); Rivas and Burghardt (2002); Seed, Emery, and Clayton (2009); and Whitehead et al. (2004).

Recent work on musicality in parrots includes Bottoni, Massa, and Boero (2003); Fitch (2009); Giret, Peron, et al. (2009); Patel et al. (2009a, 2009b); and Schachner et al. (2009).

Statistical issues relevant to the discussion of anthropomorphism versus anthropocentricism are covered in Quinn and Dunham (1983); Roughgarden (1983); and Toft and Shea (1983).

BIBLIOGRAPHY

Abbas, S. A., S. Ali, S. I. M. Halim, A. Fakhrul-Razi, R. Yunus, and T. S. Y. Choong. 2006. "Effect of thermal softening on the textural properties of palm oil fruitlets." *Journal of Food Engineering* 76(4):626–31.

Adams, M., P. R. Baverstock, D. A. Saunders, R. Schodde, and G. T. Smith. 1984. "Biochemical systematics of the Australian cockatoos (Psittaciformes:Cacatuinae)." *Australian Journal of Zoology* 32(3):363–78. doi:10.1071/zo9840363.

Akcay, E., and J. Roughgarden. 2007. "Extra-pair parentage: a new theory based on transactions in a cooperative game." *Evolutionary Ecology Research* 9(8):1223–43.

———. 2009. "The perfect family: decision making in biparental care." *Public Library of Science ONE* 4(10):e7345.

Ali, N. J., S. Farabaugh, and R. Dooling. 1993. "Recognition of contact calls by the Budgerigar (*Melopsittacus undulatus*)." *Bulletin of the Psychonomic Society* 31(5):468–70.

Allee, Warder Clyde. 1949. *Principles of animal ecology*. Philadelphia: W. B. Saunders.

Allen, C., and M. Bekoff. 1995. "Cognitive ethology and the intentionality of animal behaviour." *Mind & Language* 10(4):313–28.

Allen, G. H. 1950. "Birds as a biotic factor in the environment of pastures with particular reference to Galahs (*Cacatua roseicapilla*)." *Journal Australian Institute Agricultural Science* 16:19–25.

Allen, P., ed. 1814. *History of the expedition under the command of Captains Lewis and Clark: to the sources of the Missouri, thence across the Rocky Mountains and down the River Columbia to the Pacific Ocean*. New York: Bradford and Inskeep.

Allen, R. P., and A. Sprunt, Jr. 1936. *The Carolina Paroquet (*Conuropsis c. carolinensis*) in the Santee Swamp, South Carolina*. Report to the National Association of Audubon Societies, New York.

Amagai, S., R. J. Dooling, S. Shamma, T. L. Kidd, and B. Lohr. 1999. "Detection of modulation in spectral envelopes and linear-rippled noises by budgerigars (*Melopsittacus undulatus*)." *Journal of the Acoustical Society of America* 105(3):2029–35.

Amuno, J. B., R. Massa, and C. Dranzoa. 2007. "Abundance, movements and habitat use by African Grey Parrots (*Psittacus erithacus*) in Budongo and Mabira forest reserves, Uganda." *Ostrich* 78(2):225–31.

Andreína Pacheco, M., S. R. Beissinger, and C. Bosque. 2010. "Why grow slowly in a dangerous place? Postnatal growth, thermoregulation, and energetics of nestling Green-rumped Parrotlets (*Forpus passerinus*)." *Auk* 127(3):558–70.

Andreína Pacheco, M., Fabia U. Battistuzzi, Miguel Lentino, Roberto F. Aguilar, Sudhir Kumar, and Ananias A. Escalante. 2011. "Evolution of modern birds revealed by mitogenomics: timing the radiation and origin of major orders." *Molecular Biology and Evolution* 28(6):1927–42. doi:10.1093/molbev/msr014.

Armsworth, P. R., and J. E. Roughgarden. 2001. "An invitation to ecological economics." *Trends in Ecology & Evolution* 16(5):229–34.

Arnold, Kathryn E., Ian P. F. Owens, and N. Justin Marshall. "Fluorescent signaling in parrots." *Science* 295(5552):92.

Arrowood, P. C. 1981. "Importation and status of Canary-winged Parakeets (*Brotogeris versicolorus* P.L.S. Muller) in California." In *Conservation of New World Parrots,* edited by R. F. Pasquier, 420–25. Washington, DC: Smithsonian Institution Press.

———. 1991. "Male-male, female-female and male-female interactions within captive Canary-winged parakeet *Brotogeris v. versicolurus* flocks." *ACTA XX Congressus Internationalis Ornithologici,* 666–72.

Arrowood, P. C., and D. A. Saunders. 1991. "Concluding remarks: symposium on the ecology and social behaviour of parrots and parakeets." *ACTA XX Congressus Internationalis Ornithologici,* 697–98.

Arrowood, Patricia C. 1988. "Duetting, pair bonding and agonistic display in parakeet pairs." *Behaviour* 106:129–57.

Askins, Robert A. 2002. *Restoring North America's birds: lessons from landscape ecology.* 2nd ed. New Haven: Yale University Press.

Astuti, D., N. Azuma, H. Suzuki, and S. Higashi. 2006. "Phylogenetic relationships within parrots (Psittacidae) inferred from mitochondrial cytochrome-b gene sequences." *Zoological Science* 23(2):191–98.

Atanasov, A. T. 2007. "The near to linear allometric relationship between total metabolic energy per life span and body mass of nonpasserine birds." *Bulgarian Journal of Veterinary Medicine* 10(4):235–45.

Attrill, R. 1981. "The status and conservation of the Bahamas Amazon (*Amazona leucocephala bahamensis*)." In *Conservation of New World Parrots,* edited by R. F. Pasquier, 74–81. Washington, DC: Smithsonian Institution Press.

Audubon, J. J. 1869. *The Life of John James Audubon.* New York: G. P. Putnam & Son.

Baker, H. G., I. Baker, and S. A. Hodges. 1998. "Sugar composition of nectars and fruits consumed by birds and bats in the tropics and subtropics." *Biotropica* 30(4):559–86.

Baker, Jack, and Robert J. Whelan. 1994. "Ground Parrots and fire at Barren Grounds, New South Wales: a long-term study and an assessment of management implications." *Emu* 94(4):300–04.

Baker, Myron C. 2000. "Cultural diversification in the flight call of the Ringneck Parrot in Western Australia." *Condor* 102(4):905–10.

Balsby, T. J. S., and J. W. Bradbury. 2009. "Vocal matching by orange-fronted conures (*Aratinga canicularis*)." *Behavioral Processes* 82(2):133–39.

Baltz, A. P., and A. B. Clark. 1996. "Cere colour as a basis for extra-pair preferences of paired male budgerigars (*Melopsittacus undulatus*: Psittacidae: Aves)." *Ethology* 102(2):109–16.

———. 1997. "Extra-pair courtship behaviour of male budgerigars and the effect of an audience." *Animal Behaviour* 53:1017–24.

Bamford, Mike, and Julie Mahony. 2005. *Night Parrot* (Pezoporus occidentalis) *management plan.* East Perth, Australia: Pilbara Iron Ore, Fortesque Metal Group Ltd.

Banko, P. C., M. L. Cipollini, G. W. Breton, E. Paulk, M. Wink, and I. Izhaki. 2002. "Seed chemistry of *Sophora chrysophylla* (mamane) in relation to diet of specialist avian seed predator *Loxioides bailleui* (palila) in Hawaii." *Journal of Chemical Ecology* 28(7):1393–1410.

Barker, R.D., and W.J.M. Vestjens. 1980. "The food of Australian birds. I. Non-passerines." Clayton, Australia: CSIRO.

Barkow, J.H. 2001. "Culture and hyperculture: Why can't a cetacean be more like a (hu)man?" *Behavioral and Brain Sciences* 24(2):324–25.

Barre, N., P. Villard, N. Manceau, L. Monimeau, and C. Menard. 2006. "The birds of the Loyalty Islands (New Caledonia): census, ecological and biogeographical issues." *Revue D Ecologie-La Terre Et La Vie* 61(2):175–94.

Barre, Nicolas, Joern Theuerkauf, Ludovic Verfaille, Pierre Primot, and Maurice Saoumoe. 2010. "Exponential population increase in the endangered Ouvéa Parakeet (*Eunymphicus uvaeensis*) after community-based protection from nest poaching." *Journal of Ornithology* 151(3):695–701.

Barrett-Lennard, L.G., V.B. Deecke, H. Yurk, and J.K.B. Ford. 2001. "A sound approach to the study of culture." *Behavioral and Brain Sciences* 24(2):325–26.

Barrickman, N.L., M.L. Bastian, K. Isler, and C.P. van Schaik. 2008. "Life history costs and benefits of encephalization: a comparative test using data from long-term studies of primates in the wild." *Journal of Human Evolution* 54(5):568–90.

Bates, L.A., J.H. Poole, and R.W. Byrne. 2008. "Elephant cognition." *Current Biology* 18(13):R544–46.

Beckers, G.J.L., B.S. Nelson, and R.A. Suthers. 2004. "Vocal-tract filtering by lingual articulation in a parrot." *Current Biology* 14(17):1592–97.

Beeton, R.J.S. 1985. "The Little Corella: a seasonally adapted species." *Proceedings Ecological Society Australia* 13:53–63.

Beggs, J.R. 1988. *Energetics of Kaka in a South Island beech forest.* University of Aukland.

Beggs, J.R., and P.R. Wilson. 1987. "Energetics of the South Island Kaka (*Nestor meridionalis meridionalis*) feeding on the larvae of Kanuka Longhorn Beetles (*Ochrocydus huttoni*)." *New Zealand Journal of Ecology* 10:143–47.

———. 1991. "The Kaka *Nestor meridionalis,* a New Zealand parrot endangered by introduced wasps and mammals." *Biological Conservation* 56:23–38.

Beggs, Wayne, and Sarah Mankelow. 2002. "Kea (*Nestor notabilis*) make meals of mice (*Mus musculus*)." *Notornis* 49:50.

Beissinger, S.R. 2008. "Long-term studies of the Green-rumped Parrotlet (*Forpus passerinus*) in Venezuela: Hatching asynchrony, social system and population structure." *Ornitologia Neotropical* 19:73–83.

Beissinger, S.R., and E.H. Bucher. 1992. "Can parrots be conserved through sustainable harvesting?" *Bioscience* 42(3):164–73.

Beissinger, S.R., and N.F.R. Snyder. 1992. *New World Parrots in Crisis: Solutions from Conservation Biology.* Washington, DC: Smithsonian Institution Press.

Beissinger, S.R., and S.H. Stoleson. 1991. "Nestling mortality patterns in relation to brood size and hatching asynchrony in Green-rumped Parrotlets." *Congressus Internationalis Ornithologici* Acta XX:1727–33.

Beissinger, S.R., and J.R. Waltman. 1991. "Extraordinary clutch size and hatching asynchrony of a Neotropical parrot." *Auk* 108:863–71.

Beissinger, Steven R., and Scott H. Stoleson. 1997. "Hatching asynchrony in birds." *Trends in Ecology and Evolution* 12:112.

Bekoff, Marc, Colin Allen, and Gordon M. Burghardt. 2002. *The Cognitive Animal: empirical and theoretical perspectives on animal cognition.* Cambridge, MA: MIT Press.

Bell, H.L. 1966. "Some feeding habits of the Rainbow Lorikeet." *Emu* 66:71–72.

———. 1968. "A further note on the feeding of lorikeets." *Emu* 68:221–22.

Bellingham, M. 1987. "Red-crowned Parakeet on Burgess Island." *Notornis* 34:234–36.

Bennett, A.T.D. 2006. "Color vision and its functions." *Journal of Ornithology* 147(5):59.

Bennett, A.T.D., and M. Thery. 2007. "Avian color vision and coloration: Multidisciplinary evolutionary biology." *American Naturalist* 169(1):S1–S6.

Berg, K.S. 2007. "Great Green Macaws and the annual cycle of their food plants in Ecuador." *Journal of Field Ornithology* 78(1):1–10.

Berg, K.S., and R.R. Angel. 2006. "Seasonal roosts of Red-lored Amazons in Ecuador provide information about population size and structure." *Journal of Field Ornithology* 77(2):95–103.

Berkunsky, I., and J.C. Reboreda. 2009. "Nest-site fidelity and cavity reoccupation by Blue-fronted Parrots *Amazona aestiva* in the dry Chaco of Argentina." *Ibis* 151(1):145–50.

Berovides Alvarez, V. 1986. "Nidificación de la cotorra de Cuba (*Amazona leucocephala*) en la Isla de la Juventud." *Acad. Cienc. Biol. Cuba Ciencias Biológicas* 15:133–35.

Best, Brinley J., Niels Krabbe, Christopher T. Clarke, and Amanda L. Best. 1995. "Red-masked Parakeet *Aratinga erythrogenys* and Grey-cheeked Parakeet *Brotogeris pyrrhopterus:* Two threatened parrots from Tumbesian Ecuador and Peru?" *Bird Conservation International* 5(2–3):233–50.

Best, H.A. 1984. "The foods of Kakapo on Stewart Island as determined from their feeding sign." *New Zealand Journal of Ecology* 7:71–83.

Bianchi, C.A. 2009. "Notes on the ecology of the Yellow-Faced Parrot (*Aliopsitta xanthops*) in central Brazil." *Ornitologia Neotropical* 20(4):479–89.

Bittner, Mark. 2004. *The Wild Parakeets of Telegraph Hill.* New York: Harmony.

Blackburn, T.M., and P. Cassey. 2004. "Are introduced and re-introduced species comparable? A case study of birds." *Animal Conservation* 7:427–33.

Blackburn, T.M., P. Cassey, and R.P. Duncan. 2004. "Extinction in island endemic birds reconsidered." *Ecography* 27(1):124–28.

Blackburn, T.M., P. Cassey, R.P. Duncan, K.L. Evans, and K.J. Gaston. 2004. "Avian extinction and mammalian introductions on oceanic islands." *Science* 305(5692):1955–58.

Blackburn, T.M., O.L. Petchey, P. Cassey, and K.J. Gaston. 2005. "Functional diversity of mammalian predators and extinction in island birds." *Ecology* 86(11):2916–23.

Blackburn, Tim M., Kate E. Jones, Phillip Cassey, and Neil Losin. 2004. "The influence of spatial resolution on macroecological patterns of range size variation: A case study using parrots (Aves: Psittaciformes) of the world." *Journal of Biogeography* 31(2):285–93.

Boege, F. 2009. "Premature death in the perspective of molecular biology." *Historical Social Research-Historische Sozialforschung* 34(4):61–65.

Boles, Walter E. 1991a. "Black-light signature for birds?" *Australian Natural History* 23:752.

———. 1991b. "Glowing parrots: a need for study of hidden colors." *Birds International* 3:76–79.

———. 1993. "A new cockatoo (Psittaciformes: Cacatuidae) from the Tertiary of Riversleigh, northwestern Queensland, and an evaluation of rostral characters in the systematics of parrots." *Ibis* 135(1):8–18.

———. 1998. "A budgerigar *Melopsittacus undulatus* from the Pliocene of Riversleigh, north-western Queensland." *Emu* 98:32–35.

Boles, Walter E., N.W. Longmore, and C. Max Thompson. 1991. "Fly-by-night parrot." *Australian Natural History* 23:689–95.

———. 1994. "A recent specimen of the night parrot *Geopsittacus occidentalis*." *Emu* 94(1):37–40.

Bollen, A., and L. van Elsacker. 2004. "The feeding ecology of the lesser vasa parrot, *Coracopsis nigra*, in south-eastern Madagascar." *Ostrich* 75(3):141–46.

Bollen, A., L. Van Elsacker, and J. U. Ganzhorn. 2004. "Relations between fruits and disperser assemblages in a Malagasy littoral forest: a community-level approach." *Journal of Tropical Ecology* 20:599–612.

Bonadie, Wayne A., and Peter R. Bacon. 2000. "Year-round utilisation of fragmented palm swamp forest by Red-bellied macaws (*Ara manilata*) and Orange-winged parrots (*Amazona amazonica*) in the Nariva Swamp (Trinidad)." *Biological Conservation* 95(1):1–5.

Bond, Alan B., and Judy Diamond. 1992. "Population estimates of Kea in Arthur's Pass National Park." *Notornis* 39(3):151–60.

———. 2005. "Geographic and ontogenetic variation in the contact calls of the kea (*Nestor notabilis*)." *Behaviour* 142:1–20.

Bond, A. B., K. J. Wilson, and J. Diamond. 1991. "Sexual dimorphism in the Kea *Nestor notabilis*." *Emu* 91(1):12–19.

Bonebrake, Timothy C., and Steven R. Beissinger. 2010. "Predation and infanticide influence ideal free choice by a parrot occupying heterogeneous tropical habitats." *Oecologia (Berlin)* 163(2):385–93.

Bonilla-Ruz, C., G. Reyes-Macedo, and R. Garcia. 2007. "Observations of the military macaw (*Ara militaris*) in northern Oaxaca, Mexico." *Wilson Journal of Ornithology* 119(4):729–32.

Bonsall, M. B. 2006. "Longevity and ageing: appraising the evolutionary consequences of growing old." *Philosophical Transactions of the Royal Society B-Biological Sciences* 361(1465):119–35.

Borsari, A., and E. B. Ottoni. 2005. "Preliminary observations of tool use in captive hyacinth macaws (*Anodorhynchus hyacinthinus*)." *Animal Cognition* 8(1):48–52.

Bosque, C., and A. Pacheco. 2000. "Dietary nitrogen as a limiting nutrient in frugivorous birds." *Revista Chilena De Historia Natural* 73(3):441–50.

Bottke, W. F., D. Vokrouhlicky, and D. Nesvorny. 2007. "An asteroid breakup 160 Myr ago as the probable source of the K/T impactor." *Nature* 449(7158):48–53.

Bottoni, L., R. Massa, and D. L. Boero. 2003. "The grey parrot (*Psittacus erithacus*) as musician: an experiment with the Temperate Scale." *Ethology, Ecology & Evoluton* 15(2):133–41.

Boussekey, Marc, Jean Saint-Pie, and Olivier Morvan. 1991. "Observations on a population of Red-fronted Macaws *Ara rubrogenys* in the Rio Caine valley, central Bolivia." *Bird Conservation International* 1:335–50.

Bowles, D., D. Currey, P. Knights, and A. Michels. 1992. *Flight to extinction: the wild-caught bird trade*. Washington, DC: Animal Welfare Institute and Environmental Investigation Agency.

Bowmaker, J. K., L. A. Heath, S. E. Wilkie, and D. M. Hunt. 1997. "Visual pigments and oil droplets from six classes of photoreceptor in the retinas of birds." *Vision Research* 37:2183–84.

Boyes, R. S., and M. R. Perrin. 2009a. "Flocking dynamics and roosting behaviour of Meyer's parrot (*Poicephalus meyeri*) in the Okavango Delta, Botswana." *African Zoology* 44(2):181–93.

———. 2009b. "Generalists, specialists and opportunists: niche metrics of *Poicephalus* parrots in southern Africa." *Ostrich* 80(2):93–97.

———. 2010a. "Do Meyer's Parrots *Poicephalus meyeri* benefit pollination and seed dispersal of plants in the Okavango Delta, Botswana?" *African Journal of Ecology* 48(3):769–82.

———. 2010b. "Patterns of daily activity of Meyer's Parrot (*Poicephalus meyeri*) in the Okavango Delta, Botswana." *EMU* 110(1):54–65.

Bradbury, J. W. 2003. "Vocal communication in wild parrots." In *Animal social complexity: intelligence, culture and individualized societies*, edited by Frans B. M. de Waal and Peter L. Tyack, 293–316. Cambridge, MA: Harvard University Press.

Bradbury, Jack, and Thorsten Balsby. 2006. "Talking with the parrots of the ACG." *Psittascene* 18(3):8–11.

Bradbury, Jack W., Kathryn A. Cortopassi, and Janine R. Clemmons. 2001. "Geographical variation in the contact calls of Orange-fronted Parakeets." *Auk* 118(4):958–72.

Bradshaw, G.A., and B.L. Finlay. 2005. "Natural symmetry." *Nature* 435(7039):149.

Bradshaw, G.A., A.N. Schore, J.L. Brown, J.H. Poole, and C.J. Moss. 2005. "Elephant breakdown." *Nature* 433(7028):807.

Bradshaw, J.W.S., and R.A. Casey. 2007. "Anthropomorphism and anthropocentrism as influences in the quality of life of companion animals." *Animal Welfare* 16:149–54.

Brantley, C.G., and S.G. Platt. 2001. "Canebrake conservation in the southeastern United States." *Wildlife Society Bulletin* 29(4):1175–81.

Brauth, S.E., W.R. Liang, and W.S. Hall. 2006. "Contact-call driven and tone-driven zenk expression in the nucleus ovoidalis of the budgerigar (*Melopsittacus undulatus*)." *Neuroreport* 17(13):1407–10.

Brauth, S.E., Y.Z. Tang, W.R. Liang, and T.F. Roberts. 2003. "Contact call-driven zenk mRNA expression in the brain of the budgerigar (*Melopsittacus undulatus*)." *Molecular Brain Research* no. 117 (1):97–103.

Brauth, S.E., C.M. McHale, C.A. Brasher, and R.J. Dooling. 1987. "Auditory pathways in the budgerigar. 1: Thalamo-telencephalic projections." *Brain Behavior and Evolution* 30(3–4):174–99.

Brauth, S.E., Y.Z. Tang, W.R. Liang, and T.F. Roberts. 2003. "Contact call-driven zenk mRNA expression in the brain of the budgerigar (*Melopsittacus undulatus*)." *Molecular Brain Research* 117(1):97–103.

Brauth, Steven E., Wenru Liang, and Todd F. Roberts. 2001. "Projections of the oval nucleus of the hyperstriatum ventrale in the budgerigar: relationships with the auditory system." *Journal of Comparative Neurology* 432(4):481–511.

Brauth, Steven, Wenru Liang, F. Roberts Todd, L. Scott Lindsey, and M. Quinlan Elizabeth. 2002. "Contact call-driven Zenk protein induction and habituation in telencephalic auditory pathways in the budgerigar (*Melopsittacus undulatus*): implications for understanding vocal learning processes." *Learning & Memory (Cold Spring Harbor)* 9(2):76–88.

Brereton, J. Le Gay. 1963a. "Evolution within the Psittaciformes." *Proceedings 13th International Ornithological Congress,* 499–517.

———. 1963b. "The life cycles of three Australian parrots: some comparative and population aspects." *The Living Bird* 63:21–29.

Brereton, J. Le Gay, and R.W. Pidgeon. 1966. "The language of the Eastern Rosella." *Australian Natural History,* September, 225–29.

Brereton, R., S.A. Mallick, and S.J. Kennedy. 2004. "Foraging preferences of Swift Parrots on Tasmanian Blue-gum: tree size, flowering frequency and flowering intensity." *Emu* 104(4):377–83.

Brice, A.T., K.H. Dahl, and C.R. Grau. 1989. "Pollen digestibility by hummingbirds and psittacines." *Condor* 91(3):681–88.

Brightsmith, D.J. 2000. "Use of arboreal termitaria by nesting birds in the Peruvian Amazon." *Condor* 102:529–38.

———. 2005a. "Competition, predation and nest niche shifts among tropical cavity nesters: phylogeny and natural history evolution of parrots (Psittaciformes) and trogons (Trogoniformes)." *Journal of Avian Biology* 36(1):64–73.

———. 2005b. "Parrot nesting in southeastern Peru: seasonal patterns and keystone trees." *Wilson Bulletin* 117(3):296–305.

Brightsmith, D. J., and Romina Aramburú Muñoz-Najar. 2004. "Avian geophagy and soil characteristics in southeastern Peru." *Biotropica* 36(4):534–43.

Brightsmith, D. J., J. Hilburn, A. del Campo, J. Boyd, M. Frisius, R. Frisius, D. Janik, and F. Guillen. 2005. "The use of hand-raised psittacines for reintroduction: a case study of scarlet macaws (*Ara macao*) in Peru and Costa Rica." *Biological Conservation* 121(3):465–72.

Brightsmith, D. J., D. McDonald, D. Matsafuji, and C. A. Bailey. 2010. "Nutritional content of the diets of free-living Scarlet Macaw chicks in southeastern Peru." *Journal of Avian Medicine and Surgery* 24(1):9–23.Brightsmith, D. J., J. Taylor, and T. D. Phillips. 2008. "The roles of soil characteristics and toxin adsorption in avian geophagy." *Biotropica* 40(6):766–74.

Brillat-Savarin, Anthelme. 1826. *Physiologie du goût, ou, Méditations de gastronomie transcendante.* Paris: A. Sautelet.

Brittan-Powell, E. F., and R. J. Dooling. 2004. "Development of auditory sensitivity in budgerigars (*Melopsittacus undulatus*)." *Journal of the Acoustical Society of America* 115(6):3092–3102.

Brittan-Powell, E. F., R. J. Dooling, and S. M. Farabaugh. 1997. "Vocal development in budgerigars (*Melopsittacus undulatus*): Contact calls." *Journal of Comparative Psychology* 111(3):226–41.

Brittan-Powell, E. F., R. J. Dooling, O. N. Larsen, and J. T. Heaton. 1997. "Mechanisms of vocal production in budgerigars (*Melopsittacus undulatus*)." *Journal of the Acoustical Society of America* 101(1):578–89.

Brochier, B., D. Vangeluwe, and T. van den Berg. 2010. "Alien invasive birds." *Revue Scientifique et Technique-Office International des Epizooties* 29(2):217–25.

Brochu, C. A., C. D. Sumrall, and J. M. Theodor. 2004. "When clocks (and communities) collide: estimating divergence time from molecules and the fossil record." *Journal of Paleontology* 78(1):1–6.

Brockway, B. F. 1962. "The effects of nest-entrance positions and male vocalizations on reproduction in budgerigars." *Living Bird* 1:93–101.

———. 1964. "Ethological studies of the Budgerigar: reproductive behavior." *Behaviour* 23:294–324.

———. 1968. "Budgerigars are not determinate egg-layers." *Wilson Bulletin* 80(1):106–07.

———. 1974. "Roles of budgerigar vocalization in the integration of breeding behaviour." In *Bird Vocalizations,* edited by R. A. Hinde, 131–59. London: Cambridge University Press.

Brouwer, K. 2000. "Longevity records for Psittaciformes in captivity." *International Zoo Yearbook* 37:299–316.

Brown, David M., and Catherine A. Toft. 1999. "Molecular systematics and biogeography of the cockatoos (Psittaciformes: Cacatuidae)." *Auk* 116(1):141–57.

Brown, J. W., J. S. Rest, J. Garcia-Moreno, M. D. Sorenson, and D. P. Mindell. 2008. "Strong mitochondrial DNA support for a Cretaceous origin of modern avian lineages." *BMC Biology* 6:6.

Brown, P. 1984. "Food and feeding of the Green Rosella (*Platycercus caledonicus*)." *Tasmanian Bird Report* 13:17–24.

Brown, S. D., R. J. Dooling, and K. O'Grady. 1988. "Perceptual organization of acoustic stimuli by Budgerigars (*Melopsittacus undulatus*). III. Contact calls." *Journal of Comparative Psychology* 102(3):236–47.

Bruning, D. 1985. "Parrots for sale." *Living Bird Quarterly* 4(3):5–11.

Brush, A. H. 1990. "Metabolism of carotenoid pigments in birds." *FASEB Journal* 4(12):2969–77.

Bryant, David M. 2006. "Energetics of free-living Kakapo." *Nortornis* 53:126–37.

Bryant, Sally L. 1994. "Habitat and potential diet of the ground parrot in Tasmania." *Emu* 94(3):166–71.

Bucher, E. H., L. F. Martin, M. B. Martella, and J. L. Navarro. 1991. "Social behaviour and population dynamics of the Monk Parakeet." *ACTA XX Congressus Internationalis Ornithologici*, 681–89.

Bucher, Enrique H. 1992. "Neotropical parrots as agricultural pests." In *New World Parrots in Crisis: Solutions from Conservation Biology*, edited by Steven R. Beissinger and Noel F. R. Snyder (201–19). Washington, DC: Smithsonian Institution Press.

Bucher, Enrique H., Maria A. Bertin, and Alicia B. Santamaria. 1987. "Reproduction and molt in the Burrowing Parrot." *Wilson Bulletin* 99(1):107–09.

Bucher, T. L. 1983. "Parrot eggs, embryos and nestlings: Patterns and energetics of growth and development." *Physiological Zoology* 56(3):465–83.

Buckland, S. T., I. Rowley, and D. A. Williams. 1983. "Estimation and survival from repeated sightings of tagged galahs." *Journal of Animal Ecology* 52:563–73.

Budden, A. E., and S. R. Beissinger. 2004. "Against the odds? Nestling sex ratio variation in green-rumped parrotlets." *Behavioral Ecology* 15(4):607–13.

———. 2005. "Egg mass in an asynchronously hatching parrot: does variation offset constraints imposed by laying order?" *Oecologia* 144(2):318–26.

———. 2009. "Resource allocation varies with parental sex and brood size in the asynchronously hatching green-rumped parrotlet (*Forpus passerinus*)." *Behavioral Ecology and Sociobiology* 63(5):637–47.

Bull, J. 1971. "Monk Parakeets in the New York City Region." *Linnaean Newsletter* 25:1–2.

Bulte, E. H., and G. C. van Kooten. 1999. "Economics of antipoaching enforcement and the ivory trade ban." *American Journal of Agricultural Economics* 81(2):453–66.

Burger, J., and M. Gochfeld. 2005. "Nesting behavior and nest site selection in Monk Parakeets (*Myiopsitta monachus*) in the Pantanal of Brazil." *Acta Ethologica* 8(1):23–34.

———. 2009. "Exotic monk parakeets (*Myiopsitta monachus*) in New Jersey: nest site selection, rebuilding following removal, and their urban wildlife appeal." *Urban Ecosystems* 12(2):185–96.

Burghardt, GM. 2005. *The Genesis of Animal Play: Testing the Limits*. Boston: MIT Press.

Burish, Mark J., Hao Yuan Kueh, and Samuel S. H. Wang. 2004. "Brain architecture and social complexity in modern and ancient birds." *Brain Behavior and Evolution* 63(2):107–24.

Burton, P. J. K. 1974. "Jaw and tongue features in Psittaciformes and other orders with special reference to anatomy of Tooth-Billed Pigeon (*Didunculus strigirostris*)." *Journal of Zoology* 174 (October): 255–76.

Butler, A. B., P. R. Manger, B. I. B. Lindahl, and P. Arhem. 2005. "Evolution of the neural basis of consciousness: a bird-mammal comparison." *Bioessays* 27(9):923–36.

Butler, C. J. 2005. "Feral parrots in the continental United States and United Kingdom: past, present, and future." *Journal of Avian Medicine and Surgery* 19(2):142–49.

Butler, D. J. 2006. "The habitat, food and feeding ecology of Kakapo in Fiordland: a synopsis from the unpublished MSc thesis of Richard Gray." *Notornis* 53:55–70.

Butler, David. 1989. *Quest for the Kakapo*. Auckland: Heinemann Reed.

Byrne, R. W., and L. A. Bates. 2010. "Primate social cognition: uniquely primate, uniquely social, or just unique?" *Neuron* 65(6):815–30.

Cahill, A. J., J. S. Walker, and S. J. Marsden. 2006. "Recovery within a population of the critically endangered Citron-Crested Cockatoo *Cacatua sulphurea citrinocristata* in Indonesia after 10 years of international trade control." *Oryx* 40(2):161–67.

Cameron, Matt. 2005. "Group size and feeding rates of Glossy Black-Cockatoos in central New South Wales." *Emu* 105(4):299–304.

———. 2006. "Nesting habitat of the glossy black-cockatoo in central New South Wales." *Biological Conservation* 127(4):402–10.

————. 2007. *Cockatoos*. Collingwood, Victoria, Australia: CSIRO Publishing.

————. 2009. "The influence of climate on Glossy Black Cockatoo reproduction." *Pacific Conservation Biology* 15(1):65–71.

————. 2012. *Parrots: the Animal Answer Guide*. Baltimore: Johns Hopkins University Press.

Cameron, Matt, and R. B. Cunningham. 2006. "Habitat selection at multiple spatial scales by foraging Glossy Black-cockatoos." *Austral Ecology* 31(5):597–607.

Cannon, C. E. 1979. "Observations on the food and energy requirements of Rainbow Lorikeets *Trichoglossushaemetodus* (Aves: Psittacidae)." *Australian Wildlife Research* 6:337–346.

————. 1981. "The diet of Eastern and Pale-Headed Rosellas." *Emu* 81 (April): 101–10.

————. 1983. "Descriptions of foraging behavior of Eastern and Pale-Headed Rosellas." *Bird Behaviour* 4(2):63–70.

————. 1984a. "The diet of lorikeets *Trichoglossus* spp in the Queensland-New-South-Wales border region." *Emu* 84 (March): 16–22.

————. 1984b. "Flock size of feeding Eastern and Pale-headed Rosellas (Aves: Psittaciformes)." *Australian Wildlife Research* 11:349–55.

Cant, J. G. H. 1979. "Dispersal of *Stemmadenia donnell-smithii* by birds and monkeys." *Biotropica* 11(2):122.

Caparroz, R., C. Y. Miyaki, and A. J. Baker. 2009. "Contrasting phylogeographic patterns in mitochondrial DNA and microsatellites: Evidence of female philopatry and male-biased gene flow among regional populations of the Blue-and-Yellow Macaw (Psittaciformes: *Ara ararauna*) in Brazil." *Auk* 126(2):359–70.

Carciofi, A. C. 2008. "Protein requirements for Blue-fronted Amazon (*Amazona aestiva*) growth." *Journal of Animal Physiology and Animal Nutrition* 92(3):363–68.

Carrara, L. A., L. D. Faria, F. Q. do Amaral, and M. Rodrigues. 2007. "Eucaliptus as a roosting site for the Turquoise-fronted Parrot *Amazona aestiva* and the Yellow-faced Parrot *Salvatoria xanthops*." *Revista Brasileira De Ornitologia* 15(1):135–38.

Carrara, L. A., L. D. Faria, J. R. Matos, and P. D. Z. Antas. 2008. "Vinaceous Amazon *Amazona vinacea* (Kuhl) (Aves: Psittacidae) in the northern region of Espirito Santo state, south-eastern Brazil: rediscovery and conservation." *Revista Brasileira De Zoologia* 25(1):154–58.

Caruso, S., and F. Scelsi. 1994. "Nesting of feral Monk Parakeets, *Myiopsitta monachus*, in the town of Catania, Sicily." *Rivista Italiana di Ornitologia* 63:213–15.

Cassey, P., T. M. Blackburn, R. P. Duncan, and S. L. Chown. 2005. "Concerning invasive species: reply to Brown and Sax." *Austral Ecology* 30(4):475–80.

Cassey, P., T. M. Blackburn, R. P. Duncan, and K. J. Gaston. 2005. "Causes of exotic bird establishment across oceanic islands." *Proceedings of the Royal Society B-Biological Sciences* 272(1576):2059–63.

Cassey, Phillip, Tim M. Blackburn, Kate E. Jones, and Julie L. Lockwood. 2004. "Mistakes in the analysis of exotic species establishment: Source pool designation and correlates of introduction success among parrots (Aves: Psittaciformes) of the world." *Journal of Biogeography* 31(2):277–84.

Cassey, Phillip., Tim M. Blackburn, G. J. Russell, Kate E. Jones, and Julie L. Lockwood. 2004. "Influences on the transport and establishment of exotic bird species: an analysis of the parrots (Psittaciformes) of the world." *Global Change Biology* 10(4):417–26.

Caviedes-Vidal, E., T. J. McWhorter, S. R. Lavin, J. G. Chediack, C. R. Tracy, and W. H. Karasov. 2007. "The digestive adaptation of flying vertebrates: high intestinal paracellular absorption compensates for smaller guts." *Proceedings of the National Academy of Sciences* 104(48):19132–37.

Cemmick, David, and Dick Veitch. 1987. *Kakapo Country: the story of the world's most unusual bird.* Auckland: Hodder and Stoughton.

Chambers, Geoffrey K., and Wee Ming Boon. 2005. "Molecular systematics of Macquarie Island and Reischek's parakeets." *Notornis* 52:249–50.

Chambers, Geoffrey K., Wee Ming Boon, Thomas R. Buckley, and Rodney A. Hitchmough. 2001. "Using molecular methods to understand the Gondwanan affinities of the New Zealand biota: three case studies." *Australian Journal of Botany* 49(3):377–87.

Chan, Ken, and Dianna Mudie. 2004. "Variation in vocalisations of the ground parrot at its northern range." *Australian Journal of Zoology* 52(2):147–58.

Chapman, C. A. , L. J. Chapman, and L. LeFebvre. 1989. "Variability in parrot flock size: possible functions of communal roosts." *Condor* 91:842–47.

Chapman, T. F., and D.C. Paton. 2005. "The glossy black-cockatoo (*Calyptorhynchus lathami halmaturinus*) spends little time and energy foraging on Kangaroo Island, South Australia." *Australian Journal of Zoology* 53(3):177–83.

———. 2006. "Aspects of Drooping Sheoaks (*Allocasuarina verticillata*) that influence glossy black-cockatoo (*Calyptorhynchus lathami halmaturinus*) foraging on Kangaroo Island." *Emu* 106(2):163–68.

———. 2007. "Casuarina ecology: factors limiting cone production in the drooping sheoak, *Allocasuarina verticillata.*" *Australian Journal of Botany* 55(2):171–77.

Chiron, F., S. M. Shirley, and S. Kark. 2010. "Behind the iron curtain: socio-economic and political factors shaped exotic bird introductions into Europe." *Biological Conservation* 143(2):351–56.

Christian, Colmore S., Thomas E.Lacher, Jr., Michael P. Zamore, Thomas D. Potts, and G. Wesley Burnett. 1996. "Parrot conservation in the Lesser Antilles with some comparison to the Puerto Rican efforts." *Biological Conservation* 77(2–3):159–67.

Christiansen, Mette Bohn, and Elin Pitter. 1994. "Aspects of breeding behaviour of red-fronted macaws, *Ara rubrogenys,* in the wild." *Gerfaut* 82–83(0):51–61.

Christidis, L., Richard Schodde, D. D. Shaw, and S. F. Maynes. 1991. "Relationships among the Australo-Papuan parrots, lorikeets and cockatoos (Aves:Psittaciformes): protein evidence." *Condor* 93:302–317.

Christidis, L., D. D. Shaw, and R. Schodde. 1991. "Chromosomal evolution in parrots, lorikeets and cockatoos (Aves: Psittaciformes)." *Hereditas* 114(47–56).

Churchill, D. M., and P. Christensen. 1970. "Observations on pollen harvesting by Brush-tongued Lorikeets." *Australian Journal of Zoology* 18:427–37.

Cipollini, M. L. 2000. "Secondary metabolites of vertebrate-dispersed fruits: evidence for adaptive functions." *Revista Chilena De Historia Natural* 73(3):421–40.

Clarke, C. M. H. 1967. "Observations on population, movements and food of the Kea (*Nestornotabilis*)." *Notornis* 17:105–29.

Clarke, J.A., C. P. Tambussi, J.I. Noriega, G. M. Erickson, and R. A. Ketcham. 2005. "Definitive fossil evidence for the extant avian radiation in the Cretaceous." *Nature* 433(7023):305–08.

Clavell, J., E. Martorell, D. M. Martorell, and D. Sol. 1991. "Distribucio de la cotorreta de pit gris *Myiopsitta monachus* a Catalunya." *Butlleti del Grup Catala d'Anellament* 18:15–18w.

Clegg, S. M., and I. P. F. Owens. 2002. "The 'island rule' in birds: medium body size and its ecological explanation." *Proceedings of the Royal Society of London Series B: Biological Sciences* 269(1498):1359–65.

Cleland, J. B. 1969. "Galahs eating the corms of *Romulea.*" *Emu* 69:182–83.

Clergeau, Philippe, Alan Vergnes, and Remy Delanoue. 2009. "The Rose-ringed Parakeet, *Psittacula krameri,* introduced in Ile-de-France: distribution and diet." *Alauda* 77(2):121–32.

Clout, Mick N. 1989. "Foraging behaviour of Glossy Black Cockatoos." *Australian Wildlife Research* 16:467–73.

———. 2006. "A celebration of Kakapo: progress in the conservation of an enigmatic parrot." *Notornis* 53:1–2.

Clout, Mick N., and J. L. Craig. 1995. "The conservation of critically endangered flightless birds in New Zealand." *Ibis* 137:S181–90.

Clout, Mick N., G. P. Elliot, and B. C. Robertson. 2002. "Effects of supplementary feeding on the offspring sex ratio of Kakapo: dilemma for the conservation of a polygynous parrot." *Biological Conservation* 107:13–18.

Clout, Mick N., and Don V. Merton. 1998. "Saving the Kakapo: the conservation of the world's most peculiar parrot." *Bird Conservation International* 8(3):281–96.

Clubb, S., and K. Clubb. 1992. "Reproductive life span of macaws." In *Psittacine Aviculture: Perspectives, Techniques and Research,* edited by R. M. Schubot, K. J. Clubb, and S. L. Clubb, 25.1–25.5. Loxahatchee, FL. Aviculture Breeding and Research Center.

Coates-Estrada, Rosamond, Alejandro Estrada, and Dennis Meritt. 1993. "Foraging by parrots (*Amazona autumnalis*) on fruits of *Stemmadenia donnell smithii* (Apocynaceae) in the tropical rain forest of Los Tuxtlas, Mexico." *Journal of Tropical Ecology* 9(1):121–24.

Cochrane, E. P. 2003. "The need to be eaten: *Balanites wilsoniana* with and without elephant seed-dispersal." *Journal of Tropical Ecology* 19:579–89.

Cockle, K., G. Capuzzi, A. Bodrati, R. Clay, H. del Castillo, M. Velazquez, J. I. Areta, N. Farina, and R. Farina. 2007. "Distribution, abundance, and conservation of Vinaceous Amazons (*Amazona vinacea*) in Argentina and Paraguay." *Journal of Field Ornithology* 78(1):21–39.

Cockrem, J. F. 2002. "Reproductive biology and conservation of the endangered Kakapo (*Strigops habroptilus*) in New Zealand." 13(3):139–44.

———. 2006. "The timing of breeding in the Kakapo (*Strigops habroptilus*)." *Notornis* 53:153–59.

Cockrem, J. F., and J. R. Rounce. 1995. "Non-invasive assessment of the annual gonadal cycle in free-living Kakapo (*Strigops habroptilus*) using fecal steroid measurements." *Auk* 112(1):253–57.

Cokinos, Christopher. 2000. *Hope is the Thing with Feathers: A Personal Chronicle of Vanished Birds.* New York: J. P. Tarcher/Putnam.

Collar, N. J. 1997. "Order Psittaciformes: Family Psittacidae (Parrots)." In *Handbook of the Birds of the World,* edited by Josep del Hoyo, Andrew Elliot and Jordi Sargatal, 280–477. Barcelona: Lynx Editions.

———. 2000. "Globally threatened parrots: criteria, characteristics and cures." *International Zoo Yearbook* 37:21–35.

Conway, W. G. 1965. "Apartment-building and cliff-dwelling parrots." *Animal Kingdom,* April: 40–46.

Cooney, R. 2005. "Cambridge Conservation Forum discussion seminar on trade in wild birds." *Oryx* 39(3):246–46.

Cooney, R., and P. Jepson. 2006. "The international wild bird trade: what's wrong with blanket bans?" *Oryx* 40(1):18–23.

Cooper, A., and D. Penny. 1997. "Mass survival of birds across the Cetaceous-Tertiary boundary: molecular evidence." *Science* 275:1109–13.

Corfield, J. R. 2008. "Evolution of brain size in the palaeognath lineage, with an emphasis on new Zealand ratites." *Brain Behavior and Evolution* 71(2):87–99.

Cornelius, C., K. Cockle, N. Politi, I. Berkunsky, L. Sandoval, V. Ojeda, L. Rivera, M. Hunter, and K. Martin. 2008. "Cavity-nesting birds in Neotropical forests: cavities as a potentially limiting resource." *Ornitologia Neotropical* 19:253–68.

Cortopassi, K. A., and J. W. Bradbury. 2006. "Contact call diversity in wild Orange-fronted Parakeet pairs, *Aratinga canicularis*." *Animal Behaviour* 71:1141–54.

Cottom, Yvette, Don V. Merton, and Wouter Hendricks. 2006. "Nutrient composition of the diet of parent-raised Kakapo nestlings." *Nortornis* 53:90–99.

Cotton, P. A. 2001. "The behavior and interactions of birds visiting *Erythrina fusca* flowers in the Colombian Amazon." *Biotropica* 33(4):662–69.

Courchamp, F., T. Clutton-Brock, and B. Grenfell. 1999. "Inverse density dependence and the Allee effect." *Trends in Ecology & Evolution* 14(10):405–10.

Couzin, I. D. 2006. "Behavioral ecology: Social organization in fission-fusion societies." *Current Biology* 16(5):R169–71.

Cracraft, J. 2001. "Avian evolution, Gondwana biogeography and the Cretaceous-Tertiary mass extinction event." *Proceedings of the Royal Society of London Series B: Biological Sciences* 268(1466):459–69.

Cracraft, J., and R. Prum. 1988. "Patterns and processes of diversification: speciation and historical congruence in some Neotropical birds." *Evolution* 42(3):603–20.

Crowley, G. M., and S. T. Garnett. 2001. "Food value and tree selection by Glossy Black Cockatoos *Calyptorhynchus lathami*." *Austral Ecology* 26(1):116–26.

Cruickshank, Alick J., Jean-Pierre Gautier, and Claude Chappuis. 1993. "Vocal mimicry in wild African Grey Parrots *Psittacus erithacus*." *Ibis* 135(3):293–99.

Curnutt, John, and Stuart Pimm. 2001. "How many bird species in Hawai'i and the Central Pacific before first contact?" *Studies in Avian Biology.(22)*:15–30.

da Silva, P. A. 2007. "Seed predation by parakeets *Brotogeris chiriri* (Psittacidae) in *Chorisia speciosa* (Bombacaceae)." *Revista Brasileira De Ornitologia* 15(1):127–29.

da Silva, Paulo Antonio. 2005. "Seed predation by Red-shouldered Macaw (*Diopsittaca nobilis*, Psittacidae) in an exotic plant (*Melia azedurach*, Meliaccae) in the state of Sao Paulo, Brazil." *Revista Brasileira De Ornitologia* 13(2):183–85.

Dahlin, C. R., and T. F. Wright. 2009. "Duets in Yellow-naped Amazons: Variation in syntax, note composition and phonology at different levels of social organization." *Ethology* 115(9):857–71.

Dangoisse, Gersende. 2009. "A study of the population of Monk Parakeets (*Myiopsitta monachus*) in Brussels." *Aves* 46(2):57–69.

Davis, A. R. 1997. "Influence of floral visitation on nectar-sugar composition and nectary surface changes in *Eucalyptus*." *Apidologie* 28(1):27–42.

de Kloet, R. S., and S. R. de Kloet. 2005. "The evolution of the spindlin gene in birds: sequence analysis of an intron of the spindlin W and Z gene reveals four major divisions of the Psittaciformes." *Molecular Phylogenetics and Evolution* 36(3):706–21.

de las Pozas, G., and H. Gonzalez. 1984a. "Comportamiento de una pareja de cotorras (Aves: Psittacidae) durante un período de la nidificacion." *Miscelanea Zoologica* 18:2.

———. 1984b. "Dismunción de los sitios de nidificación de cotorra y catey (Aves: Psittacidae) por la tala de palmas en ciénaga de Zapata, Cuba." *Miscelanea Zoologica* 18:4.

de Moura, L. N., J. M. E. Vielliard, and M. L. da Silva. 2010. "Seasonal fluctuation of the Orange-winged Amazon at a roosting site in Amazonia." *Wilson Journal of Ornithology* 122(1):88–94.

De Schaetzen, P., and J. P. Jacob. 1985. "Installation d'une colonie de perriches jeune-veuve (*Myiopsitta monachus*) a Bruxelles." *Aves* 22:127–30.

Deecke, V. B., J. K. B. Ford, and P. Spong. 2000. "Dialect change in resident killer whales: implications for vocal learning and cultural transmission." *Animal Behaviour* 60: 629–38.

del Hoyo, J., A. Elliot, and J. Sargatal. 1997. *Handbook of the Birds of the World. Vol. 4: Sandgrouse to cuckoos*. Barcelona: Lynx Editions.

Del Rio, C. M., J. E. Schondube, T. J. McWhorter, and L. G. Herrera. 2001. "Intake responses in nectar feeding birds: digestive and metabolic causes, osmoregulatory consequences, and coevolutionary effects." *American Zoologist* 41(4):902–15.

Dent, M. L., E. F. Brittan-Powell, R. J. Dooling, and A. Pierce. 1997. "Perception of synthetic |ba|-|wa| speech continuum by budgerigars (*Melopsittacus undulatus*)." *Journal of the Acoustical Society of America* 102(3):1891–97.

Dent, M. L., O. N. Larsen, and R. J. Dooling. 1997. "Free-field binaural unmasking in budgerigars (*Melopsittacus undulatus*)." *Behavioral Neuroscience* 111(3):590–98.

Dent, Micheal L., and Robert J. Dooling. 2004. "The precedence effect in three species of birds (*Melopsittacus undulatus, Serinus canaria*, and *Taeniopygia guttata*)." *Journal of Comparative Psychology* 118(3):325–31.

Desenne, P. 1995. "Preliminary study of the diet of 15 species of psittacids in an evergreen forest, Tawadu river basin, El Caura forest reserve, state of Bolivar, Venezuela." In *Biología y conservacion de los psitácidos de Venezuela*, edited by G. Morales, I. Novo, D. Bigio, B. Luy, and F. Rojas-Suarez. Cambridge: Birdlife International.

Desenne, Philip, and S. Strahl. 1991. "Trade and the conservation status of the family Psittacidae in Venezuela." *Bird Conservation International* 1:153–69.

Diamond, J., and A. B. Bond. 1991. "Social behavior and the ontogeny of foraging in the Kea (*Nestornotabilis*)." *Ethology* 88:128–44.

———. 1999. *Kea: bird of paradox*. Berkeley: University of California Press.

———. 2003. "A comparative analysis of social play in birds." *Behaviour* 140:1091–15.

———. 2004. "Social play in kaka (*Nestor meridionalis*) with comparisons to kea (*Nestor notabilis*)." *Behaviour* 141:777–98.

Diamond, J., D. Eason, C. Reid, and A. B. Bond. 2006. "Social play in kakapo (*Strigops habroptilus*) with comparisons to kea (*Nestor notabilis*) and kaka (*Nestor meridionalis*)." *Behaviour* 143:1397–1423.

Díaz, S., and T. Kitzberger. 2006. "High Nothofagus flower consumption and pollen emptying in the southern South American Austral Parakeet (*Enicognathus ferrugineus*)." *Austral Ecology* 31(6):759–66.

———. 2012. "Food resources and reproductive output of the Austral Parakeet (*Enicognathus ferrugineus*) in forests of northern Patagonia." *Emu* 112:234–43.

Díaz, S., and S. Peris. 2011. "Consumption of Larvae by the Austral Parakeet (*Enicognathus ferrugineus*)." *Wilson Journal of Ornithology* 123:168–71.

Dilger, William C. 1960. "The comparative ethology of the African parrot genus *Agapornis*." *Zeitschrift für Tierpsychologie* 17(6):649–85.

———. 1962. "The behavior of lovebirds." *Scientific American* 206:88–98.

Dold, C. 1992. "Exotic birds, at risk in wild, may be banned as imports to U.S." *New York Times*, October 20, 1992, B7.

Dooling, R. J. 1986. "Perception of vocal signals by budgerigars (*Melopsittacus undulatus*)." *Experimental Biology* 45:195–218.

Dooling, R. J., S. D. Brown, G. M. Klump, and K. Okanoya. 1992. "Auditory-perception of conspecific and heterospecific vocalizations in birds: evidence for special processes." *Journal of Comparative Psychology* 106(1):20–28.

Dooling, R. J., S. D. Brown, T. J. Park, K. Okanoya, and S. D. Soli. 1987. "Perceptual organization of acoustic stimuli by Budgerigars (*Melopsittacus undulatus*). 1. Pure tones." *Journal of Comparative Psychology* 101(2):139–49.

Dooling, R.J., M.L. Dent, M.R. Leek, and O. Gleich. 2001. "Masking by harmonic complexes in birds: behavioral thresholds and cochlear responses." *Hearing Research* 152(1–2):159–72.

Dooling, R.J., B.F. Gephart, P.H. Price, C. McHale, and S.E. Brauth. 1987. "Effects of deafening on the contact call of the Budgerigar, *Melopsittacus undulatus*." *Animal Behaviour* 35:1264–66.

Dooling, R.J., M.R. Leek, O. Gleich, and M.L. Dent. 2002. "Auditory temporal resolution in birds: discrimination of harmonic complexes." *Journal of the Acoustical Society of America* 112(2):748–59.

Dooling, R.J., T.J. Park, S.D. Brown, and K. Okanoya. 1990. "Perception of species-specific vocalizations by isolate-reared budgerigars (*Melopsittacusundulatus*)." *International Journal of Comparative Psychology* 4(1):57–78.

Dooling, R.J., T.J. Park, S.D. Brown, K. Okanoya, and S.D. Soli. 1987. "Perceptual organization of acoustic stimuli by Budgerigars (*Melopsittacus undulatus*). 2. Vocal Signals." *Journal of Comparative Psychology* 101(4):367–81.

Dooling, R.J., B.M. Ryals, M.L. Dent, and T.L. Reid. 2006. "Perception of complex sounds in budgerigars (*Melopsittacus undulatus*) with temporary hearing loss." *Journal of the Acoustical Society of America* 119(4):2524–32.

Dooling, R.J., and J.C. Saunders. 1975. "Hearing in the parakeet (*Melopsittacus undulatus*): absolute threshholds, critical ratios, frequency difference limens and vocalizations." *Journal of Comparative and Physiological Psychology* 88(1):1–20.

Dooling, R.J., and M.H. Searcy. 1981. "Amplitude modulation thresholds for the parakeet (*Melopsittacus undulatus*)." *Journal of Comparative Psychology* 143:383–88.

———. 1985. "Nonsimultaneous auditory masking in the Budgerigar (*Melopsittacus undulatus*)." *Journal of Comparative Psychology* 99(2):226–30.

Downing, J.D., K. Okanoya, and R.J. Dooling. 1988. "Auditory short-term memory in the Budgerigar (*Melopsittacus undulatus*)." *Animal Learning & Behavior* 16(2):153–156.

Drechsler, Martin. 1998. "Spatial conservation management of the Orange-Bellied Parrot *Neophema chrysogaster*." *Biological Conservation* 84(3):283–92.

Drechsler, Martin, Mark A. Brugman, and Peter W. Menkhorst. 1998. "Uncertainty in population dynamics and its consequences for the management of the Orange-Bellied Parrot *Neophema chrysogaster*." *Biological Conservation* 84(3):269–81.

Duffy, D.C. 1990. "Can the parrot trade help save the rainforest?" *Watchbird*, August/September: 13–15.

Dumser, Frances M. 1987. *The effects of nest presence and group size on the behavior of Quaker Parakeets (Myiopsitta monachus)*. Chicago: University of Illinois.

Durand, Sarah E., James T. Ehaton, Stuart A. Amateau, and Steven E. Brauth. 1997. "Vocal control pathways through the anterior forebrain of a parrot (*Melopsittacus undulatus*)." *Journal of Comparative Neurology* 377(2):179–206.

Durand, Sarah E., Wenru Liang, and Steven E. Brauth. 1998. "Methionine enkephalin immunoreactivity in the brain of the budgerigar (*Melopsittacus undulatus*): similarities and differences with respect to oscine songbirds." *Journal of Comparative Neurology* 393(2):145–68.

Dyck, J. 1971a. "Structure and colour-production of the blue barbs of *Agapornis roseicollis* and *Cotinga maynana*." *Zeitschrift für Zellforschung und Mikroskopische Anatomie* 115:17–29.

———. 1971b. "Structure and spectral reflectance of green and blue feathers of the Rose-faced Lovebird (*Agapornisroseicollis*)." *Biologiske Skrifter* 18(2):1–67.

———. 1977. "Feather ultrastructure of Pesquet's Parrot *Psittrichas fulgidus*." *Ibis* 119(3):364–66.

Dyke, G.J. 2001. "The evolutionary radiation of modern birds: systematics and patterns of diversification." *Geological Journal* 36(3–4):305–15.

———. 2003. "'Big bang' for Tertiary birds?" *Trends in Ecology & Evolution* 18(9):441–42.

Dyke, G. J., and J. H. Cooper. 2000. "A new psittaciform bird from the London clay (Lower Eocene) of England." *Palaeontology* 43:271–85.

Dyke, G. J., and G. Mayr. 1999. "Did parrots exist in the Cretaceous period?" *Nature* 399:317–18.

Dyke, G. J., R. L. Nudds, and M. J. Benton. 2007. "Modern avian radiation across the Cretaceous-Paleogene boundary." *Auk* 124(1):339–41.

Eason, Daryl K., Graeme P. Elliott, Don V. Merton, Paul W. Jansen, Grant A. Harper, and Ron J. Moorhouse. 2006. "Breeding biology of Kakapo (*Strigops habroptilus*) on offshore island sanctuaries, 1990–2002." *Nortornis* 53:27–36.

Eason, Daryl K., and Ron J. Moorhouse. 2006. "Hand-rearing Kakapo (*Strigops habroptilus*), 1997–2005." *Nortornis* 53:116–25.

Eberhard, Jessica R. 1998a. "Breeding biology of the monk parakeet." *Wilson Bulletin* 110(4):463–73.

———. 1998b. "Evolution of nest-building behavior in *Agapornis* parrots." *Auk* 115(2):455–64.

Eberhard, Jessica R., and E. Bermingham. 2005. "Phylogeny and comparative biogeography of *Pionopsitta* parrots and *Pteroglossus* toucans." *Molecular Phylogenetics and Evolution* 36(2):288–304.

Eberhard, Jessica R., Timothy F. Wright, and Eldredge Bermingham. 2001. "Duplication and concerted evolution of the mitochondrial control region in the parrot genus *Amazona*." *Molecular Biology and Evolution* 18(7):1330–42.

Eda-Fujiwara, Hiroko, and Hiroshi Okumura. 1992. "The temporal pattern of vocalizations in the budgerigar *Melopsittacus undulatus*." *Journal of the Yamashina Institute for Ornithology* 24:18–31.

Eda-Fujiwara, Hiroko, R. Satoh, J. J. Bolhuis, and T. Kimura. 2003. "Neuronal activation in female budgerigars is localized and related to male song complexity." *European Journal of Neuroscience* 17(1):149–54.

Eda-Fujiwara, Hiroko, Aiko Watanabe, and Takeji Kimura. 2002. "The role of learned acoustic structure in the song of the budgerigar, *Melopsittacus undulatus*." *Journal of the Yamashina Institute for Ornithology* 34:9–15.

Einoder, Luke, and Alastair Richardson. 2006. "An ecomorphological study of the raptorial digital tendon locking mechanism." *Ibis* 148(3):515–25.

Ekstrom, J. M. M., T. Burke, L. Randrianaina, and T. R. Birkhead. 2007. "Unusual sex roles in a highly promiscuous parrot: the Greater Vasa Parrot Caracopsis vasa." *Ibis* 149(2):313–20.

Elliott, Graeme P. 2006. "A simulation of the future of Kakapo." *Nortornis* 53:164–72.

Elliott, Graeme P., P. J. Dilks, and C. F. J. O'Donnell. 1996. "The ecology of yellow-crowned parakeets (*Cyanoramphus auriceps*) in Nothofagus forest in Fiordland, New Zealand." *New Zealand Journal of Zoology* 23(3):249–65.

Elliott, Graeme P., Daryl K. Eason, Paul W. Jansen, Don V. Merton, Grant A. Harper, and Ron J. Moorhouse. 2006. "Productivity of Kakapo (*Strigops habroptilus*) on offshore island refuges." *Nortornis* 53:123–42.

Elliott, Graeme P., D. Merton, and Paul Jansen. 2001. "Intensive management of a critically endangered species: the kakapo." *Biological Conservation* 99:121–33.

Elphick, Chris, S., David L. Roberts, and J. Michael Reed. 2010. "Estimated dates of recent extinctions for North American and Hawaiian birds." *Biological Conservation* 143(3):617–24.

Emery, N. J. 2006. "Cognitive ornithology: the evolution of avian intelligence." *Philosophical Transactions of the Royal Society B: Biological Sciences* 361(1465):23–43.

Emison, W. B., and D. G. Nicholls. 1992. "Notes on the feeding patterns of the long-billed corella, sulphur-crested cockatoo and galah in southeastern Australia." *South Australian Ornithologist* 31:117–21.

Emison, W. B., C. N. White, and W. D. Caldow. 1995. "Presumptive renesting of red-tailed black cockatoos in south-eastern Australia." *Emu* 95:141–44.

Engebretson, M. 2006. "The welfare and suitability of parrots as companion animals: a review." *Animal Welfare* 15(3):263–76.

Engeman, R., D. Whisson, J. Quinn, F. Cano, P. Quinones, and T. H. White. 2006. "Monitoring invasive mammalian predator populations sharing habitat with the critically endangered Puerto Rican Parrot *Amazona vittata*." *Oryx* 40(1):95–102.

Enkerlin-Hoeflich, E. C., and K. M. Hogan. 1997. "Red-crowned Parrot (*Amazona virdigenalis*)." In *The Birds of North America Online*, edited by A. Poole. Ithaca, NY: Cornell Lab of Ornithology. http://libezp.nmsu.edu:2333/bna/species/292.

Ericson, P. G. P., C. L. Anderson, T. Britton, A. Elzanowski, U.S. Johansson, M. Kallersjo, J. I. Ohlson, T. J. Parsons, D. Zuccon, and G. Mayr. 2006. "Diversification of Neoaves: integration of molecular sequence data and fossils." *Biology Letters* 2(4):543-U1.

Ericson, P. G. P., C. L. Anderson, and G. Mayr. 2007. "Hangin' on to our rocks 'n clocks: a reply to Brown et al." *Biology Letters* 3(3):260–61.

Eriksson, O. 2008. "Evolution of seed size and biotic seed dispersal in angiosperms: Paleoecological and neoecological evidence." *International Journal of Plant Sciences* 169(7):863–70.

Evans, B. E. I., J. Ashley, and S. J. Marsden. 2005. "Abundance, habitat use, and movements of Blue-winged Macaws (*Primolius maracana*) and other parrots in and around an Atlantic Forest Reserve." *Wilson Bulletin* 117(2):154–64.

Evans, Christopher S. 2002. "Cracking the code: communication and cognition in birds." In *The Cognitive Animal: empirical and theoretical perspectives on animal cognition*, edited by Marc Bekoff, Colin Allen, and Gordon M. Burghardt, 315–22. Cambridge, MA: MIT Press.

Fairfax, R. J., and R. J. Fensham. 2000. "The effect of exotic pasture development on floristic diversity in central Queensland, Australia." *Biological Conservation* 94:11–21.

Farabaugh, Susan M., M. L. Dent, and R. J. Dooling. 1998. "Hearing and vocalizations of wild-caught Australian budgerigars (*Melopsittacus undulatus*)." *Journal of Comparative Psychology* 112(1):74–81.

Farabaugh, Susan M., and Robert J. Dooling. 1996. "Acoustic communication in parrots: laboratory and field studies of Budgerigars, *Melopsittacus undulatus*." In *Ecology and Evolution of Acoustic Communication in Birds*, edited by D. E. Kroodsma and E. H. Miller (97–117). Ithaca, NY: Cornell University Press.

Farabaugh, Susan M., A. Linzenbold, and R. J. Dooling. 1994. "Vocal plasticity in Budgerigars (*Melopsittacus undulatus*): evidence for social factors in the learning of contact calls." *Journal of Comparative Psychology* 108(1):81–92.

Faria, P. J., N. M. R. Guedes, C. Yamashita, P. Martuscelli, and C. Y. Miyaki. 2008. "Genetic variation and population structure of the endangered Hyacinth Macaw (*Anodorhynchus hyacinthinus*): implications for conservation." *Biodiversity and Conservation* 17(4):765–79.

Farrimond, Melissa, Mick N. Clout, and Graeme P. Elliott. 2006. "Home range size of Kakapo (*Strigops habroptilus*) on Codfish Island." *Notornis* 53:150–52.

Farrimond, Melissa, and Graeme Elliott. 2006. "Growth and fledging of Kakapo." *Notornis* 53:112–115.

Feduccia, A. 1999. *The Origin and Evolution of Birds*. 2nd ed. New Haven: Yale University Press.

———. 2003a. "'Big bang' for tertiary birds?" *Trends in Ecology & Evolution* 18(4):172–76.

———. 2003b. "Response to Dyke, and van Tuinen et al.: 'Big bang' for Tertiary birds?" *Trends in Ecology & Evolution* 18(9):443–44.

Feenders, G., M. Liedvogel, M. Rivas, M. Zapka, H. Horita, E. Hara, K. Wada, H. Mouritsen, and E. D. Jarvis. 2008. "Molecular mapping of movement-associated areas in the avian brain: A motor theory for vocal learning origin." *PLoS ONE* 3(3).

Fensham, R. J., J. C. McCosker, and M. J. Cox. 1998. "Estimating clearance of Acacia-dominated ecosystems in central Queensland using land-system mapping data." *Australian Journal of Botany* 46:305–19.

Fernandez-Juricic, Esteban, and Monica B. Martella. 2000. "Guttural calls of Blue-fronted Amazons: structure, context, and their possible role in short range communication." *Wilson Bulletin* 112(1):35–43.

Fernandez-Juricic, Esteban, Monica B. Martella, and Eugenia V. Alvarez. 1998. "Vocalizations of the Blue-fronted Amazon (*Amazona aestiva*) in the Chancani Reserve, Cordoba, Argentina." *Wilson Bulletin* 110(3):352–61.

Filardi, C. E., and J. Tewksbury. 2005. "Ground-foraging palm cockatoos (*Probosciger aterrimus*) in lowland New Guinea: fruit flesh as a directed deterrent to seed predation?" *Journal of Tropical Ecology* 21:355–61.

Finger, E., D. Burkhardt, and J. Dyck. 1992. "Avian plumage colors: origin of UV reflection in a black parrot." *Naturwissenschaften* 79(4):187–88.

Fischer, C. 2004. "The complex interactions of markets for endangered species products." *Journal of Environmental Economics and Management* 48:926–53.

Fisher, C.T. 1986. "A type specimen of the Paradise Parrot *Psephotus pulcherrimus* (Gould, 1845)." *Australian Zoologist* 22(3):10–12.

Fitch, WT. 2009. "Biology of music: another one bites the dust." *Current Biology* 19(10):R403–04.

Fitzpatrick, J. W., M. Lammertink, M. D. Luneau Jr., T. W. Gallagher, B. R. Harrison, G. M. Sparling, K. V. Rosenberg, et al. 2006. "Clarifications about current research on the status of Ivory-billed Woodpecker (*Campephilus principalis*) in Arkansas." *Auk* 123(2):587–93.

Fleming, P. A., S. Xie, K. Napier, T. J. McWhorter, and S. W. Nicolson. 2008. "Nectar concentration affects sugar preferences in two Australian honeyeaters and a lorikeet." *Functional Ecology* 22(4):599–605.

Fleming, P. J. S., A. Gilmour, and J. A. Thompson. 2002. "Chronology and spatial distribution of cockatoo damage to two sunflower hybrids in south-eastern Australia, and the influence of plant morphology on damage." *Agriculture Ecosystems & Environment* 91:127–37.

Foote, A. D., R. M. Griffin, D. Howitt, L. Larsson, P. J. O. Miller, and A. R. Hoelzel. 2006. "Killer whales are capable of vocal learning." *Biology Letters* 2(4):509–12.

Ford, H. A., D. C. Paton, and N. Forde. 1979. "Birds as pollinators of Australian plants." *New Zealand Journal of Botany* 17:509–19.

Ford, Julian. 1980. "Morphological and ecological divergence and convergence in isolated populations of the Red-Tailed Black Cockatoo." *Emu* 80:130–20.

Forshaw, Joseph M. 1970. "Early record of the Night Parrot in New South Wales." *Emu* 70:34.

———. 1981. "The Norfolk Island Parakeet (*Cyanoramphus novaezelandiea cooki*): a threatened population status and management options." In *Conservation of New World Parrots*, edited by R.F. Pasquier, 446–461. Washington, DC: Smithsonian Institution Press.

———. 1989. *Parrots of the World*. Melbourne: Landsdowne Editions.

———. 2006. *Parrots of the World: an Identification Guide*. Princeton: Princeton University Press.

Forshaw, J. M., P. J. Fullagar, and J. I. Harris. 1976. "Specimens of the Night Parrot throughout the world." *Emu* 76:122–26.

———. 2010. *Parrots of the world (Princeton Field Guides)*. Princeton: Princeton University Press.

Fouts, Roger S., Mary Lee A. Jensvold, and Deborah H. Fouts. 2002. "Chimpanzee signing: Darwinian realities and Cartesian delusions." In *The Cognitive Animal: empirical and theoretical perspectives on animal cognition*, edited by Marc Bekoff, Colin Allen, and Gordon M. Burghardt, 285–91. Cambridge, MA: MIT Press.

Francisco, Mercival Roberto, Vitor de Oliveira Lunardi, and Mauro Galetti. 2002. "Massive seed predation of *Pseudobombax grandiflorum* (Bombaceae) by parakeets *Brotogeris versicolurus* (Psittacidae) in a forest fragment in Brazil." *Biotropica* 34:613–15.

Francisco, Mercival Roberto, Vitor de Oliveira Lunardi, P. R. Guimaraes, and Mauro Galetti. 2008. "Factors affecting seed predation of *Eriotheca gracipiles* (Bombacaceae) by parakeets in a cerrado fragment." *Acta Oecologica-International Journal of Ecology* 33(2):240–45.

Franke, E., S. Jackson, and S. Nicolson. 1998. "Nectar sugar preferences and absorption in a generalist African frugivore, the Cape White-eye *Zosterops pallidus*." *Ibis* 140(3):501–06.

Frankel, T. L., and D. Avram. 2001. "Protein requirements of rainbow lorikeets, *Trichoglossus haematodus*." *Australian Journal of Zoology* 49(4):435–43.

Franklin, J., and D. W. Steadman. 1991. "The potential for conservation of Polynesian birds through habitat mapping and species translocation." *Conservation Biology* 5(4):506–21.

Freeland, David B. 1973. "Some food preferences and aggressive behavior by Monk Parakeets." *Wilson Bulletin* 85(3):332–34.

Freeland, W. J., and L. R. Saladin. 1989. "Choice of mixed diets by herbivores: the idiosyncratic effects of plant secondary compounds." *Biochemical Systematics and Ecology* 17(6):493–97.

Frémont, J. C. 1845. *Report of the exploring expedition to the Rocky Mountains in the year 1842, and the Oregon and North California in the Years 1843-'44*. Washington, DC: Gales and Seaton.

French, A. R., and T. B. Smith. 2005. "Importance of body size in determining dominance hierarchies among diverse tropical frugivores." *Biotropica* 37(1):96–101.

Frynta, Daniel, Silvie Liskova, Sebastian Bueltmann, and Hynek Burda. 2010. "Being attractive brings advantages: the case of parrot species in captivity." *PLoS One* 5(9):e12568.

Fule, P. Z., J. Villanueva-Diaz, and M. Ramos-Gomez. 2005. "Fire regime in a conservation reserve in Chihuahua, Mexico." *Canadian Journal of Forest Research-Revue Canadienne De Recherche Forestiere* 35(2):320–30.

Fuller, Errol. 2001. *Extinct Birds*. 2nd ed. Ithaca: Cornell University Press.

Gajdon, G. K., N. Fijn, and L. Huber. 2006. "Limited spread of innovation in a wild parrot, the Kea (*Nestor notabilis*)." *Animal Cognition* 9(3):173–81.

Galetti, Mauro. 1993. "Diet of the Scaly-headed Parrot (*Pionus maximiliani*) in a semideciduous forest in Southeastern Brazil." *Biotropica* 25(4):419–25.

———. 1997. "Seasonal abundance and feeding ecology of parrots and parakeets in a lowland Atlantic forest of Brazil." *Ararajuba* 5(2):115–26.

Galetti, Mauro, and M. Rodrigues. 1992. "Comparative seed predation on pods by parrots in brazil." *Biotropica* 24(2):222–24.

Garner, J. P. 2005. "Stereotypies and other abnormal repetitive behaviors: Potential impact on validity, reliability, and replicability of scientific outcomes." *Ilar Journal* 46(2): 106–17.

Garner, J. R., C. L. Meehan, T. R. Famula, and J. A. Mench. 2006. "Genetic, environmental, and neighbor effects on the severity of stereotypies and feather picking in Orange-winged Amazon parrots (*Amazona amazonica*): an epidemiological study." *Applied Animal Behaviour Science* 96(1–2):153–68.

Garner, J. P., C. L. Meehan, and J. A. Mench. 2003. "Stereotypies in caged parrots, schizophrenia and autism: evidence for a common mechanism." *Behavioural Brain Research* 145(1–2):125–34.

Garnett, Stephen, and Gabriel Crowley. 1995. "Feeding ecology of Hooded Parrots *Psephotus dissimilis* during the early wet season." *Emu* 95(1):54–61.

————. 1997. *"Report to the Queensland Department of Environment on the feasibility of conducting field research on the Palm Cockatoo at Crater Mountain Wildlife Management Area in Paupua New Guinea."* July 31, 1997.

Garnett, Stephen, Gabriel Crowley, Ray Duncan, Niel Baker, and Patrick Doherty. 1993. "Notes on live night parrot sightings in north-western Queensland." *Emu* 93(4):292–96.

Garnett, Stephen, and David B. Lindenmayer. 2011. "Conservation science must engender hope to succeed." *Trends in Ecology & Evolution* 26(2):59–60. doi:10.1016/j.tree.2010 .11.009.

Garnett, Stephen, L. P. Pedler, and G. M. Crowley. 1999. "The breeding biology of the Glossy Black-Cockatoo *Calyptorhynchus lathami* on Kangaroo Island, South Australia." *Emu* 99:262–279.

Garnetzke-Stollmann, K., and D. Franck. 1988. "Long-lasting sibling relationships as a means to acquire reproductive ability in the Spectacled Parrotlet (*Forpus conspicillatus*)." Meeting on Current Topics in Avian Biology, Bonn

————. 1991. "Socialisation tactics of the Spectacled Parrotlet (*Forpus conspicillatus*)." *Behaviour* 119(1–2):1–29.

Garrod, A. J. 1872. "Note on the tongue of the Psittacine genus *Nestor*." *Proceedings of the Zoological Society of London* 1872:787–89.

Gartrell, Brett D. 2000. "The nutritional, morphologic, and physiologic bases of nectarivory in Australian birds." *Journal of Avian Medicine and Surgery* 14(2):85–94.

Gartrell, Brett D., and S. M. Jones. 2001. "Eucalyptus pollen grain emptying by two Australian nectarivorous psittacines." *Journal of Avian Biology* 32(3):224–30.

Gartrell, Brett D., Susan M. Jones, Raymond N. Brereton, and Lee B. Astheimer. 2000. "Morphological adaptations to nectarivory of the alimentary tract of the Swift Parrot *Lathamus discolor*." *Emu* 100(4):274–79.

Gavrilov, L. A., and N. S. Gavrilova. 2004. "The reliability-engineering approach to the problem of biological aging." *Annals of the New York Academy of Sciences* 1019:509–12.

Gerischer, Bernd-Henning, and Bruno A. Walthier. 2003. "Behavioural observations of the blue lorikieet (*Vini peruviana*) on Rangiroa atoll, Tuamotu Archipelago, French Polynesia." *Nortonis* 50:54–58.

Ghalambor, Cameron K., and Thomas E. Martin. 2001. "Fecundity-survival trade-offs and parental risk-taking in birds." *Science* 292:494–97.

Gibb, G. C., O. Kardailsky, R. T. Kimball, E. L. Braun, and D. Penny. 2007. "Mitochondrial genomes and avian phylogeny: Complex characters and resolvability without explosive radiations." *Molecular Biology and Evolution* 24(1):269–80.

Gibbs, J. P., M. L. Hunter Jr., and S. M. Melvin. 1993. "Snag availability and communities of cavity nesting birds in tropical versus temperate forests." *Biotropica* 1993(2):235–41.

Gibson, L. 2007. "Dealing with uncertain absences in habitat modelling: a case study of a rare ground-dwelling parrot." *Diversity and Distributions* 13:704–13.

Gilardi, James D. 2006. "To ban or not to ban: a reply to Jorge Rabinovich." *Oryx* 40(3): 264–65.

Gilardi, James D., Sean S. Duffey, Charles A. Munn, and Lisa A. Tell. 1999. "Biochemical functions of geophagy in parrots: detoxification of dietary toxins and cytoprotective effects." *Journal of Chemical Ecology* 25(4):897–922.

Gilardi, James D., and Charles A. Munn. 1998. "Patterns of activity, flocking, and habitat use in parrots of the Peruvian Amazon." *Condor* 100(4):641–53.

Gilardi, James D., and Catherine A. Toft. 2012. "Parrots eat nutritious foods despite toxins." *PLoS ONE* 7(6):e38293.

Giret, N., M. Monbureau, M. Kreutzer, and D. Bovet. 2009. "Conspecific discrimination in an object-choice task in African grey parrots (*Psittacus erithacus*)." *Behavioral Processes* 82(1):75–77.

Giret, N., F. Peron, J. Lindova, L. Tichotova, L. Nagle, M. Kreutzer, F. Tymr, and D. Bovet. 2010. "Referential learning of French and Czech labels in African grey parrots (*Psittacus erithacus*): Different methods yield contrasting results." *Behavioural Processes* 85(2):90–98.

Giret, N., F. Peron, L. Nagle, M. Kreutzer, and D. Bovet. 2009. "Spontaneous categorization of vocal imitations in African grey parrots (Psittacus erithacus)." *Behavioural Processes* 82(3):244–48.

Gnam, Rosemarie. 1990. "Conservation of the Bahama Parrot." *American Birds* 44:32–36.

———. 1991a. "Natural history of the Bahama Parrot *Amazona leucocephala bahamensis*on Abaco Island." Paper read at the 4th Symposium on the Natural History of the Bahamas, San Salvador, Bahamas.

———. 1991b. "Nesting behaviour of the Bahama parrot *Amazona leucocephala bahamensis* on Abaco Island, Bahamas." *ACTA XX Congressus Internationalis Ornithologici*, 673–81.

———. 1991c. "Underground parrots: nesting habits of the Bahaman Parrot." *American Zoologist* 31:11A.

Gnam, Rosemarie, and R. F. Rockwell. 1991. "Reproductive potential and output of the Bahama Parrot *Amazona leucocephala bahamensis*." *Ibis* 133:400–05.

Gochfeld, M. 1973. "Ecologic aspects of ectopic populations of Monk Parakeets (*Myiopsitta monachus*) and possible agricultural consequences." *Journal of Agriculture of the University of Puerto Rico* 57:262–70.

Goldingay, R. L., and J. R. Stevens. 2009. "Use of artificial tree hollows by Australian birds and bats." *Wildlife Research* 36(2):81–97.

Goldsmith, T. H., and B. K. Butler. 2003. "The roles of receptor noise and cone oil droplets in the photopic spectral sensitivity of the budgerigar, *Melopsittacus undulatus*." *Journal of Comparative Physiology A: Neuroethology Sensory Neural and Behavioral Physiology* 189(2):135–42.

———. 2005. "Color vision of the budgerigar (*Melopsittacus undulatus*): hue matches, tetrachromacy, and intensity discrimination." *Journal of Comparative Physiology A: Neuroethology Sensory Neural and Behavioral Physiology* 191(10):933–51.

Gonçalves da Silva, A., J. R. Eberhard, T. F. Wright, M. L. Avery, and M. A. Russello. 2010. "Genetic evidence for high propagule pressure and long-distance dispersal in Monk Parakeet (*Myiopsitta monachus*) invasive populations." *Molecular Ecology* 19(16):3336–50.

González, J. A. 2003. "Harvesting, local trade, and conservation of parrots in the northeastern Peruvian Amazon." *Biological Conservation* 114:437–46.

Gonzalez-Lagos, C., D. Sol, and S. M. Reader. 2010. "Large-brained mammals live longer." *Journal of Evolutionary Biology* 23(5):1064–74.

Goodland, Robert J. A. 1987. "Protection of the Chilean Conure or Burrowing Parrot." *Environmental Conservation* 14:180.

Graves, Gary R., and D. U. Restrepo. 1989. "A new allopatric taxon in the *Hapalopsittaca amazonina* (Psitticidae) superspecies from Colombia." *Wilson Bulletin* 101(3):369–76.

Green, R. M., and J. W Swift. 1965. "Rosellas as insect eaters." *Emu* 65(1):75.

Greenberg, R. 1981. "The abundance and seasonality of forest canopy birds on Barro-Colorado Island, Panama." *Biotropica* 13(4):241–51.

Greene, T. C. 1998. "Foraging ecology of the Red-Crowned Parakeet (*Cyanoramphus novaezelandiae novaezelandiae*) and Yellow-Crowned Parakeet (*C. auriceps auriceps*) on Little Barrier Island, Hauraki Gulf, New Zealand." 22(2):161–71.

Greene, Terry C. 1999. "Aspects of the ecology of Antipodes Island Parakeet (*Cyanoramphus unicolor*) and Reischek's Parakeet (*C. novaezelandiae hochstetteri*) on Antipodes Island, October-November 1995." *Notornis* 46:301–10.

Griffin, Donald R. 2001. *Animal Minds: beyond cognition to consciousness*. 2nd ed. Chicago: University of Chicago Press.

Griggio, M., V. Zanollo, and H. Hoi. 2010. "UV plumage color is an honest signal of quality in male budgerigars." *Ecological Research* 25(1):77–82.

Groombridge, Jim J., Carl G. Jones, Richard A. Nichols, Mark Carlton, and Michael W. Bruford. 2004. "Molecular phylogeny and morphological change in the *Psittacula* parakeets." *Molecular Phylogenetics and Evolution* 31(1):96–108.

Grzelewski, D. 2002. "Going to extremes: without the extraordinary dedication of a few conservationists, New Zealand's Kakapo would likely have gone the way of the dodo." *New Zealand Geographic* 33(7):90–95.

Guedes, N. M. R. 1994. "A história das araras azuis." *ONATI Revista Tecnico-cientifica e cultural do CESUP* 1(1):57–61.

———. 1995. "Alguns aspectos sobre o comportamento reprodutivo da arara-azul (*Anodorhynchus hyacinthinus*) e a necessidade de manejo para a conservação da espécie." *Anais de Etologia* 13:274–92.

———. 2004. "Management and conservation of the large macaws in the wild." *Ornitologia Neotropical* 15:279–83.

Gueneau, P., F. Pizani, A. Porco Giambra, and C. Bosque. 2006. "Presence of nitrogen-fixing bacteria in the crops of the Hoatzin and Green-rumped Parrotlet." *Journal of Ornithology* 147(5):176–176.

Guerra, J. E., J. Cruz-Nieto, S. G. Ortiz-Maciel, and T. F. Wright. 2008. "Limited geographic variation in the vocalizations of the endangered Thick-billed Parrot: Implications for conservation strategies." *Condor* 110(4):639–47.

Guittar, J. L., F. Dear, and C. Vaughan. 2009. "Scarlet Macaw (*Ara macao*, Psittaciformes: Psittacidae) nest characteristics in the Osa Peninsula Conservation Area (ACOSA), Costa Rica." *Revista de Biologia Tropical* 57(1–2):387–93.

Güntert, M., and V. Ziswiler. 1972. "Konvergenzen in der struktur von zunge und verdauungsstrakt nektarfressender papageien." *Revue Suisse de Zoologie* 79:1016–26.

Hackett, S. J., et al. 2008. "A phylogenomic study of birds reveals their evolutionary history." *Science* 320(5884):1763–1768. doi:10.1126/science.1157704.

Haesler, Sebastian, Kazuhiro Wada, A. Nshdejan, Edward E. Morrisey, Thierry Lints, Eric D. Jarvis, and Constance Scharff. 2004. "FoxP2 expression in avian vocal learners and non-learners." *Journal of Neuroscience* 24(13):3164–75.

Halliday, T. R. 1980. "Extinction of the Passenger Pigeon *Ectopistes migratorius* and its relevance to contemporary conservation." *Biological Conservation* 17(2):157–62.

Hardy, J. W. 1963. "Epigamic and reproductive behavior of the Orange-fronted Parrakeet." *Condor* 65(3):168–99.

Hardy, John William. 1965. "Flock social behavior of the Orange-fronted Parakeet." *Condor* 67:140–56.

Harper, Grant A., Graeme P. Elliott, Daryl K. Eason, and Ron J. Moorhouse. 2006. "What triggers nesting of Kakapo (*Strigops habroptilus*)?" *Notornis* 53:160–64.

Harper, Grant A., and Joanne Joice. 2006. "Agonistic display and social interaction between female Kakapo (*Strigops habroptilus*)." *Notornis* 53:191–97.

Harrison, C. J. O. 1971. "Nest building behavior of Quaker Parrots *Myiopsitta monachus*." *Ibis* 115,124–28.

Harrison, G. L., P. A. McLenachan, M. J. Phillips, K. E. Slack, A. Cooper, and D. Penny. 2004. "Four new avian mitochondrial genomes help get to basic evolutionary questions in the late Cretaceous." *Molecular Biology and Evolution* 21(6):974–83.

Hasebe, Makoto, and Donald C. Franklin. 2004. "Food sources of the Rainbow Lorikeet *Trichoglossus haematodus* during the early wet season on the urban fringe of Darwin, Northern Australia." *Corella* 28:68–74.

Haugaasen, T. 2008. "Seed predation of *Couratari guianensis* (Lecythidaceae) by macaws in central Amazonia, Brazil." *Ornitologia Neotropical* 19(3):321–28.

Franziska Hausmann, Kathryn E. Arnold, N. Justin Marshall, and Ian P. F. Owens. 2003. "Ultraviolet signals in birds are special." *Proceedings of the Royal Society of London, Series B: Biological Sciences* 270(1510):61–67.

Heatherbell, C. 1992. "Anting by an Orange-fronted Parakeet." *Notornis* 39:131–32.

Heaton, J. T., and S. E. Brauth. 2000a. "Effects of lesions of the central nucleus of the anterior archistriatum on contact call and warble song production in the budgerigar (Melopsittacus undulatus)." *Neurobiology of Learning and Memory* 73(3):207–42.

———. 2000b. "Telencephalic nuclei control late but not early nestling calls in the budgerigar." *Behavioural Brain Research* 109(1):129–35.

Heinrich, Bernd. 1989. *Ravens in Winter*. New York: Summit Books.

Heinsohn, R. S. 2007. "Genetic evidence for cooperative polyandry in reverse dichromatic *Eclectus* parrots." *Animal Behaviour* 74:1047–54.

———. 2008a. "The ecological basis of unusual sex roles in reverse-dichromatic eclectus parrots." *Animal Behaviour* 76:97–103.

———. 2008b. "Ecology and evolution of the enigmatic eclectus parrot (*Eclectus roratus*)." *Journal of Avian Medicine and Surgery* 22(2):146–50.

Heinsohn, R. S., and S. Legge. 2003. "Breeding biology of the reverse-dichromatic, co-operative parrot *Eclectus roratus*." *Journal of Zoology* 259:197–208.

Heinsohn, R. S., S. Legge, and J. A. Endler. 2005. "Extreme reversed sexual dichromatism in a bird without sex role reversal." *Science* 309(5734):617–19.Heinsohn, R. S., S. Legge, and S. Barry. 1997. "Extreme bias in sex allocation in *Eclectus* parrots." *Proceedings of the Royal Society of London Series B* 264:1325–29.

Heinsohn, R. S., S. Murphy, and S. Legge. 2003. "Overlap and competition for nest holes among Eclectus Parrots, Palm Cockatoos and Sulfur-Crested Cockatoos." *Australian Journal of Zoology* 51:81–94.

Hesse, Alan J., and Giles E. Duffield. 2000. "The status and conservation of the Blue-throated Macaw *Ara glaucogularis*." *Bird Conservation International* 10(3):255–75.

Heyes, C. 2003. "Four routes of cognitive evolution." *Psychological Review* 110(4):713–27.

Heyes, C., and A. Saggerson. 2002. "Testing for imitative and nonimitative social learning in the budgerigar using a two-object/two-action test." *Animal Behaviour* 64:851–59.

Higgins, M. L. 1979. "Intensity of seed predation on *Brosimum utile* by Mealy Parrots (*Amazona farinosa*)." *Biotropica* 11(1):80.

Higgins, P. J. 1999. *Handbook of Australian, New Zealand and Antarctic birds, Vol. 4. Parrots to dollarbird*. Oxford: Oxford University Press.

Hile, A. G., N. T. Burley, C. B. Coopersmith, V. S. Foster, and G. F. Striedter. 2005. "Effects of male vocal learning on female behavior in the budgerigar, *Melopsittacus undulatus*." *Ethology* 111(10):901–23.

Hile, A. G., T. K. Plummer, and G. F. Striedter. 2000. "Male vocal imitation produces call convergence during pair bonding in budgerigars, *Melopsittacus undulatus*." *Animal Behaviour* 59:1209–18.

Hile, A. G., and G. F. Striedter. 2000. "Call convergence within groups of female budgerigars (*Melopsittacus undulatus*)." *Ethology* 106(12):1105–14.

Hingston, A. B., B. D. Gartrell, and G. Pinchbeck. 2004. "How specialized is the plant-pollinator association between *Eucalyptus globulus* ssp *globulus* and the swift parrot *Lathamus discolor*?" *Austral Ecology* 29(6):624–30.

Hingston, A. B., B. M. Potts, and P. B. McQuillan. 2004. "The swift parrot, *Lathamus discolor* (Psittacidae), social bees (Apidae) and native insects as pollinators of *Eucalyptus globulus* ssp *globulus* (Myrtaceae)." *Australian Journal of Botany* 52(3):371–79.

Hofmann, R. R. 1989. "Evolutionary steps of ecophysiological adaptation and diversification of ruminants: a comparative view of their digestive systems." *Oecologia* 78:443–57.

Holmes, D. J., R. Flückiger, and S. N. Austad. 2001. "Comparative biology of aging in birds: an update." *Experimental Gerontology* 36:869–83.

Homberger, D. G. 1981. "Functional morphology and evolution of the feeding apparatus in parrots with special reference to the Pesquet's Parrot, *Psittrichas fulgidus* (Lesson)." In *Conservation of New World Parrots,* edited by R. F. Pasquier, 462–471. Washington, DC: Smithsonian Institution Press.

———. 2002. "Function, ecology and evolution of two morphological bill types of Psittaciformes." *Integrative and Comparative Biology* 42(6):1245–45.

———. 2003. "The comparative biomechanics of a prey-predator relationship: the adaptive morphologies of the feeding apparatus of Australian black cockatoos and their foods as a basis for the reconstruction of the evolutionary history of the Psittaciformes." In *Vertebrate Biomechanics and Evolution,* edited by V. L. Bels, J.-P. Gasc, and A. Casinos. Oxford: BIOS Scientific.

Homberger, D. G., and A. H. Brush. 1986. "Functional-morphological and biochemical correlations of the keratinized structures in the African Grey Parrot, *Psittacus erithacus* (Aves)." *Zoomorphology* 106(2):103–14.

Hopper, S. D. 1980. "Pollen and nectar feeding by Purple-Crowned Lorikeets on *Eucalyptus occidentalis*." *Emu* 80(OCT):239–40.

Hopper, S. D., and A. A. Burbidge. 1979. "Feeding behavior of a Purple-crowned Lorikeet on flowers of *Eucalyptus buprestium*." *Emu* 79:40–42.

Hoppes, S., and P. Gray. 2010. "Parrot rescue organizations and sanctuaries: a growing presence in 2010." *Journal of Exotic Pet Medicine* 19(2):133–39.

Hotchner, A. E. 1966. *Papa Hemingway: a personal memoir.* New York: Random House.

Houston, D., K. McInnes, G. Elliott, D. Eason, R. Moorhouse, and J. Cockrem. 2007. "The use of a nutritional supplement to improve egg production in the endangered kakapo." *Biological Conservation* 138(1–2):248–55.

Hrabar, H. D. K., and M. Perrin. 2002. "The effect of bill structure on seed selection by granivorous birds." *African Zoology* 37(1):67–80.

Huber, L., and G. K. Gajdon. 2006. "Technical intelligence in animals: the kea model." *Animal Cognition* 9(4):295–305.

Hudon, J., and A. H. Brush. 1992. "Identification of carotenoid pigments in birds." In *Methods in Enzymology, Vol. 213. Carotenoids, Part A: Chemistry, separation, quantitation, and antioxidation,* edited by L Packer. San Diego, CA: Academic Press.

Hurley, Timothy. 2001. "Parrot invasion worries Maui." *Honolulu Advertiser,* January 8, 2001.

Hutching, G. 1997. "A good year for the Kakapo." *Forest and Bird* 154(2078):5.

Hyman, Jeremy, and Stephen Pruett-Jones. 1995. "Natural history of the Monk Parakeet in Hyde Park, Chicago." *Wilson Bulletin* 107(3):510–17.

Imboden, C. 1992. "The wild bird trade." *World Birdwatch* 14(1):6–7.

Imboden, Christoph, Peter Jones, and Ian Aktinson. 1995. "Review of the Kakapo recovery programme." Wellington: New Zealand Department of Conservation.

Iñigo-Elias, E. E., and M. A. Ramos. 1991. "The psittacine trade in Mexico." In *Neotropical Wildlife Use and Conservation*, edited by J. G. Robinson and K. H. Redford, 380–92. Chicago: University of Chicago Press.

Iriarte, J. A., G. A. Lobos, and F. M. Jaksic. 2005. "Invasive vertebrate species in Chile and their control and monitoring by governmental agencies." *Revista Chilena De Historia Natural* 78(1):143–54.

Ishii, S. 1999. "Application of modern endocrine methods to conservation biology." *Ostrich* 70(1):33–38.

Isler, K., and C. van Schaik. 2006. "Costs of encephalization: the energy trade-off hypothesis tested on birds." *Journal of Human Evolution* 51(3):228–43.

———. 2008. "Life history pace and brain size: from correlation to causation." *Folia Primatologica* 79(5):342–43.

———. 2009. "Why are there so few smart mammals (but so many smart birds)?" *Biology Letters* 5(1):125–29.

Iwaniuk, A. N., D. H. Clayton, and D. R. W. Wylie. 2006. "Echolocation, vocal learning, auditory localization and the relative size of the avian auditory midbrain nucleus (MLd)." *Behavioural Brain Research* 167(2):305–17.

Iwaniuk, A. N., K. M. Dean, and J. E. Nelson. 2004. "A mosaic pattern characterizes the evolution of the avian brain." *Proceedings of the Royal Society of London Series B-Biological Sciences* 271:S148–51.

———. 2005. "Interspecific allometry of the brain and brain regions in parrots (Psittaciformes): Comparisons with other birds and primates." 65(1):40–59.

Iwaniuk, A. N., and P. L. Hurd. 2005. "The evolution of cerebrotypes in birds." *Brain Behavior and Evolution* 65(4):215–30.

Iwaniuk, A. N., P. L. Hurd, and D. R. W. Wylie. 2006. "Comparative morphology of the avian cerebellum: I. Degree of foliation." *Brain Behavior and Evolution* 68(1):45–62.

———. 2007. "Comparative morphology of the avian cerebellum: II. Size of folia." *Brain Behavior and Evolution* 69(3):196–219.

Iwaniuk, A. N., and J. E. Nelson. 2002. "Can endocranial volume be used as an estimate of brain size in birds?" *Canadian Journal of Zoology* [print] 80(1):16–23.

———. 2003. "Developmental differences are correlated with relative brain size in birds: a comparative analysis." *Canadian Journal of Zoology-Revue Canadienne De Zoologie* 81(12):1913–28.

Iwaniuk, A. N., J. E. Nelson, H. E. James, and S. L. Olson. 2004. "A comparative test of the correlated evolution of flightlessness and relative brain size in birds." *Journal of Zoology* 263:317–27.

Iwaniuk, A. N., J. E. Nelson, and S. O'Leary. 2001. "Big brains need big protection: The 'interhemispheric septum' in the psittaciform braincase." *Journal of Morphology* [print] 248(3):245.

Iyer, P., and J. Roughgarden. 2008. "Gametic conflict versus contact in the evolution of anisogamy." *Theoretical Population Biology* 73(4):461–72.

Izhaki, I. 1993. "Influence of nonprotein nitrogen on estimation of protein from total nitrogen in fleshy fruits." *Journal of Chemical Ecology* 19(11):2605–15.

———. 2002. "Emodin: a secondary metabolite with multiple ecological functions in higher plants." *New Phytologist* 155(2):205–17.

Izhaki, I., and U. N. Safriel. 1989. "Why are there so few exclusively frugivorous birds? Experiments on fruit digestibility." *Oikos* 54(1):23–32.

———. 1990. "Weight losses due to exclusive fruit diet: interpretation and evolutionary implications—Reply." *Oikos* 57(1):140–42.

Jacobs, M. D., and J. S. Walker. 1999. "Density estimates of birds inhabiting fragments of cloud forest in southern Ecuador." *Bird Conservation International* 9(1):73–79.

Jahelkova, H., I. Horacek, and T. Bartonicka. 2008. "The advertisement song of *Pipistrellus nathusii* (Chiroptera, Vespertilionidae): a complex message containing acoustic signatures of individuals." *Acta Chiropterologica* 10(1):103–26.

James, Frances. 1990. "The selling of wild birds: out of control?" *The Living Bird* 9:9–15.

Jansen, Paul W. 2006. "Kakapo recovery: the basis of decision-making." *Notornis* 53:184–90.

Janzen, D. H. 1981. "Ficus ovalis seed predation by an Orange-chinned Parakeet (*Brotogeris jugularis*) in Costa Rica." *Auk* 98(4):841–44.

———. 1983. "*Ara macao* (Lapa, Scarlet Macaw)." In *Costa Rican natural history*, edited by D. H. Janzen, 547–48. Chicago: University of Chicago Press.

Jarry, Guy. 2003. "Trade and traffic of birds." *Bulletin de la Société Zoologique de France* 129:103–10.

Jarvis, E. D. 2006. "Selection for and against vocal learning in birds and mammals." *Ornithological Science* 5(1): 5–14.

Jarvis, E. D., O. Gunturkun, L. Bruce, A. Csillag, H. Karten, W. Kuenzel, L. Medina, et al. 2005. "Avian brains and a new understanding of vertebrate brain evolution." *Nature Reviews Neuroscience* 6(2):151–59.

Jarvis, E. D., and Claudio V. Mello. 2000. "Molecular mapping of brain areas involved in parrot vocal communication." *Journal of Comparative Neurology* 419(1):1–31.

Jarvis, E. D., V. A. Smith, K. Wada, M. V. Rivas, M. McElroy, T. V. Smulders, P. Carninci, et al. 2002. "A framework for integrating the songbird brain." *Journal of Comparative Physiology A: Neuroethology Sensory Neural and Behavioral Physiology* 188(11–2):961–80.

Jennings, S., J. D. Reynolds, and S. C. Mills. 1998. "Life history correlates of responses to fisheries exploitation." *Proceedings of the Royal Society of London Series B* 265(1393):333–39.

Johnson, M. A., M. T. Tomas, and N. M. R. Guedes. 1997. "On the Hyacinth Macaw's nesting tree: density of young manduvis around adult trees under three different management conditions in the Pantanal wetland, Brazil." *Ararajuba* 5:185–88.

Johnson, P. N. 1976. "Vegetation associated with Kakapo *Strigops habroptilus* in Sinbad Gully Fiordland New Zealand." *New Zealand Journal of Botany* 14:151–59.

Jones, Darryl. 1987. "Feeding ecology of the Cockatiel, *Nymphicus hollandicus*, in a grain-growing area." *Australian Wildlife Research* 14:105–15.

———. 2008. "Feeding birds in our towns and cities: a global research opportunity." *Journal of Avian Biology* 39(3):265–71.

Jordan, O. C., and C. A. Munn. 1993. "First observations of the Blue-Throated Macaw in Bolivia." *Wilson Bulletin* 105:694–95.

Jordano, P. 1983. "Fig-seed predation and dispersal by birds." *Biotropica* 15(1):38–41.

———. 1987. "Avian fruit removal: effects of fruit variation, crop size, and insect damage." *Ecology* 68(6):1711–23.

———. 1989. "Pre-dispersal biology of *Pistacia lentiscus* (Anacardiaceae): cumulative effects on seed removal by birds." *Oikos* 55(3):375–86.

Joseph, L. 1982a. "The Red-tailed Black Cockatoo in Southeastern Australia." *Emu* 82:42–45.

———. 1982b. "The Glossy Black-Cockatoo on Kangaroo Island." *Emu* 82:46–49.

Joseph, L., W. B. Emison, and W. M. Bren. 1991. "Critical assessment of the conservation status of the Red-tailed Black-Cockatoos in South-eastern Australia with special reference to nesting requirements." *Emu* 91:46–50.

Joseph, L., A. Toon, E. E. Schirtzinger, and T. F. Wright. 2011. "Molecular systematics of two enigmatic genera *Psittacella* and *Pezoporus* illuminate the ecological radiation of Australo-Papuan parrots (Aves: Psittaciformes)." *Molecular Phylogenetics and Evolution* 59:675–84.

Joseph, L., A. Toon, E. E. Schirtzinger, T. F. Wright, and R. Schodde. 2012. "A revised nomenclature and classification for family-group taxa of parrots (Psittaciformes)." *Zootaxa* (3205):26–40.

Joseph, L., and T. Wilke. 2006. "Molecular resolution of population history, systematics and historical biogeography of the Australian ringneck parrots *Barnardius*: are we there yet?" *Emu* 106(1):49–62.

Jumars, P. A. 2000. "Animal guts as nonideal chemical reactors: partial mixing and axial variation in absorption kinetics." *American Naturalist* 155(4):544–55.

Juniper, A. T., and C. Yamashita. 1991. "The habitat and status of Spix's Macaw *Cyanopsitta spixii.*" *Bird Conservation International* 1:1–9.

Juniper, T., and M. Parr. 1998. *Parrots: a guide to parrots of the world.* New Haven, CT: Yale University Press.

Juniper, Tony. 1991. "Last of a kind: a lone Spix's Macaw survives in the wild." *Birds International* 3:10–16.

———. 2002. *Spix's Macaw: the race to save the world's rarest bird.* London: Fourth Estate.

Juniper, Tony, and Carlos Yamashita. 1990. "The conservation of Spix's Macaw." *Oryx* 24(4):224–28.

Jurisevic, M. A. 2003. "Convergent characteristics of begging vocalisations in Australian birds." *Lundiana* 4(1):25–33.

Jurisevic, M. A., and K. J. Sanderson. 1994. "Alarm vocalizations in Australian birds: convergent characteristics and phylogenetic differences." *Emu* 94:67–77.

Juste, B. J. 1996. "Trade in the Gray Parrot *Psittacus erithacus* on the island of Principe (Sao Tome and Principe, Central Africa): initial assessment of the activity and its impact." *Biological Conservation* 76(2):101–04.

Kako, E. 2001. "The promise of an ecological, evolutionary approach to culture and language." *Behavioral and Brain Sciences* 24(2):338.

Kanesada, A., H. Eda-Fujiwara, R. Satoh, and T. Miyamoto. 2005. "Vocal learning during pair bonding in the Budgerigar, *Melopsittacus undulatus.*" *Zoological Science* 22(12):1513–13.

Karasov, W. H., and S. J. Cork. 1994. "Glucose absorption by a nectarivorous bird: the passive pathway is paramount." *American Journal of Physiology* 267(1):G18–26.

———. 1996. "Test of a reactor-based digestion optimization model for nectar-eating rainbow lorikeets." *Physiological Zoology* 69(1):117–38.

Karasov, W. H., and C. Martinez del Rio. 2007. *Physiological Ecology: how animals process energy, nutrients and toxins.* Princeton, NJ: Princeton University Press.

Karl, B. J., and H. A. Best. 1982. "Feral cats on Stewart Island: their foods, and their effects on Kakapo." *New Zealand Journal of Zoology* 9(2):287–93.

Kearvell, J. C., J. R. Young, and A. D. Grant. 2002. "Comparative ecology of sympatric orange-fronted parakeets (*Cyanoramphus malherbi*) and yellow-crowned parakeets (*C. auriceps*), South Island, New Zealand." *New Zealand Journal of Ecology* 26(2):139–48.

Kennedy, S. J., and A. E. Overs. 2001. "Foraging ecology and habitat use of the Swift Parrot on the south-western slopes of New South Wales." *Corella* 25:68–74.

Khaleghizadeh, Abolghasem. 2004. "On the diet and population of the Alexandrine Parakeet, *Psittacula eupatria,* in the urban environment of Tehran, Iran." *Zoology in the Middle East* 32:27–32.

Khan, Hammad Ahmad, Mirza Azhar Beg, and Akbar Ali Khan. 2004. "Breeding habitats of the Rose-ringed Parakeet (*Psittacula krameri*) in the cultivations of Central Punjab." *Pakistan Journal of Zoology* 36(2):133–138.

Kinnaird, Margaret F., Timothy G. O'Brien, Frank R. Lambert, and David Purmiasa. 2000. "Project Kakatua Seram: current status and conservation needs of the Seram Cockatoo, *Cacatua moluccensis*."

———. 2003. "Density and distribution of the endemic Seram cockatoo *Cacatua moluccensis* in relation to land use patterns." *Biological Conservation* 109(2):227–235.

Kirchman, J. J., E. E. Schirtzinger, and Wright T. F. 2012. "Phylogenetic relationships of the extinct Carolina Parakeet (*Conuropsis carolinensis*) inferred from DNA sequence data." *Auk* 129:197–204.

Kirk, Edwin J., R. G. Powlesland, and Susan C. Cork. 1993. "Anatomy of the mandibles, tongue and alimentary tract of kakapo, with some comparative information from kea and kaka." *Notornis* 40(1):55–63.

Kirkwood, T. B. L. 2002. "Evolution of ageing." *Mechanisms of Ageing and Development* 123(7):737–45.

———. 2008. "Understanding ageing from an evolutionary perspective." *Journal of Internal Medicien* 263(2):117–27.

———. 2011. "Systems biology of ageing and longevity." *Philosophical Transactions of the Royal Society B: Biological Sciences* 366(1561):64–70. doi:10.1098/rstb.2010.0275.

Kleeman, P. M., and J. D. Gilardi. 2005. "Geographical variation of St Lucia Parrot flight vocalizations." *Condor* 107(1):62–68.

Knights, P., and D. Currey. 1990. "Will Europe ban wild-bird imports?" *Defenders* (Nov./Dec.): 20–25.

Koenig, S. E. 2001. "The breeding biology of Black-billed Parrot *Amazona agilis* and Yellow-billed Parrot *Amazona collaria* in Cockpit Country, Jamaica." *Bird Conservation International* 11(3):205–225.

Koenig, S. E., J. M. Wunderle, and E. C. Enkerlin-Hoeflich. 2007. "Vines and canopy contact: a route for snake predation on parrot nests." *Bird Conservation International* 17(1):79–91.

Kondo, N., E. I. Izawa, and S. Watanabe. 2010. "Perceptual mechanism for vocal individual recognition in jungle crows (*Corvus macrorhynchos*): contact call signature and discrimination." *Behaviour* 147(8):1051–1072.

Konishi, M. 2006. "Behavioral guides for sensory neurophysiology." *Journal of Comparative Physiology A: Neuroethology Sensory Neural and Behavioral Physiology* 192(6):671–76.

Koutsos, E. A., K. D. Matson, and K. C. Klasing. 2001. "Nutrition of birds in the order Psittaciformes: a review." *Journal of Avian Medicine and Surgery* 15(4):257–75.

Kozlowski, C. P., and R. E. Ricklefs. 2010. "Egg size and yolk steroids vary across the laying order in cockatiel clutches: a strategy for reinforcing brood hierarchies?" *General and Comparative Endocrinology* 168(3):460–65.

Krabbe, N. 2000. "Overview of conservation priorities for parrots in the Andean region with special consideration for Yellow-eared Parrot *Ognorhynchus icterotis*." *International Zoo Yearbook* 37:283–88.

Krebs, E. A. 1998. "Breeding biology of crimson rosellas (*Platycercus elegans*) on Black Mountain, Australian Capital Territory." *Australian Journal of Zoology* 46(2):119–136.

———. 1999. "Last but not least: nestling growth and survival in asynchronously hatching crimson rosellas." *Journal of Animal Ecology* 68:266–281.

———. 2001. "Begging and food distribution in crimson rosella (*Platycercus elegans*) broods: why don't hungry chicks beg more?" *Behavioral Ecology and Sociobiology* 50(1):20–30.

Krebs, E. A., R. B. Cunningham, and C. F. Donnelly. 1999. "Complex patterns of food allocation in asynchronously hatching broods of crimson rosellas." *Animal Behaviour* 57:753–763.

Krebs, E. A., and R. D. Magrath. 2000. "Food allocation in crimson rosella broods: parents differ in their responses to chick hunger." *Animal Behavior* 59:739–751.

Krukenberg, C. F. W. 1882. "Die Federfarbstoffe der Psittaciden." *Vergleichedn-physiologische Studien Reihe 2. Abtlg* 2:29–36.

Kuehler, C., A. Lieberman, A. Varney, P. Unitt, R. M. Sulpice, J. Azua, and B. Tehevini. 1997. "Translocation of Ultramarine Lories *Vini ultramarina* in the Marquesas Islands: Ua Huka to Fatu Hiva." *Bird Conservation International* 7(1):69–79.

Kumschick, S., and W. Nentwig. 2010. "Some alien birds have as severe an impact as the most effectual alien mammals in Europe." *Biological Conservation* 143(11):2757–62.

Kundu, S., C. G. Jones, R. P. Prys-Jones, and J. J. Groombridge. 2012. "The evolution of the Indian Ocean parrots (Psittaciformes): Extinction, adaptive radiation and eustacy." *Molecular Phylogenetics and Evolution* 62(1):296–305. doi:10.1016/j.ympev.2011.09.025.

Lachlan, R. F., and P. J. B. Slater. 1999. "The maintenance of vocal learning by gene-culture interaction: the cultural trap hypothesis." *Proceedings of the Royal Society of London Series B: Biological Sciences* 266(1420):701–706.

Lambert, Frank R. 1993. "Trade, status and management of three parrots in the North Moluccas, Indonesia: White Cockatoo *Cacatua alba*, Chattering Lory *Lorius garrulus* and Violet-eared Lory *Eos squamata*." *Bird Conservation International* 3(2):145–68.

Langen, Marieke, Martien J. H. Kas, Wouter G. Staal, Herman van Engeland, and Sarah Durston. 2011. "The neurobiology of repetitive behavior: of mice . . . " *Neuroscience & Biobehavioral Reviews* 35(3):345–55. doi:10.1016/j.neubiorev.2010.02.004.

Langer, P. 1988. *The Mammalian Herbivore Stomach: comparative anatomy, function and evolution.* New York: G. Fischer.

Lanning, D. V. 1991. "Distribution and breeding biology of the Red-fronted Macaw." *Wilson Bulletin* 103(3):357–65.

Lanning, D. V., and J. T. Shiflett. 1983. "Nesting ecology of Thick-billed Parrots." *Condor* (85):66–73.

Larsen, O. N., R. J. Dooling, and A. Michelsen. 2006. "The role of pressure difference reception in the directional hearing of budgerigars (*Melopsittacus undulatus*)." *Journal of Comparative Physiology A: Neuroethology Sensory Neural and Behavioral Physiology* 192(10):1063–72.

Lavenex, P. B. 2000. "Lesions in the budgerigar vocal control nucleus NLc affect production, but not memory, of English words and natural vocalizations." *Journal of Comparative Neurology* 421(4):437–60.

Lavenex, Pamela Banta. 1999. "Vocal production mechanisms in the budgerigar (*Melopsittacus undulatus*): the presence and implications of amplitude modulation." *Journal of the Acoustical Society of America* 106(1):491–505.

Lebas, N. R. 2006. "Female finery is not for males." *Trends in Ecology & Evolution* 21(4):170–73.

Lee, A. T. K., S. Kumar, D. J. Brightsmith, and S. J. Marsden. 2010. "Parrot claylick distribution in South America: do patterns of 'where' help answer the question 'why'?" *Ecography* 33(3):503–13.

Lee, Jessica, Hugh Finn, and Michael Calver. 2010. "Mine-site revegetation monitoring detects feeding by threatened black-cockatoos within 8 years." *Ecological Management & Restoration* 11(2):141–43.

Lee, W. G., J. R. Wood, and G. M. Rogers. 2010. "Legacy of avian-dominated plant-herbivore systems in New Zealand." *New Zealand Journal of Ecology* 34(1):28–47.

Leech, Tara J., Andrew M. Gormley, and Philip J. Seddon. 2008. "Estimating the minimum viable population size of Kaka (*Nestor meridionalis*), a potential surrogate species in New Zealand lowland forest." *Biological Conservation* 141(3):681–91.

Leek, M. R., M. L. Dent, and R. J. Dooling. 2000. "Masking by harmonic complexes in budgerigars (*Melopsittacus undulatus*)." *Journal of the Acoustical Society of America* 107(3):1737–44.

Leeton, Peter R. J., Leslie Christidis, Michael Westerman, and Walter E. Boles. 1994. "Molecular phylogenetic affinities of the night parrot (*Geopsittacus occidentalis*) and the ground parrot (*Pezoporus wallicus*)." *Auk* 111(4):833–43.

Lefebvre, L., A. Gaxiola, S. Dawson, S. Timmermans, L. Rosza, and P. Kabai. 1998. "Feeding innovations and forebrain size in Australasian birds." *Behaviour* 135:1077–97.

Legge, S., R. Heinsohn, and S. Garnett. 2004. "Availability of nest hollows and breeding population size of eclectus parrots, *Eclectus roratus*, on Cape York Peninsula, Australia." *Wildlife Research* 31(2):149–61.

Leite, K. C. E., G. H. F. Seixas, I. Berkunsky, R. G. Collevatti, and R. Caparroz. 2008. "Population genetic structure of the Blue-fronted Amazon (*Amazona aestiva*, Psittacidae: Aves) based on nuclear microsatellite loci: implications for conservation." *Genetics and Molecular Research* 7(3):819–29.

Lengagne, T., T. Aubin, P. Jouventin, and J. Lauga. 2000. "Perceptual salience of individually distinctive features in the calls of adult king penguins." *Journal of the Acoustical Society of America* 107(1):508–16.

Leopold, Aldo. 1966. *The Sand County Almanac*. 2nd ed. New York: Oxford University Press.

Leslie, D. 2005. "Is the Superb Parrot *Polytelis swainsonii* population in Cuba State Forest limited by hollow or food availability?" *Corella* 29:77–87.

Levey, D. J., and C. M. Del Rio. 2001. "It takes guts (and more) to eat fruit: lessons from avian nutritional ecology." *Auk* 118(4):819–31.

Lewis, M. H., Y. Tanimura, L. W. Lee, and J. W. Bodfish. 2007. "Animal models of restricted repetitive behavior in autism." *Behavioural Brain Research* 176(1):66–74.

Lieberman, B. S. 2003. "Paleobiogeography: The relevance of fossils to biogeography." *Annual Review of Ecology Evolution and Systematics* 34:51–69.

Lin, Ruey-Shing, and Pei-Fen Lee. 2006. "Status of feral populations of exotic cockatoos (Genus *Cacatua*) in Taiwan." *Taiwania* 51:188–94.

Lindsey, G. D., W. J. Arendt, and J. Kalina. 1994. "Survival and causes of mortality in juvenile Puerto Rican parrots." *Journal of Field Ornithology* 65(1):76–82.

Lindsey, G. D., W. J. Arendt, J. Kalina, and G. W. Pendleton. 1991. "Home range and movements of juvenile Puerto Rican Parrots." *Journal of Wildlife Management* 55(2):318–22.

Livezey, B. C. 1992. "Morphological corollaries and ecological implications of flightlessness in the Kakapo (Psittaciformes: *Strigops habroptilus*)." *Journal of Morphology* 213:105–45.

Lloyd, B. D., and R. G. Powlesland. 1994. "The decline of Kakapo *Strigops habroptilus* and attempts at conservation by translocation." *Biological Conservation* 69(1):75–85.

Lohr, Bernard, Timothy F. Wright, and Robert J. Dooling. 2003. "Detection and discrimination of natural calls in masking noise by birds: estimating the active space of a signal." *Animal Behaviour* 65(4):763–77.

Long, J. L. 1981. *Introduced birds of the world: the worldwide history, distribution and influence of birds introduced to new environments*. New York: Universe.

———. 1984. "The diets of three species of parrots in the South of Western Australia." *Australian Wildlife Research* 11:357–71.

———. 1985. "Damage to cultivated fruit by parrots in the south of Western Australia." *Australian Wildlife Research* 12:75–80.

Long, J. L., and P. R. Mawson. 1994. "Diet of regent parrots (*Polytelis anthopeplus*) in the south-west of Western Australia." *Western Australian Naturalist* 19(4):293–99.

Lorenz, Konrad Z. 1952. *King Solomon's Ring: new light on animal ways.* London: Methuen & Co.

Lotz, C. N., and J. E. Schondube. 2006. "Sugar preferences in nectar- and fruit-eating birds: behavioral patterns and physiological causes." *Biotropica* 38(1):3–15.

Lowry, H., and A. Lill. 2007. "Ecological factors facilitating city-dwelling in red-rumped parrots." *Wildlife Research* 34:624–31.

Loyn, R. H., B. A. Lane, C. Chandler, and G. W. Carr. 1986. "Ecology of Orange-bellied Parrots *Neophema chrysogaster* at their main remnant wintering site." *Emu* 86(4):195–206.

Lynn, S. K., and I. M. Pepperberg. 2001. "Culture: in the beak of the beholder?" *Behavioral and Brain Sciences* 24(2):341–42.

Mack, A. L., and D. D. Wright. 1998. "The Vulturine Parrot, *Psittrichas fulgidus*, a threatened New Guinea endemic: notes on its biology and conservation." *Bird Conservation International* 8(2):185–94.

Mac Nally, Ralph, and Gregory Horrocks. 2000. "Landscape-scale conservation of an endangered migrant: the Swift Parrot (*Lathamus discolor*) in its winter range." *Biological Conservation* 92(3):335–43.

Magrath, R. D. 1994. "Footedness in the Glossy Black Cockatoo: some observations and a review of the literature with a note on the husking of *Allocasuarina* cones by this species." *Corella* 18:21–24.

Magrath, R., and A. Lill. 1983. "The use of time and energy by the Crimson Rosella in a temperate wet forest in winter." *Australian Journal of Zoology* 31:903–12.

———. 1985. "Age-related differences in behaviour and ecology in Crimson Rosellas, *Platycercus elegans,* during the non-breeding season." *Australian Wildlife Research* 12:299–306.

Major, R. E., and H. Parsons. 2010. "What do museum specimens tell us about the impact of urbanisation? A comparison of the recent and historical bird communities of Sydney." *Emu* 110(1):92–103.

Manabe, K. 1997. "Vocal plasticity in budgerigars: various modifications of vocalization by operant conditioning." *Biomedical Research-Tokyo* 18:125–32.

———. 2008. "Vocal learning in Budgerigars (*Melopsittacus undulatus*): effects of an acoustic reference on vocal matching." *Journal of the Acoustical Society of America* 123(3):1729–36.

Manabe, K., and R. J. Dooling. 1997. "Control of vocal production in budgerigars (*Melopsittacus undulatus*): selective reinforcement, call differentiation, and stimulus control." *Behavioural Processes* 41(2):117–32.

Manabe, K., T. Kawashima, and J. E. R. Staddon. 1995. "Differential localization in budgerigars: towards an experimental analysis of naming." *Journal of the Experimental Analysis of Behavior* 63(1):111–26.

Manabe, K., E. I. Sadr, and R. J. Dooling. 1998. "Control of vocal intensity in budgerigars (*Melopsittacus undulatus*): differential reinforcement of vocal intensity and the Lombard effect." *Journal of the Acoustical Society of America* 103(2):1190–98.

Manabe, K., J. E. R. Staddon, and J. M. Cleaveland. 1997. "Control of vocal repertoire by reward in budgerigars (*Melopsittacus undulatus*)." *Journal of Comparative Psychology* 111(1):50–62.

Manning, A. D., D. B. Lindenmayer, and S. C. Barry. 2004. "The conservation implications of bird reproduction in the agricultural 'matrix': a case study of the vulnerable Superb Parrot of south-eastern Australia." *Biological Conservation* 120(3):363–74.

Marini, Miguel Angelo, Morgane Barbet-Massin, Jaime Martinez, Nemora P. Prestes, and Frederic Jiguet. 2010. "Applying ecological niche modelling to plan conservation actions for the Red-spectacled Amazon (*Amazona pretrei*)." *Biological Conservation* 143(1):102–12.

Marino, L., R. C. Connor, R. E. Fordyce, L. M. Herman, P. R. Hof, L. Lefebvre, D. Lusseau, et al. 2007. "Cetaceans have complex brains for complex cognition." *PLoS Biology* 5(5):e139.

Maron, M., P. K. Dunn, C. A. McAlpine, and A. Apan. 2010. "Can offsets really compensate for habitat removal? The case of the endangered Red-tailed Black-Cockatoo." *Journal of Applied Ecology* 47(2):348–55.

Maron, M., and A. Lill. 2004. "Discrimination among potential buloke (*Allocasuarina luehmannii*) feeding trees by the endangered south-eastern red-tailed black-cockatoo (*Calyptorhynchus banksii graptogyne*)." *Wildlife Research* 31(3):311–17.

Marsden, Stuart J., and Martin T. Jones. 1997. "The nesting requirements of the parrots and hornbill of Sumba, Indonesia." *Biological Conservation* 82(3):279–87.

Marsden, Stuart, J., and John D. Pilgrim. 2003. "Factors influencing the abundance of parrots and hornbills in pristine and disturbed forests on New Britain, PNG." *Ibis* 145(1):45–53.

Marsden, Stuart, J., John D. Pilgrim, and Roger Wilkinson. 2001. "Status, abundance and habitat use of Blue-eyed Cockatoo *Cacatua ophthalmica* on New Britain, Papua New Guinea." *Bird Conservation International* 11(3):151–60.

Marsden, Stuart J., Mark Whiffin, Lisa Sadgrove, and Paulo Guimaraes Jr. 2000. "Parrot populations and habitat use in and around two lowland Atlantic forest reserves, Brazil." *Biological Conservation* 96(2):209–17.

Marsh, K. J., I. R. Wallis, R. L. Andrew, and W. J. Foley. 2006. "The detoxification limitation hypothesis: Where did it come from and where is it going?" *Journal of Chemical Ecology* 32(6):1247–66.

Marsh, K. J., I. R. Wallis, S. McLean, J. S. Sorensen, and W. J. Foley. 2006. "Conflicting demands on detoxification pathways influence how common brushtail possums choose their diets." *Ecology* 87(8):2103–12.

Martella, M. B., and E. H. Bucher. 1990. "Vocalizations of the Monk Parakeet." *Bird Behavior* 8:101–10.

Martin, L. F. 1989. *Caracteristicas del sistema social cooperativo de la Cotorras Myiopsitta monachus.* Unversidad Nacional de Cordoba, Argentina.

Martin, L. F., and Enrique H. Bucher. 1993. "Natal dispersal and first breeding age in Monk Parakeets." *The Auk* 110:930–33.

Martin, Thomas E. 1993. "Evolutionary determinants of clutch size in cavity-nesting birds: nest predation or limited breeding opportunities?" *American Naturalist* 142:937–46.

———. 2002. "A new view of avian life-history evolution tested on an incubation paradox." *Proceedings of the Royal Society of London, Series B: Biological Sciences* 269:309–16.

Martuscelli, Paulo. 1994. "Maroon-bellied Conures feed on gall-forming homopteran larvae." *Wilson Bulletin* 106(4):769–770.

Masello, J. F., M. L. Pagnossin, C. Sommer, and P. Quillfeldt. 2006. "Population size, provisioning frequency, flock size and foraging range at the largest known colony of Psittaciformes: the Burrowing Parrots of the north-eastern Patagonian coastal cliffs." *Emu* 106(1):69–79.

Masello, J. F., and P. Quillfeldt. 2003a. "Body size, body conditions and ornamental feathers of Burrowing Parrots: variation between years and sexes, assortative mating and influences on breeding success." *Emu* 103:1–13.

———. 2003b. *Hatching asynchrony and brood reduction in burrowing parrots during the La Niña of 1998.* Institut für Ökoloigie, Friedrich-Shiller-Universität Jena, Germany.

Masello, J. F., R. G. Choconi, M. Helmer, T. Kremberg, T. Lubjuhn, and P. Quillfeldt. 2009. "Do leucocytes reflect condition in nestling burrowing parrots *Cyanoliseus patagonus* in the wild?" *Comparative Biochemistry and Physiology A: Molecular & Integrative Physiology* 152(2):176–81.

Masello, Juan F., Maria Lujn Pagnossin, Thomas Lubjuhn, and Petra Quillfeldt. 2004. "Ornamental non-carotenoid red feathers of wild burrowing parrots." *Ecological Research* 19(4):421–32.

Masello, Juan F., and Petra Quillfeldt. 2002. "Chick growth and breeding success of the Burrowing Parrot." *Condor* 104(3):574–86.

Masello, Juan F., Anna Sramkova, Petra Quillfeldt, Thomas Epplen Joerg, and Thomas Lubjuhn. 2002. "Genetic monogamy in burrowing parrots *Cyanoliseus patagonus?*" *Journal of Avian Biology* 33(1):99–103.

Masin, S., R. Massa, and L. Bottoni. 2004. "Evidence of tutoring in the development of subsong in newly-fledged Meyer's Parrots *Poicephalus meyeri.*" 76(2):231–36.

Mason, G., R. Clubb, N. Latham, and S. Vickery. 2007. "Why and how should we use environmental enrichment to tackle stereotypic behaviour?" *Applied Animal Behaviour Science* 102(3–4):163–88.

Mason, G. J. 2010. "Species differences in responses to captivity: stress, welfare and the comparative method." *Trends in Ecology & Evolution* 25(12):713–21.

Massa, R., V. Galanti, and L. Bottoni. 1996. "Mate choice and reproductive success in the domesticated budgerigar, *Melopsittacus undulatus.*" *Italian Journal of Zoology* 63:243–46.

Matuzak, G. D., M. Bernadette Bezy, and Donald J. Brightsmith. 2008. "Foraging ecology of parrots in a modified landscape: seasonal trends and introduced species." *Wilson Journal of Ornithology* 120(2):353–65.

Mawson, P. R., and J. L. Long. 1994. "Size and age parameters of nest trees used by four species of parrot and one species of cockatoo in south-west Australia." *Emu* 94(3):149–55.

———. 1996. "Changes in the status and distribution of four species of parrot in the south of Western Australia during 1970–90." *Pacific Conservation Biology* 2(2):191–99.

May, Diana L. 1996. "Behaviour of Grey Parrots in the rainforest of the Central African Republic." *Psittascene* 8(3):8–9.

Mayr, Ernst, and Jared Diamond. 2001. *The Birds of Northern Melanesia: speciation, ecology, & biogeography.* Oxford: Oxford University Press.

Mayr, G. 2002. "On the osteology and phylogenetic affinities of the Pseudasturidae: Lower Eocene stem-group representatives of parrots (Aves, Psittaciformes)." *Zoological Journal of the Linnean Society* 136(4):715–29.

———. 2008a. "Avian higher-level phylogeny: well-supported clades and what we can learn from a phylogenetic analysis of 2954 morphological characters." *Journal of Zoological Systematics and Evolutionary Research* 46(1):63–72.

———. 2008b. "The phylogenetic affinities of the parrot taxa Agapornis, Loriculus and Melopsittacus (Aves: Psittaciformes): hypotarsal morphology supports the results of molecular analyses." *Emu* 108:23–27.

———. 2009. *Paleogene fossil birds.* Heidelberg: Springer.

———. 2010. "Parrot interrelationships: morphology and the new molecular phylogenies." *EMU* 110(4):348–57. doi:10.1071/mu10035.

———. 2014. "Frontiers in palaeontology: the origins of crown group birds—molecules and fossils." *Palaeontology (Oxford)* 57(2):231–42. doi:10.1111/pala.12103.

Mayr, G., and J. Clarke. 2003. "The deep divergences of neornithine birds: a phylogenetic analysis of morphological characters." *Cladistics* 19(6):527–53.

Mayr, G., and M. Daniels. 1998. "Eocene parrots from Messel (Hessen, Germany) and the London Clay of Walton-on-the-Naze (Essex, England)." *Senckenbergiana Lethaea* 78(1–2):157–77.

Mayr, G., and U. B. Gohlich. 2004. "A new parrot from the Miocene of Germany, with comments on the variation of hypotarsus morphology in some Psittaciformes." *Belgian Journal of Zoology* 134(1):47–54.

Mbatha, K., C. T. Downs, and M. Penning. 2002. "Nectar passage and gut morphology in the Malachite Sunbird and the Black-capped Lory: implications for feeding in nectarivores." *Ostrich* 73(3–4):138–42.

McCormack, Gerald. 1997. "The status of Cook Islands birds." *World Birdwatch* 19:13–16.

———. 2006. Rimatara Lorikeet reintroduction programme. http://cookislands.bishopmuseum. org/showarticle.asp?id = 24.

McCormack, Gerald, and Judith Künzlé. 1996. "The 'Ura or Rimatara Lorikeet Vini kuhlii: its former range, present status, and conservation priorities." *Bird Conservation International* 6:325–34.

McCormack, John E., Michael G. Harvey, Brant C. Faircloth, Nicholas G. Crawford, Travis C. Glenn, and Robb T. Brumfield. 2013. "A phylogeny of birds based on over 1,500 loci collected by target enrichment and high-throughput sequencing." *PLoS ONE* 8(1):e54848. doi:10.1371/journal.pone.0054848.

McDiarmid, R. W., R. E. Ricklefs, and M. S. Foster. 1977. "Dispersal of *Stemmadenia donnellsmithii* (Apocynaceae) by birds." *Biotropica* 9(1):9–25.

McDonald, D. 2003. "Feeding ecology and nutrition of Australian lorikeets." *Seminars in Avian and Exotic Pet Medicine* 12:195–204.

McElreath, R., R. Boyd, and P. J. Richerson. 2003. "Shared norms and the evolution of ethnic markers." *Current Anthropology* 44(1):122–29.

McElreath, R., M. Lubell, P. J. Richerson, T. M. Waring, W. Baum, E. Edsten, C. Efferson, and B. Paciotti. 2005. "Applying evolutionary models to the laboratory study of social learning." *Evolution and Human Behavior* 26(6):483–508.

McFarland, D. C. 1991a. "The biology of the Ground Parrot, *Pezoporus wallicus,* in Queensland. II. Spacing, calling and breeding behaviour." *Wildlife Research* 18:185–97.

———. 1991b. "The biology of the Ground Parrot, *Pezoporus wallicus,* in Queensland. III. Distribution and abundance." *Wildlife Research* 18:199–213.

———. 1991c. "Flush behaviour, catchability and morphometrics of the Ground Parrot Pezoporus wallicus in south-eastern Queensland." *Corella* 15:143–49.

———. 1991d. "The biology of the Ground Parrot, *Pezoporus wallicus,* in Queensland. I. Microhabitat use, activity cycle and diet." *Wildlife Research* 18:169–184.

McGraw, K. J. 2005. "The antioxidant function of many animal pigments: are there consistent health benefits of sexually selected colourants?" *Animal Behaviour* 69:757–64.

McGraw, K. J., and M. C. Nogare. 2005. "Distribution of unique red feather pigments in parrots." *Biology Letters* 1(1):38–43.

McGraw, K. J., K. Wakamatsu, S. Ito, P. M. Nolan, P. Jouventin, F. S. Dobson, R. E. Austic, R. J. Safran, L. M. Siefferman, G. E. Hill, and R. Parker. 2004. "You can't judge a pigment by its color: carotenoid and melanin content of yellow and brown feathers in swallows, bluebirds, penguins, and domestic chickens." *Condor* 106(2):390–95.

McInnes, R. S., and P. B. Carne. 1978. "Predation of Cossid Moth larvae by Yellow-Tailed Black Cockatoos causing losses in plantations of *Eucalyptus grandis* in north coastal New South Wales." *Australian Wildlife Research* 5:101–21.

McKinley, D. 1978. "Clutch size, laying date, and incubation period in Carolina Parakeet." *Bird-Banding* 49(3):223–33.

———. 1980a. "The balance of decimating factors and recruitment in extinction of the Carolina Parakeet. Part I." *Indiana Audobon Quarterly* 58:8–18.

———. 1980b. "The balance of decimating factors and recruitment in extinction of the Carolina Parakeet. Part II." *Indiana Audobon Quarterly* 58:50–61.

———. 1980c. "The balance of decimating factors and recruitment in extinction of the Carolina Parakeet. Part III." *Indiana Audobon Quarterly* 58:103–14.

McNab, B. K. 2003. "Metabolism: ecology shapes bird bioenergetics." *Nature* 426(6967):620–21.

McNab, B. K., and H. I. Ellis. 2006. "Flightless rails endemic to islands have lower energy expenditures and clutch sizes than flighted rails on islands and continents." *Comparative Biochemistry and Physiology A: Molecular & Integrative Physiology* 145(3):295–311.

McNab, Brian K., and Charles A. Salisbury. 1995. "Energetics of New Zealand's temperate parrots." *New Zealand Journal of Zoology* 22(3):339–49.

McWhorter, T. J., and M. V. Lopez-Calleja. 2000. "The integration of diet, physiology, and ecology of nectar-feeding birds." *Revista Chilena De Historia Natural* 73(3):451–60.

McWhorter, T. J., D. R. Powers, and C. M. del Rio. 2003. "Are hummingbirds facultatively ammonotelic? Nitrogen excretion and requirements as a function of body size." *Physiological and Biochemical Zoology* 76(5):731–43.

McWilliams, S. R., D. Afik, and S. Secor. 1997. "Patterns and processes in the vertebrate digestive system." *Trends in Ecology & Evolution* 12(11):420–22.

Medway, David G. 2005. "Feeding association of tui (*Prosthemadera n. novaeseelandiae*) with North Island kaka (*Nestor meridionalis septentrionalis*)." *Nortonis* 52:111–12.

Mee, A., R. Denny, K. Fairclough, D. M. Pullan, and W. Boyd-Wallis. 2005. "Observations of parrots at a geophagy site in Bolivia." *Biota Neotropica* 5(2):321–24.

Meehan, C. L., J. P. Garner, and J. A. Mench. 2004. "Environmental enrichment and development of cage stereotypy in Orange-winged Amazon parrots (*Amazona amazonica*)." *Developmental Psychobiology* 44(4):209–18.

Mello, C. V. 2002. "Mapping vocal communication pathways in birds with inducible gene expression." *Journal of Comparative Physiology A: Neuroethology Sensory Neural and Behavioral Physiology* 188(11–2):943–59.

———. 2004. "Identification and analysis of vocal communication pathways in birds through inducible gene expression." *Anais da Academia Brasileira de Ciencias* 76(2):243–46.

Mello, C. V., T. A. F. Velho, and R. Pinaud. 2004. "A window on song auditory processing and perception." *Behavioral Neurobiology of Birdsong* 1016:263–81.

Menkhorst, P.W., and A.C. Isles. 1981. "The Night Parrot *Geopsittacus occidentalis*: evidence of its occurence in north-western Victoria during the 1950s." *Emu* 81:239–40.

Meredith, C.W., A.M. Gilmore, and A.C. Isles. 1984. "The Ground Parrot (*Pezoporus wallicus*) in south-eastern Australia: a fire adapted species?" *Australian Journal of Ecology* 9:367–80.

Merton, D., and R. Empson. 1989. "But it doesn't look like a parrot!" *Birds International* 1(1):60–72.

Merton, D.V., R. B. Morris, and I. A. E. Atkinson. 1984. "Lek behaviour in a parrot: the Kakapo *Strigops habroptilus* of New Zealand." *Ibis* 126:277–83.

Messing, H. 2008. "Why do parrots have the ability to mimic?" *Scientific American* 298(3):102.

Meyers, J. Michael. 1996. "New nesting area of Puerto Rican Parrots." *Wilson Bulletin* 108(1):164–66.

Miller, G. F. 2000. "Evolution of human music through sexual selection." In *The Origins of Music*, edited by N. L. Wallin, B. Merker, and S. Brown, 329–60. Cambridge, MA: MIT Press.

Miller, Hillary C, David M. Lambert, Craig D. Millar, and Bruce C. Robertson. 2003. "Minisatellite DNA profiling detects lineages and parentage in the endangered Kakapo (*Strigops habroptilus*) despite low microsatellite DNA variation." *Conservation Genetics* 4:265–74.

Milton, Katharine. 1980. *The Foraging Strategy of Howler Monkeys: a study in primate economics*. New York: Columbia University Press.

———. 2000. "Quo vadis? Tactics of food search and group movement in primates and other animals." In *On the Move: how and why animals travel in groups*, edited by S. Boinski and P. Garber, 375–418. Chicago: University of Chicago Press.

Minot, Ed, Bruce Robertson, and Hillary Miller. 2000. "Molecular ecology of Kakapo (*Strigops habroptilus*)." *Notornis* 47:172.

Mishra, R. M. 2007. "Role of birds in the seed dispersal of *Zizyphus oenoplia* (Mill.) in a tropical deciduous forest of Central India." *Asian Journal of Environmental Science* 2(1/2):14–17.

Mithen, S. 1996. "Anthropomorphism and the evolution of cognition." *Journal of the Royal Anthropological Institute* 2(4):717–19.

Miyaki, Christina Yumi, Sergio Russo Matioli, Terry Burke, and Anita Wajntal. 1998. "Parrot evolution and paleogeographical events: mitochondrial DNA evidence." *Molecular Biology and Evolution* 15(5):544–51.

Mlikovsky, Jiri. 1998. "A new parrot (Aves: Psittacidae) from the early Miocene of the Czech Republic." *Acta Societatis Zoologicae Bohemicae* 62(4):335–41.

Moegenburg, Susan M., and Douglas J. Levey. 2003. "Do frugivores respond to fruit harvest? An experimental study of short-term responses." *Ecology (Washington DC)* 84(10):2600–12.

Monroe, Burt L, Jr., and Charles G. Sibley. 1993. *A World Checklist of Birds*. New Haven: Yale University Press.

Montagna, E., and B. B. Torres. 2008. "Expanding ecological possibilities." *Biochemistry and Molecular Biology Education* 36(2):99–105.

Monterrubio, Tiberio, Ernesto Enkerlin-Hoeflich, and B. Hamilton Robert. 2002. "Productivity and nesting success of Thick-billed Parrots." *Condor* 104(4):788–94.

Monterrubio-Rico, T., and E. Enkerlin-Hoeflich. 2004. "Present use and characteristics of Thick-billed Parrot nest sites in northwestern Mexico." *Journal of Field Ornithology* 75(1):96–103.

Monterrubio-Rico, T.C., J. M. Ortega-Rodriguez, M. C. Marin-Togo, A. Salinas-Melgoza, and K. Renton. 2009. "Nesting habitat of the Lilac-crowned Parrot in a modified landscape in Mexico." *Biotropica* 41(3):361–68.

Moore, B. R. 2004. "The evolution of learning." *Biological Reviews* 79(2):301–35.

Moore, N. L. 1994. *Affiliative behavior of captive pair bonded Orange-winged Amazon parrots*. University of California, Davis.

Moorhouse, R. J. 1997. "The diet of the North Island kaka (*Nestor meridionalis septentrionalis*) on Kapiti Island." *New Zealand Journal of Ecology* 21(2):141–52.

Moorhouse, R. J., and R. G. Powlesland. 1991. "Aspects of the ecology of Kakapo *Strigops habroptilus* liberated on Little Barrier Island (Hauturu), New Zealand." *Biological Conservation* 56:349–65.

Moorhouse, Ron, Terry Greene, Peter Dilks, Ralph Powlesland, Les Moran, Genevieve Taylor, Alan Jones, et al. 2003. "Control of introduced mammalian predators improves kaka *Nestor meridionalis* breeding success: reversing the decline of a threatened New Zealand parrot." *Biological Conservation* 110(1):33–44.

Moorhouse, Ron J., Mick J. Sibley, Brian D. Lloyd, and Terry C. Greene. 1999. "Sexual dimorphism in the North Island Kaka *Nestor meridionalis septentrionalis*: selection for enhanced male provisioning ability?" *Ibis* 141(4):644–51.

Moravec, M. L., G. F. Striedter, and N. T. Burley. 2006. "Assortative pairing based on contact call similarity in budgerigars, *Melopsittacus undulatus*." *Ethology* 112(11):1108–16.

———. 2010. "'Virtual parrots' confirm mating preferences of female Budgerigars." *Ethology* 116(10):961–71.

Morbey, Volanda E., Ronald C. Ydenberg, Hugh A. Knechtel, and Anne Hafenist. 1999. "Parental provisioning, nestling departure decisions and prefledging mass recession in Cassin's auklets." *Animal Behavior* 57:873–81.

Morowitz, H., and E. Smith. 2007. "Energy flow and the organization of life." *Complexity* 13(1):51–59.

Mourer-Chauviré, Cécile 1992. "Une nouvelle famille de Perroquets (Aves, Psittaciformes) dans l'Eocène supérieur des Phosphorites du Quercy." *Geobios, Mémoire Spécial* 14:169–77.

Moyle, B. J., M. Vardon, and R. Noske. 1997. "Harvesting Black Cockatoos in the Northern Territory: catastrophe or conservation?" *Australian Biologist* 10:84–93.

Munn, C. A. 1988. "Macaw biology in Manu National Park, Peru." *Parrotletter* 1:18–21.

———. 1992. "Macaw biology and ecotourism, or 'when a bird in the bush is worth two in the hand'." In *New World Parrots in Crisis: solutions from conservation biology,* edited by S. R. Beissinger and N. F. R. Snyder, 47–72. Washington, DC: Smithsonian Institution Press.

Munn, Charles A., Jorgen B. Thomsen, and Carlos Yamashita. 1990. "The Hyacinth Macaw." In *Audubon Wildlife Report 1989/90,* edited by W. J. Chandler, 404–19. New York: Academic.

Munoz, A. R., and R. Real. 2006. "Assessing the potential range expansion of the exotic Monk Parakeet in Spain." *Diversity and Distributions* 12(6):656–65.

Munshi-South, J., and G. S. Wilkinson. 2006. "Diet influences life span in parrots (Psittaciformes)." *Auk* 123(1):108–18.

———. 2010. "Bats and birds: Exceptional longevity despite high metabolic rates." *Ageing Research Reviews* 9(1):12–19.

Murgui, Enrique, and Anna Valentin. 2003. "Relacion entre las caracteristicas del paisaje urbano y la comunidad de aves introducidas en la cuidad de Valencia (España)." *Ardeola* 50(2):201–14.

Murphy, S. A., and S. M. Legge. 2007. "The gradual loss and episodic creation of Palm Cockatoo (*Probosciger aterrimus*) nest-trees in a fire- and cyclone-prone habitat." *Emu* 107(1):1–6.

Murphy, S., S. Legge, and R. Heinsohn. 2003. "The breeding biology of palm cockatoos (*Probosciger aterrimus*): a case of a slow life history." 261:327–39.

Murphy, Stephen A., Leo Joseph, Allan H. Burbidge, and Jeremy Austin. 2011. "A cryptic and critically endangered species revealed by mitochondrial DNA analyses: the Western Ground Parrot." *Conservation Genetics* 12(2):595–600. doi:10.1007/s10592-010-0161-1.

Myers, M. C., and C. Vaughan. 2004. "Movement and behavior of scarlet macaws (*Ara macao*) during the post-fledging dependence period: implications for *in situ* versus *ex situ* management." 118(3):411–20.

Nair, S., R. Balakrishnan, C. S. Seelamantula, and R. Sukumar. 2009. "Vocalizations of wild Asian elephants (*Elephas maximus*): Structural classification and social context." *Journal of the Acoustical Society Of America* 126(5):2768–78.

Napier, K. R., C. Purchase, T. J. McWhorter, S. W. Nicolson, and P. A. Fleming. 2008. "The sweet life: diet sugar concentration influences paracellular glucose absorption." *Biology Letters* 4(5):530–33.

Ndithia, H., and M. R. Perrin. 2006. "The spatial ecology of the Rosy-faced Lovebird *Agapornis roseicollis* in Namibia." *Ostrich* 77(1–2):52–57.

Ndithia, H., M. R. Perrin, and M. Waltert. 2007. "Breeding biology and nest site characteristics of the Rosy-faced Lovebird *Agapornis roseicollis* in Namibia." *Ostrich* 78(1):13–20.

Neidermyer, W. J., and J. J. Hickey. 1977. "The Monk Parakeet in the United States, 1970–75." *American Birds* 31(3):273–78.

Neill, D. 2010. "A proposal in relation to a genetic control of lifespan in mammals." *Ageing Research Reviews* 9(4):437–46.

Neuhauser, Markus. 2003. "Further evidence for Emlen's hypothesis from two parrot species." *New Zealand Journal of Zoology* 30(3):221–25.

Nicolas, G., C. Fraigneau, and T. Aubin. 2004. "Variation in the behavioral responses of Budgerigars *Melopsittacus undulatus* to an alarm call in relation to sex and season." *Anais Da Academia Brasileira De Ciencias* 76(2):359–364.

Nicolson, SW, and PA Fleming. 2003. "Nectar as food for birds: the physiological consequences of drinking dilute sugar solutions." *Plant Systematics and Evolution* 238(1–4):139–53.

Nixon, A. J. 1994. "Feeding ecology of hybridizing parakeets on Mangere Island, Chatham Islands." *Notornis Supplement* 41:5–18.

Nores, M. 2009. "Use of active monk parakeet nests by common pigeons and response by the host." *Wilson Journal of Ornithology* 121(4):812–15.

Nowotny, Dr. 1898. "The breeding of the Carolina Parakeet in captivity." *Auk* 15:28–32.

Nunes, M. F. C., and M. Galetti. 2007. "Use of forest fragments by Blue-winged Macaws (*Primolius maracana*) within a fragmented landscape." *Biodiversity and Conservation* 16(4):953–67.

O'Donnell, C. F. J. 1993. "More sap feeding by the Kaka." *Notornis* 40:70–80.

O'Donnell, C. F. J., and P. J. Dilks. 1986. "Forest birds in South Westland; status, distribution and habitat use." *Occasional Publication* 10. Wellington: New Zealand Wildlife Service.

———. 1989. "Sap-feeding by the Kaka *Nestor meridionalis* in South Westland New Zealand." *Notornis* 36(1):65–71.

———. 1994. "Foods and foraging of forest birds in temperate rainforest, South Westland, New Zealand." *New Zealand Journal of Ecology* 18(2):87–107.

Okanoya, K. 2007. "Language evolution and an emergent property." *Current Opinion in Neurobiology* 17(2):271–76.

Okanoya, K., and R. J. Dooling. 1987. "Hearing in passerine and psittacine birds: a comparative study of absolute and masked auditory thresholds." *Journal of Comparative Psychology* 101(1):7–15.

———. 1990a. "Detection of auditory sinusoids of fixed and uncertain frequency by Budgerigars (*Melopsittacus undulatus*) and Zebra Finches (*Poephila guttata*)." *Hearing Research* 50(1–2):175–84.

———. 1990b. "Detection of gaps in noise by budgerigars (*Melopsittacusundulatus*) and zebra finches (*Poephilaguttata*)." *Hearing Research* 50:185–92.

———. 1991. "Perception of distance calls by Budgerigars (*Melopsittacus undulatus*) and Zebra Finches (*Poephila guttata*): assessing species-specific advantages." *Journal of Comparative Psychology* 105(1):60–72.

Oliverieri, A., and L. Pearson. 1992. "Monk Parakeets in Bridgeport Connecticut." *Connecticut Warbler* 12:104–11.

Olson, S. L., and E. J. M. Lopez. 2008. "New evidence of *Ara autochthones* from an archeological site in Puerto Rico: a valid species of West Indian macaw of unknown geographical origin (Aves: Psittacidae)." *Caribbean Journal of Science* 44(2):215–22.

Oren, D.C., and F. Novaes. 1986. "Observations on the Golden Parakeet *Aratinga guarouba* in northern Brazil." *Biological Conservation* 36:329–37.

Orensanz, J.M., C.M. Hand, A.M. Parma, J. Valero, and R. Hilborn. 2004. "Precaution in the harvest of Methuselah's clams: the difficulty of getting timely feedback from slow-paced dynamics." *Canadian Journal of Fisheries and Aquatic Sciences* 61(8):1355–72.

Ortega-Rodriguez, J. M., and T. C. Monterrubio-Rico. 2008. "Geographic characteristics of the Lilac-crowned Parrot (*Amazona finschi*) nest sites on coastal Michoacan, Mexico." *Ornitologia Neotropical* 19(3):427–39.

Ortiz-Catedral, L., and D. H. Brunton. 2008. "Clutch parameters and reproductive success of a translocated population of red-crowned parakeet (*Cyanoramphus novaezelandiae*)." *Australian Journal of Zoology* 56(6):389–93.

———. 2009. "Nesting sites and nesting success of reintroduced Red-crowned Parakeets (*Cyanoramphus novaezelandiae*) on Tiritiri Matangi Island, New Zealand." *New Zealand Journal of Zoology* 36(1):1–10.

Ortiz-Maciel, S.G., C. Hori-Ochoa, and E. Enkerlin-Hoeflich. 2010. "Maroon-fronted Parrot (*Rhynchopsitta terrisi*) breeding home range and habitat selection in the northern Sierra Madre Oriental, Mexico." *Wilson Journal of Ornithology* 122(3):513–17.

Pacheco, M.A., M.A. García-Amado, C. Bosque, and M.G. Domínguez-Bello. 2004. "Bacteria in the crop of the seed-eating Green-rumped Parrotlet." *Condor* 106(1):139–43.

Pacheco, M.A., M. Lentino, C. Mata, S. Barreto, and M. Araque. 2008. "Microflora in the crop of adult Dusky-billed Parrotlets (Forpus modestus)." *Journal of Ornithology* 149(4):621–28.

Pain, D.J., T.L.F. Martins, M. Boussekey, S.H. Diaz, C.T. Downs, J.M.M. Ekstrom, S. Garnett, et al. 2006. "Impact of protection on nest take and nesting success of parrots in Africa, Asia and Australasia." *Animal Conservation* 9(3):322–30.

Paranhos, S.J., C.B. de Araujo, and L.O. Marcondes-Machado. 2007. "Feeding behavior of the Yellow-chevroned Parakeet at the northeast of the state of Sao Paulo State, Brazil." *Revista Brasileira De Ornitologia* 15(1):95–101.

Park, T.J., and R.J. Dooling. 1985. "Perception of species-specific contact calls by Budgerigars (*Melopsittacus undulatus*)." *Journal of Comparative Psychology* 99(4):391–402.

Parrado-Rosselli, Angela, Jaime Cavelier, and Arthur van Dulmen. 2002. "Effect of fruit size on primary seed dispersal of five canopy tree species of the Colombian Amazon." *Selbyana* 23(2):245–57.

Partridge, L. 2010. "The new biology of ageing." *Philosophical Transactions of the Royal Society B: Biological Sciences* 365(1537):147–54.

Patel, A.D., J.R. Iversen, M.R. Bregman, and I. Schulz. 2009a. "Experimental evidence for synchronization to a musical beat in a nonhuman animal." *Current Biology* 19(10):827–30.

———. 2009b. "Studying synchronization to a musical beat in nonhuman animals." *Neuroscience and Music III: Disorders and Plasticity* 1169:459–69.

Patterson, D.K., and I.M. Pepperberg. 1994. "A comparative study of human and parrot phonation: acoustic and articulatory correlates of vowels." *Journal of the Acoustical Society of America* 96(2):634–48.

———. 1998. "Acoustic and articulatory correlates of stop consonants in a parrot and a human subject." *Journal of the Acoustical Society of America* 103(4):2197–2215.

Pearn, S.M., A.T.D. Bennett, and I.C. Cuthill. 2001. "Ultraviolet vision, fluorescence and mate choice in a parrot, the budgerigar *Melopsittacus undulatus*." *Proceedings of the Royal Society Biological Sciences Series B* 268(1482):2273–79.

———. 2002. "The role of ultraviolet-A reflectance and ultraviolet-A induced florescence in the appearance of budgerigar plumage: insights from spectroflurometry and reflectance spectrometry." *Proceedings of the Royal Society Biological Sciences Series B* 270:859–65.

———. 2003. "The role of ultraviolet-A reflectance and ultraviolet-A-induced fluorescence in budgerigar mate choice." *Ethology* 109(12):961–70.

Pell, A.S., and C.R. Tidemann. 1997. "The impact of two exotic hollow-nesting birds on two native parrots in savannah and woodland in eastern Australia." *Biological Conservation* 79(2–3):145–53.

Pepper, J.W. 1997. "A survey of the south Australian glossy black-cockatoo (*Calyptorhynchus lathami halmaturinus*) and its habitat." *Wildlife Research* 24(2):209–23.

Pepper, J.W., T.D. Male, and G.E. Roberts. 2000. "Foraging ecology of the South Australian glossy black-cockatoo (*Calyptorhynchus lathami halmaturinus*)." *Austral Ecology* 25(1):16–24.

Pepperberg, I.M. 1981. "Functional vocalizations by an African Grey Parrot (*Psittacus erithacus*)." *Zeitschrift für Tierpsychologie* 55:139–60.

———. 1983. "Cognition in the African Grey Parrot: preliminary evidence for auditory vocal comprehension of the class concept." *Animal Learning & Behavior* 11(2):179–85.

———. 1984. "Cognition and categorization in the African Grey Parrot: auditory vocal comprehension of the class concept." *Bulletin of the British Psychological Society* 37 (February): A36.

———. 1985. "Social modeling theory: a possible framework for understanding avian vocal learning." *Auk* 102:854–64.

———. 1986. "Sensitive periods, social interaction, and song acquisition: the dialectics of dialects." *Behavioral and Brain Sciences* 9(4):756–57.

———. 1987a. "Acquisition of the same different concept by an African Gray Parrot (*Psittacus erithacus*): Learning with respect to categories of color, shape, and material." *Animal Learning & Behavior* 15(4):423–32.

———. 1987b. "Evidence for conceptual quantitative abilities in the African Gray Parrot: labeling of cardinal sets." *Ethology* 75(1):37–61.

———. 1988a. "Comprehension of absence by an African Grey Parrot: learning with respect to questions of same different." *Journal of the Experimental Analysis of Behavior* 50(3):553–64.

———. 1988b. "An interactive modeling technique for acquisition of communication skills: separation of labeling and requesting in a psittacine subject." *Applied Psycholinguistics* 9(1):59–76.

———. 1988c. "Nature nurture reflux." *Behavioral and Brain Sciences* 11(4):645–46.

———. 1988d. "Studying numerical competence: a trip through linguistic Wonderland." *Behavioral and Brain Sciences* 11(4):595–96.

———. 1990a. "Referential mapping: a technique for attaching functional significance to the innovative utterances of an African Gray Parrot (Psittacus erithacus)." *Applied Psycholinguistics* 11(1):23–44.

———. 1990b. "Cognition in an African Gray Parrot (*Psittacus erithacus*): further evidence for comprehension of categories and labels." *Journal of Comparative Psychology* 104(1):41–52.

———. 1991. "Relative size discrimination by an African Grey Parrot." *Bulletin of the Psychonomic Society* 29(6):531.

———. 1992. "Proficient performance of a conjunctive, recursive task by an African Grey Parrot (*Psittacus erithacus*)." *Journal of Comparative Psychology* 106(3):295–305.

———. 1993a. "Kommunikation zwischen Mensch und Vogel: Eine Fallstudie zu den kognitiven Fähigkeiten eines Papageis." *Zeitschrift für Semiotek* 15(1–2):41–67.

———. 1993b. "A review of the effects of social interaction on vocal learning in African grey parrots (*Psittacus erithacus*)." *Netherlands Journal of Zoology* 43(1–2):104–24.

———. 1994a. "Numerical competence in an African gray parrot (*Psittacus erithacus*)." *Journal of Comparative Psychology* 108(1):36–44.

———. 1994b. "Vocal learning in grey parrots (*Psittacus erithacus*): effect of social interaction, reference, and context." *Auk* 111(2):300–13.

———. 1997. "Social influences on the acquisition of human-based codes in parrots and nonhuman primates." In *Social influences on vocal development*, 157–177. Cambridge: Cambridge University Press.

———. 1998. "To see or not to see, that is the question: designing experiments to test perspective-taking in nonhumans." *Behavioral and Brain Sciences* 21(1):128–29.

———. 1999. *The Alex Studies: cognitive and communicative abilities of grey parrots.* Cambridge, MA: Harvard University Press.

———. 2001a. "The conundrum of correlation and causation." *Behavioral and Brain Sciences* 24(6):1073–74.

———. 2001b. "Avian cognitive abilities." *Bird Behavior* 14:51–70.

———. 2002a. "Cognitive and communicative abilities of grey parrots." *Current Directions in Psychological Science* 11(3):83–87.

———. 2002b. "The value of the Piagetian framework for comparative cognitive studies." *Animal Cognition* 5(3):177–82.

———. 2002c. "In search of King Solomon's ring: cognitive and communicative studies of Grey parrots (*Psittacus erithacus*)." *Brain Behavior and Evolution* 59(1–2):54–67.

———. 2002d. "Cognitive and communicative abilities of grey parrots." In *The Cognitive Animal: empirical and theoretical perspectives on animal cognition*, edited by Marc Bekoff, Colin Allen, and Gordon M. Burghardt, 247–253. Cambridge, MA: MIT Press.

———. 2004a. "Cognitive and communicative capacities of Grey parrots: implications for the enrichment of many species." *Animal Welfare* 13:S203–08.

———. 2004b. "'Insightful' string-pulling in Grey parrots (*Psittacus erithacus*) is affected by vocal competence." *Animal Cognition* 7(4):263–66.

———. 2005. "An avian parallel to primate mirror neurons and language evolution?" *Behavioral and Brain Sciences* 28(2):141.

———. 2006a. "Cognitive and communicative abilities of grey parrots." *Applied Animal Behaviour Science* 100(1–2):77–86.

———. 2006b. "Grey parrot (*Psittacus erithacus*) numerical abilities: addition and further experiments on a zero-like concept." *Journal of Comparative Psychology* 120(1):1–11.

———. 2006c. "Grey parrot numerical competence: a review." *Animal Cognition* 9(4):377–91.

———. 2006d. "Ordinality and inferential abilities of a grey parrot (Psittacus erithacus)." *Journal of Comparative Psychology* 120(3):205–16.

———. 2007a. "Grey parrots do not always 'parrot': the roles of imitation and phonological awareness in the creation of new labels from existing vocalizations." *Language Sciences* 29(1):1–13.

———. 2007b. "Individual differences in grey parrots (*Psittacus erithacus*): effects of training." *Journal of Ornithology* 148:S161–68.

Pepperberg, I. M., K. J. Brese, and B. J. Harris. 1991. "Solitary sound play during acquisition of English vocalizations by an African Grey parrot (*Psittacuserithacus*): possible parallels with children's monologue speech." *Applied Psycholinguistics* 12(2):151–78.

Pepperberg, I. M., and M. V. Brezinsky. 1991. "Acquisition of a relative class concept by an African Gray Parrot (*Psittacus erithacus*): discriminations based on relative size." *Journal of Comparitive Psychology* 105(3):286–94.

Pepperberg, I. M., and M. S. Funk. 1990. "Object permanence in four species of psittacine birds: an African Grey Parrot (*Psittacus erithacus*), an Illiger Mini Macaw (*Ara maracana*), a parakeet (*Melopsittacus undulatus*) and a cockatiel (*Nymphicus hollandicus*)." *Animal Learning and Behavior* 18(1):97–108.

Pepperberg, I. M., Sean E. Garcia, Eric C. Jackson, and Sharon Marconi. 1995. "Mirror use by African grey parrots (*Psittacus erithacus*)." *Journal of Comparative Psychology* 109(2):182–95.

Pepperberg, I. M., Lisa I. Gardiner, and Lori J. Luttrell. 1999. "Limited contextual vocal learning in the grey parrot (*Psittacus erithacus*): the effect of interactive co-viewers on videotaped instruction." *Journal of Comparative Psychology* 113(2):158–72.

Pepperberg, I. M., and J. D. Gordon. 2005. "Number comprehension by a grey parrot (*Psittacus erithacus*), including a zero-like concept." *Journal of Comparative Psychology* 119(2):197–209.

Pepperberg, I. M., Kirk S. Howell, Pamela Banta, A., Patterson Dianne K., and Melissa Meister. 1998. "Measurement of grey parrot (*Psittacus erithacus*) trachea via magnetic resonance imaging, dissection, and electron beam computed tomography." *Journal of Morphology* 238(1):81–91.

Pepperberg, I. M., and F. A. Kozak. 1986. "Object permanence in the African Grey Parrot (*Psittacus erithacus*)." *Animal Learning and Behavior* 14(3):322–30.

Pepperberg, I. M., and Spencer K. Lynn. 2000. "Possible levels of animal consciousness with reference to grey parrots (*Psittacus erithacus*)." *American Zoologist* 40(6):893–901.

Pepperberg, I. M., and Mary A. McLaughlin. 1996. "Effect of avian-human joint attention on allospecific vocal learning by grey parrots (*Psittacus erithacus*)." *Journal of Comparative Psychology* 110(3):286–97.

Pepperberg, I. M., James R. Naughton, and Pamela A. Banta. 1998. "Allospecific vocal learning by grey parrots (*Psittacus erithacus*): a failure of videotaped instruction under certain conditions." *Behavioural Processes* 42(2–3):139–58.

Pepperberg, I. M., and D. M. Neapolitan. 1988. "2nd language-acquisition: a framework for studying the importance of input and interaction in exceptional song acquisition." *Ethology* 77(2):150–68.

Pepperberg, I. M., Robert M. Sandefer, Dawn A. Noel, and Clare P. Ellsworth. 2000. "Vocal learning in the grey parrot (*Psittacus erithacus*): effects of species identity and number of trainers." *Journal of Comparative Psychology* 114(4):371–80.

Pepperberg, I. M., and H. R. Shive. 2001. "Simultaneous development of vocal and physical object combinations by a grey parrot (*Psittacus erithacus*): bottle caps, lids, and labels." *Journal of Comparative Psychology* 115(4):376–84.

Pepperberg, I. M., and Sarah E. Wilcox. 2000. "Evidence for a form of mutual exclusivity during label acquisition by grey parrots (*Psittacus erithacus*)?" *Journal of Comparative Psychology* 114(3):219–31.

Pepperberg, I. M., Mark R. Willner, and Lauren B. Gravitz. 1997. "Development of Piagetian object permanence in a grey parrot (*Psittacus erithacus*)." *Journal of Comparative Psychology* 111(1):63–75.

Perez, J. Juan, and Luis E. Eguiarte. 1989. "Situacion actual de tres especies del genero *Amazona* (*A. ochrocephala, A. viridigenalis* and *A. autumnalis*) en el noreste de Mexico." *Vida Sylvestre Neotropical* 2(1):63–67.

Peris, Salvador J., and Roxana M. Aramburu. 1995. "Reproductive phenology and breeding success of the Monk Parakeet (*Myiopsitta monachus monachus*) in Argentina." *Studies on Neotropical Fauna & Environment* 30(2):115–19.

Perrin, M. R. 2005. "A review of the taxonomic status and biology of the Cape Parrot *Poicephalus robustus,* with reference to the Brown-necked Parrot *P. fusciollis fusciollis* and the Grey-headed Parrot *P. f. suahelicus.*" *Ostrich* 76(3–4):195–205.

Philip, S. A., M. M. Oommen, and K. V. Baiju. 2010. "Linear sex ratio change in the clutch sequence of *Melopsittacus undulatus.*" *Current Science* 98(11):1520–23.

Pidgeon, R. 1981. "Calls of the galah *Cacatua roseicapilla* and some comparisons with four other species of Australian parrots." *Emu* 81:158–68.

Pillay, K., D. A. Dawson, G. J. Horsburgh, M. R. Perrin, T. Burke, and T. D. Taylor. 2010. "Twenty-two polymorphic microsatellite loci aimed at detecting illegal trade in the Cape Parrot, *Poicephalus robustus* (Psittacidae, Aves)." *Molecular Ecology Resources* 10(1):142–49.

Pini, E., A. Bertelli, R. Stradi, and M. Falchi. 2004. "Biological activity of parrodienes, a new class of polyunsaturated linear aldehydes similar to carotenoids." *Drugs under Experimental and Clinical Research* 30(5–6):203–06.

Pithon, J. A., and C. Dytham. 1999a. "Breeding performance of Ring-necked Parakeets *Psittacula krameri* in small introduced populations in southeast England." *Bird Study* 46:342–47.

———. 2002. "Distribution and population development of introduced Ring-necked Parakeets *Psittacula krameri* in Britain between 1983 and 1998." *Bird Study* 49:110–17.

Pithon, J.A., and C. Dytham. 1999b. "Census of the British Ring-necked Parakeet *Psittacula krameri* population by simultaneous counts of roosts." *Bird Study* 46:112–15.

Pizo, M.A., I. Simão, and M. Galetti. 1997. "Daily variation in activity and flock size of two parakeet species from southeastern Brazil." *Wilson Bulletin* 109(2):343–48.

Pizo, Maroc Aurelio, Isaac Simão, and Mauro Galetti. 1995. "Diet and flock size of sympatric parrots in the Atlantic forest of Brazil." *Ornithologia Neotropical* 6:87–95.

Plischke, Andreas, Petra Quillfeldt, Thomas Lubjuhn, Santiago Merino, and Juan Masello, F. 2010. "Leucocytes in adult burrowing parrots *Cyanoliseus patagonus* in the wild: variation between contrasting breeding seasons, gender, and individual condition." *Journal of Ornithology* 151(2):347–54.

Plummer, Thane K., and Georg F. Striedter. 2000. "Auditory responses in the vocal motor system of budgerigars." *Journal of Neurobiology* 42(1):79–94.

———. 2002. "Brain lesions that impair vocal imitation in adult budgerigars." *Journal of Neurobiology* 53(3):413–28.

Poe, S., and A.L. Chubb. 2004. "Birds in a bush: five genes indicate explosive evolution of avian orders." *Evolution* 58(2):404–15.

Pomarede, M. 1988. "Le drame des perroquets-siffleurs ou Vinis." *Le Journal Des Oiseaux* 221:7–10.

Powell, G.V.N., P. Wright, C. Guindon, I. Alemán, and R. Bjork. 1999. *Results and recommendations for the conservation of the Great Green Macaw (Ara ambigua) in Costa Rica.* San José, Costa Rica: Centro Científico Tropical.

Powell, L.L., T.U. Powell, G.V.N. Powell, and D.J. Brightsmith. 2009. "Parrots take it with a grain of salt: available sodium content may drive collpa (clay lick) selection in southeastern Peru." *Biotropica* 41(3):279–82.

Power, D.M. 1966. "Agonistic behavior and vocalizations of Orange-chinned Parakeets in captivity." *Condor* 68:562–81.

———. 1967. "Epigamic and reproductive behavior of Orange-chinned Parakeets in captivity." *Condor* 69:28–41.

Powlesland, R.G., and B.D. Lloyd. 1994. "Use of supplementary feeding to induce breeding in free-living Kakapo *Strigops habroptilus* in New Zealand." *Biological Conservation* 69(1):97–106.

Powlesland, R.G., B.D. Lloyd, H.A. Best, and D.V. Merton. 1992. "Breeding biology of the Kakapo *Strigops habroptilus* on Stewart Island, New Zealand." *Ibis* 134:361–73.

Powlesland, R.G., B.D. Lloyd, and A.D. Grant. 1986. "Breeding of Kakapo of Stewart-Island 1985." *New Zealand Journal of Ecology* 9:161.

Powlesland, R.G., D.V. Merton, and J.F. Cockrem. 2006. "A parrot apart: the natural history of the Kakapo (*Strigops habroptilus*), and the context of its conservation management." *Notornis* 53:3–26.

Powlesland, R.G., A. Roberts, B.D. Lloyd, and D.V. Merton. 1995. "Number, fate, and distribution of Kakapo (*Strigops habroptilus*) found on Stewart Island, New Zealand, 1979–92." *New Zealand Journal of Zoology* 22(3):239–48.

Pranty, Bill. 2009. "Nesting substrates of Monk Parakeets (*Myiopsitta monachus*) in Florida." *Florida Field Naturalist* 37(2):51–57.

Pranty, Bill, Daria Feinstein, and Karen Lee. 2010. "Natural history of Blue-and-yellow Macaws (*Ara ararauna*) in Miami-Dade County, Florida." *Florida Field Naturalist* 38(2):55–62.

Pratt, R.C., G.C. Gibb, M. Morgan-Richards, M.J. Phillips, M.D. Hendy, and D. Penny. 2009. "Toward resolving deep Neoaves phylogeny: data, signal enhancement, and priors." *Molecular Biology and Evolution* 26(2):313–26.

Pratt, T.K., and E.W. Stiles. 1985. "The influence of fruit size and structure on composition of frugivore assemblages in New Guinea." *Biotropica* 17(4):314–21.

Prestwich, A. A. 1966. "Records of breeding the Carolina Parakeet (*Conuropsis carolinensis*) in captivity." *Avicultural Magazine* 72:20–22.

Prum, R. O., S. Andersson, and R. H. Torres. 2003. "Coherent scattering of ultraviolet light by avian feather barbs." *Auk* 120(1):163–70.

Prum, R. O., R. Torres, S. Williamson, and J. Dyck. 1999. "Two-dimensional Fourier analysis of the spongy medullary keratin of structurally coloured feather barbs." *Proceedings of the Royal Society of London Series B: Biological Sciences* 266(1414):13–22.

Pryor, G. S., D. J. Levey, and E. S. Dierenfeld. 2001. "Protein requirements of a specialized frugivore, Pesquet's Parrot (*Psittrichas fulgidus*)." *Auk* 118(4):1080–88.

Pryor, Gregory S. 2003. "Protein requirements of three species of parrots with distinct dietary specializations." *Zoo Biology* 22(2):163–77.

Quammen, David. 1996. *The Song of the Dodo: island biogeography in an age of extinction*. New York: Scribner.

Quinn, James F., and Arthur E. Dunham. 1983. "On hypothesis testing in ecology and evolution." *American Naturalist* 122(5):602–17.

Ragusa-Netto, J. 2007. "Nectar, fleshy fruits and the abundance of parrots at a gallery forest in the southern Pantanal (Brazil)." *Studies on Neotropical Fauna and Environment* 42(2):93–99.

———. 2008. "Yellow-chevroned Parakeet (*Brotogeris chiriri*) abundance and canopy foraging at a dry forest in western Brazil." *Studies on Neotropical Fauna and Environment* 43(2):99–105.

Ramos-Fernandez, G. 2005. "Vocal communication in a fission-fusion society: do spider monkeys stay in touch with close associates?" *International Journal of Primatology* 26(5):1077–92.

Ramzan, M., and H. S. Toor. 1972. "Studies on damage to guava fruit due to Roseringed Parakeet, *Psittacula krameri* (Scopoli), at Ludhiana (Pb.)." *The Punjab Horticultural Journal* 12(2&3):144–45.

———. 1973. "Damage to maize crop by Roseringed Parakeet, *Psittacula krameri* (Scopoli) in the Punjab." *Journal, Bombay Natural Hist. Society* 70(1):201–04.

Raubenheimer, David, and Stephen J. Simpson. 2006. "The challenge of supplementary feeding: can geometric analysis help save the Kakapo?" *Nortornis* 53:100–11.

Reed, M. A., and S. C. Tidemann. 1994. "Nesting sites of the Hooded Parrot *Psephotus dissimilis* in the Northern Territory." *Emu* 94:225–29.

Reillo, Paul R., Stephen Durand, and Karen A. McGovern. 1999. "First sighting of eggs and chicks of the red-necked Amazon parrot (*Amazona arausiaca*) using an intra-cavity video probe." *Zoo Biology* 18(1):63–70.

Reiner, A., D. J. Perkel, L. L. Bruce, A. B. Butler, A. Csillag, W. Kuenzel, L. Medina, et al. 2004. "Revised nomenclature for avian telencephalon and some related brainstem nuclei." *Journal of Comparative Neurology* 473(3):377–414.

Reiner, A., D. J. Perkel, C. V. Mello, and E. D. Jarvis. 2004. "Songbirds and the revised avian brain nomenclature." In *Behavioral Neurobiology of Birdsong*, 77–108.

Remsen, J. V., Jr., Erin E. Schirtzinger, Anna Ferraroni, Luís Fábio Silveira, and Timothy F. Wright. 2013. "DNA-sequence data require revision of the parrot genus Aratinga (Aves: Psittacidae)." *Zootaxa* 3641(3):296–300.

Rendall, Drew, and Michael J. Owren. 2002. "Animal vocal communication: say what?" In *The Cognitive Animal: empirical and theoretical perspectives on animal cognition*, edited by Marc Bekoff, Colin Allen, and Gordon M. Burghardt, 307–13. Cambridge, MA: MIT Press.

Renton, K. 2001. "Lilac-crowned parrot diet and food resource availability: resource tracking by a parrot seed predator." *Condor* 103(1):62–69.

———. 2002a. "Seasonal variation in occurrence of macaws along a rainforest river." *Journal of Field Ornithology* 73(1):15–19.

———. 2002b. "Influence of environmental variability on the growth of Lilac-crowned Parrot nestlings." *Ibis* 144(2):331–39.

———. 2004. "Agonistic interactions of nesting and nonbreedlng Macaws." *Condor* 106(2):354–62.

———. 2006. "Diet of adult and nestling Scarlet Macaws in southwest Belize, Central America." *Biotropica* 38(2):280–83.

Renton, K., and D. J. Brightsmith. 2009. "Cavity use and reproductive success of nesting macaws in lowland forest of southeast Peru." *Journal of Field Ornithology* 80(1):1–8.

Renton, K., and A. Salinas-Melgoza. 1999. "Nesting behavior of the Lilac-crowned Parrot." *Wilson Bulletin* 111(4):488–93.

Ribas, C. C., R. Gaban-Lima, C. Y. Miyaki, and J. Cracraft. 2005. "Historical biogeography and diversification within the Neotropical parrot genus *Pionopsitta* (Aves: Psittacidae)." *Journal of Biogeography* 32(8):1409–27.

Ribas, C. C., L. Joseph, and C. Y. Miyaki. 2006. "Molecular systematics and patterns of diversification in *Pyrrhura* (Psittacidae), with special reference to the *Picta leucotis* complex." *Auk* 123(3):660–80.

Ribas, C. C., and C. Y. Miyaki. 2003. "Molecular systematics in *Aratinga* parakeets: species limits and historical biogeography in the 'solstitialis' group, and the systematic position of *Nandayus nenday*." *Molecular Phylogenetics and Evolution* 30(3):663–75.

Ribas, C. C., C. Y. Miyaki, and J. Cracraft. 2009. "Phylogenetic relationships, diversification and biogeography in Neotropical *Brotogeris* parakeets." *Journal of Biogeography* 36(9):1712–29.

Ribas, C. C., E. S. Tavares, C. Yoshihara, and C. Y. Miyaki. 2007. "Phylogeny and biogeography of Yellow-headed and Blue-fronted Parrots (*Amazona ochrocephala* and *Amazona aestiva*) with special reference to the South American taxa." *Ibis* 149(3):564–74.

Richardson, K. C., and R. D. Wooller. 1990. "Adaptations of the alimentary tracts of some Australian lorikeets to a diet of pollen and nectar." *Australian Journal of Zoology* 38:581–86.

Ricklefs, R. E. 2000. "Intrinsic aging-related mortality in birds." *Journal of Avian Biology* 31(2):103–11.

———. 2006. "Embryo development and ageing in birds and mammals." *Proceedings of the Royal Society B: Biological Sciences* 273(1597):2077–82.

———. 2010a. "Embryo growth rates in birds and mammals." *Functional Ecology* 24(3):588–96.

———. 2010b. "Insights from comparative analyses of aging in birds and mammals." *Aging Cell* 9(2):273–84.

———. 2010c. "Life-history connections to rates of aging in terrestrial vertebrates." *Proceedings of the National Academy of Sciences of the United States of America* 107(22):10314–19.

Ridgely, R. S. 1989. "Hyacinth Macaws in the wild." *Birds International* 1(1):8–17.

Rios-Munoz, C. A., and A. G. Navarro-Siguenza. 2009. "Effects of land use change on the hypothetical habitat availability for Mexican parrots." *Ornithologia Neotropical* 20(4):491–509.

Rivas, Jesús, and Gordon M. Burghardt. 2002. "Crotalomorphism: a metaphor for understanding anthropomorphism by omission." In *The cognitive animal: empirical and theoretical perspectives on animal cognition,* edited by Marc Bekoff, Colin Allen, and Gordon M. Burghardt, 9–17. Cambridge, MA: MIT Press.

Rivera, L., R. R. Llanos, N. Politi, B. Hennessey, and E. H. Bucher. 2010. "The near threatened Tucuman Parrot *Amazona tucumana* in Bolivia: insights for a global assessment." *Oryx* 44(1):110–13.

Rivera-Ortiz, F. A., A.M.C. Contreras-Gonzalez, C. A. S. Soberanes-Gonzalez, A. Valiente-Banuet, and M. D. Arizmendi. 2008. "Seasonal abundance and breeding chronology of the

Military Macaw (*Ara militaris*) in a semi-arid region of Central Mexico." *Ornitologia Neotropical* 19(2):255–63.

Robbins, R. L. 2000. "Vocal communication in free-ranging African wild dogs (*Lycaon pictus*)." *Behaviour* 137:1271–98.

Roberts, D. L., C. S. Elphick, and J. M. Reed. 2010. "Identifying anomalous reports of putatively extinct species and why it matters." *Conservation Biology* 24(1):189–96.

Roberts, T. F., S. E. Brauth, and W. S. Hall. 2001. "Distribution of iron in the parrot brain: conserved (pallidal) and derived (nigral) labeling patterns." *Brain Research* 921(1–2):138–49.

Roberts, T. F., K. K. Cookson, K. J. Heaton, W. S. Hall, and S. E. Brauth. 2001. "Distribution of tyrosine hydroxylase-containing neurons and fibers in the brain of the budgerigar (*Melopsittacus undulatus*): general patterns and labeling in vocal control nuclei." *Journal of Comparative Neurology* 429(3):436–54.

Roberts, T. F., W. S. Hall, and S. E. Brauth. 2000. "Chemical anatomy of the avian basal forebrain: a histochemical, and cytoarchitectural study in a parrot (*Melopsittacus undulatus*)." *Society for Neuroscience Abstracts* 26(1–2): Abstract No. 462 12.

———. 2002. "Organization of the avian basal forebrain: chemical anatomy in the parrot (*Melopsittacus undulatus*)." *Journal of Comparative Neurology* 454(4):383–408.

Robertson, B.C., G. P. Elliott, D. K. Eason, M. N. Clout, and N. J. Gemell. 2006. "Sex allocation theory aids species conservation." *Biology Letters* 2(2):229–31.

Robertson, B.C., C. D. Millar, E. O. Minot, D. V. Merton, and D. M. Lambert. 2000. "Sexing the critically endangered Kakapo *Strigops habroptilus*." 100:336–39.

Robertson, Bruce C. 2006. "The role of genetics in Kakapo recovery." *Notornis* 53:173–83.

Robinet, O., V. Bretagnolle, and M. Clout. 2003. "Activity patterns, habitat use, foraging behaviour and food selection of the Ouvea Parakeet (*Eunymphicus cornutus uvaeensis*)." *Emu* 103(1):71–80.

Robinet, O., and M. Salas. 1999. "Reproductive biology of the endangered Ouvea Parakeet *Eunymphicus cornutus uvaeensis*." *Ibis* 141(4):660–69.

Robinson, Angus. 1965. "Feeding notes on the white-tailed black cockatoo." *Western Australian Naturalist* 9:169–70.

Robinson, Tim, and David Paull. 2009. "Comparative evaluation of suburban bushland as foraging habitat for the Glossy Black-Cockatoo." *Corella* 33(1):7–12.

Rodríguez Castillo, A.M., and J. R. Eberhard. 2006. "Reproductive behavior of the Yellow-crowned Parrot (*Amazona ochrocephala*) in western Panama." *Wilson Journal of Ornithology* 118(2):225–36.

Rodriguez, J. P., F. Rojas-Suarez, and C. J. Sharpe. 2004. "Setting priorities for the conservation of Venezuela's threatened birds." *Oryx* 38(4):373–82.

Rodríguez-Estrella, R., E. Mata, and L. Rivera. 1992. "Ecological notes on the Green Parakeet of Isla Socorro, Mexico." *The Condor* 94:523–25.

Rodriguez-Ferraro, A. 2007. "Natural history and population status of the yellow-shouldered parrot on La Blanquilla Island, Venezuela." *Wilson Journal of Ornithology* 119:602–09.

Roet, E.C., D. S. Mack, and N. Duplaix. 1981. "Psittacines imported by the United States (October 1979-June 1980)." In *Conservation of New World Parrots*, edited by R. F. Pasquier, 21–56. Washington, DC: Smithsonian Institution Press.

Rogers, L. J., and H. McCulloch. 1981. "Pair-bonding in the galah *Cacatua roseicapilla*." *Bird Behaviour* 3:80–92.

Romer, L. 2000. "Management of the double-eyed or red-browed fig parrot *Cyclopsitta diophthalma macleayana* at Currumbin Sanctuary, Queensland." *International Zoo Yearbook* 37:152–58.

Roth, Paul. 1984. "Reparticao do habitat psitacideos simpatricos no sul da Amazonia." *Acta Amazonica* 14:175–221.

Roudybush, T. E., and C. R. Grau. 1986. "Food and water interrelations and the protein require-ment for growth of an altricial bird, the cockatiel (*Nymphicus hollandicus*)." *Journal of Nutrition* 116 (4):552–59.

———. 1991. "Cockatiel (*Nymphicus hollandicus*) nutrition." *Journal of Nutrition* 121:S206.

Roughgarden, J. 1983. "Competition and theory in community ecology." *American Naturalist* 122(5):583–601.

Roughgarden, J., and F. Smith. 1996. "Why fisheries collapse and what to do about it." *Proceedings of the National Academy of Sciences* 93(10):5078–83.

Rowley, I. 1980a. "Social organisation and the use of créches in the galah, *Cacatua roseicapilla*." Paper read at ACTA XVII International Ornithological Congress.

———. 1997. "Order Psittaciformes: Family Cacatuidae (Cockatoos)." In *Handbook of the birds of the world,* edited by Josep del Hoyo, Andrew Elliot, and Jordi Sargatal, 246–79. Barcelona: Lynx Editions.

Rowley, Ian. 1980b. "Parent-offspring recognition in a cockatoo, the galah, *Cacatua roseicapilla*." *Australian Journal of Zoology* 28:445–56.

———. 1983. "Mortality and dispersal of juvenile Galahs, *Cacatua roseicapilla*, in the Western Australia wheatbelt." *Australian Wildlife Research* 10:329–42.

———. 1990. *Behavioural ecology of the Galah, Eolophus roseicapillus, in the wheatbelt of Western Australia.* Chipping Norton: Surrey Beatty & Sons.

Rowley, Ian, and Graeme Chapman. 1991. "The breeding biology, food, social organisation, demography and conservation of the Major Mitchell or Pink Cockatoo,*Cacatua leadbeateri,* on the margin of the Western Australia wheatbelt." *Australian Journal of Zoology* 39:211–61.

Runde, David E., and William C. Pitt. 2007a. "Maui's Mitred Parakeets (*Aratinga mitrata*), Part 1." *'Elepaio* 68(1):1–4.

———. 2007b. "Maui's Mitred Parakeets (*Aratinga mitrata*), Part 2." *'Elepaio* 68(2):1–2.

Russello, M. A., and G. Amato. 2004. "A molecular phylogeny of *Amazona:* implications for Neotropical parrot biogeography, taxonomy, and conservation." *Molecular Phylogenetics and Evolution* 30:421–37.

Russello, M. A., C. Stahala, D. Lalonde, K. L. Schmidt, and G. Amato. 2010. "Cryptic diversity and conservation units in the Bahama Parrot." *Conservation Genetics* 11(5):1809–21.

Salaman, Paul. 1999. "New flocks of Yellow-eared Parrot *Ognorhynchus icterotis* found in Colom-bia." *Bird Conservation International* 9:284.

Salaman, Paul G. W., F. Gary Stiles, Clara Isabel Bohorquez, R. Mauricio Alvarez, Ana Maria Umana, Thomas M. Donegan, and Andres M. Cuervo. 2002. "New and noteworthy bird records from the east slope of the Andes of Colombia." *Caldasia* 24(1):157–89.

Salinas-Melgoza, A., and K. Renton. 2007. "Postfledging survival and development of juvenile lilac-crowned parrots." *Journal of Wildlife Management* 71(1):43–50.

Salinas-Melgoza, A., and T. F. Wright. 2012. "Evidence for vocal learning and limited dispersal as dual mechanisms for dialect maintenance in a parrot." *PLoS ONE* 7(11):e48667. doi:10.1371/journal.pone.0048667.

Sanchez-Martinez, T. C., and K. Renton. 2009. "Availability and selection of arboreal termitaria as nest-sites by Orange-fronted Parakeets *Aratinga canicularis* in conserved and modified landscapes in Mexico." *IBIS* 151(2):311–20.

Sandercock, B. K., and S. R. Beissinger. 2002. "Estimating rates of population change for a neotropical parrot with ratio, mark-recapture and matrix methods." *Journal of Applied Statistics* 29(1–4):589–607.

Sandercock, B. K., S. R. Beissinger, S. H. Stoleson, R. R. Melland, and C. R. Hughes. 2000. "Survival rates of a Neotropical parrot: Implications for latitudinal comparisons of avian demography." *Ecology (Washington D C)* 81(5):1351–70.

Santharam, V. 2004. "Woodpecker holes used for nesting by secondary cavity-nesters in the Western Ghats, India." *Journal of the Bombay Natural History Society* 101(1):158–59.

Santos, S. I. C. O., L. De Neve, J. T. Lumeij, and M. I. Förschler. 2007. "Strong effects of various incidence and observation angles on spectrometric assessment of plumage colouration in birds." *Behavioral Ecology and Sociobiology* 61(9):1499–1506.

Santos, S. I. C. O., B. Elward, and J. T. Lumeij. 2006. "Sexual dichromatism in the blue-fronted amazon parrot (*Amazona aestiva*) revealed by multiple-angle spectrometry." *Journal of Avian Medicine and Surgery* 20(1):8–14.

Sanvito, S., F. Galimberti, and E. H. Miller. 2007. "Observational evidences of vocal learning in southern elephant seals: a longitudinal study." *Ethology* 113(2):137–46.

Sanz, V. 2008. "Multiscale and multivariate analyses for assessing psittacine nesting susceptibility to predation: an example from Yellow-shouldered Parrots (*Amazona barbadensis*)." *Ornitologia Neotropical* 19:123–34.

Sanz, V., and A. Grajal. 1998. "Successful reintroduction of captive-raised yellow-shouldered Amazon parrots on Margarita Island, Venezuela." *Conservation Biology* 12(2):430–41.

Sanz, V., and A. Rodriguez-Ferraro. 2006. "Reproductive parameters and productivity of the Yellow-shouldered Parrot on Margarita Island, Venezuela: a long-term study." *Condor* 108(1):178–92.

Sanz, V., A. Rodriguez-Ferraro, M. Albornoz, and C. Bertsch. 2003. "Use of artificial nests by the Yellow-shouldered Parrot (*Amazona barbadensis*)." *Ornitologia Neotropical* 14(3):345–51.

Saunders, D. A. 1977a. "Breeding of the Long-billed Corella at Coomallo Creek, WA." *Emu* 77:223–27.

———. 1977b. "The effect of agricultural clearing on the breeding success of the White-tailed Black Cockatoo." *Emu* 77:180–84.

———. 1979. "The availability of tree hollows for use as nest sites by White-tailed Black Cockatoos." *Australian Wildlife Research* 6:205–16.

———. 1980. "Food and movements of the short-billed form of the White-tailed Black Cockatoo." *Australian Wildlife Research* 7:257–69.

———. 1982. "The breeding behaviour and biology of the short-billed form of the White-tailed Black Cockatoo *Calyptorhynchus funereus*." *Ibis* 124(4):422–55.

———. 1983. "Vocal repertoire and individual vocal recognition in the Short-billed White-tailed Black Cockatoo, *Calyptorhynchus funereus latirostris* Carnaby." *Australian Wildlife Research* 10:527–36.

———. 1986. "Breeding season, nesting success, and nestling growth in Carnaby's Cockatoo, *Calyptorhynchus funereus latirostris*, over 16 years at Coomallo Creek, and a method for assessing the viability of populations in other areas." *Australian Wildlife Research* 13:261–73.

———. 1990. "Problems of survival in an extensively cultivated landscape: the case of Carnaby's Cockatoo *Calyptorhynchus funereus latirostris*." *Biological Conservation* 54:277–90.

Saunders, D. A., and R. Dawson. 2009. "Update on longevity and movements of Carnaby's Black Cockatoo." *Pacific Conservation Biology* 15(1):72–74.

Saunders, D. A, and J. A. Ingram. 1987. "Factors affecting survival of breeding populations of Carnaby's cockatoo *Calyptorhynchus funerus latirostris* in remnants of native vegetation." In *Nature Conservation: the role of remnants of native vegetation.*, edited by D. A. Saunders, G. W. Arnold, A. A. Burbidge, and A. J. M. Hopkins, 249–58. Chipping Norton, NSW: Surrey Beatty and Sons with CSIRO & CALM.

Saunders, D. A., and G. T. Smith. 1981. "Egg dimensions and egg weight loss during incubation in five species of cockatoo, and the use of measurements to determine the stage of incubation of birds' eggs." *Australian Wildlife Research* 8:411–19.

Saunders, D. A., G. T. Smith, and N. A. Campbell. 1984. "The relationship between body weight, egg weight, incubation period, nestling period and nest site in the Psittaciformes, Falconiformes, Strigiformes and Columbiformes." *Australian Journal of Zoology* 32:57–65.

Saunders, D. A., G. T. Smith, and I. Rowley. 1982. "The availability and dimensions of tree hollows that provide nest sites for cockatoos (Psittaciformes) in western Australia." *Australian Wildlife Research* 9:541–56.

Saunders, Debra L., and Robert Heinsohn. 2008. "Winter habitat use by the endangered, migratory Swift Parrot (*Lathamus discolor*) in New South Wales." *Emu* 108(1):81–89.

Sazima, I. 1989. "Peach-fronted Parakeet feeding on winged termites." *Wilson Bulletin* 101(4):656–57.

———. 2008. "The parakeet *Brotogeris tirica* feeds on and disperses the fruits of the palm *Syagrus romanzoffiana* in southeastern Brazil." *Biota Neotropica* 8:231–34.

Scarl, J. C. 2009. "Heightened responsiveness to female-initiated displays in an Australian cockatoo, the Galah (*Eolophus roseicapillus*)." *Behaviour* 146:1313–30.

Scarl, J. C. , and J. W. Bradbury. 2009. "Rapid vocal convergence in an Australian cockatoo, the Galah (*Eolophus roseicapillus*)." *Animal Behaviour* 77:1019–26.

Schachner, Adena, Timothy F. Brady, Irene M. Pepperberg, and Marc D. Hauser. 2009. "Spontaneous motor entrainment to music in multiple vocal mimicking species." *Current Biology* 19(10):831–36. doi:10.1016/j.cub.2009.03.061.

Schaefer, H. M., V. Schmidt, and F. Bairlein. 2003. "Discrimination abilities for nutrients: which difference matters for choosy birds and why?" *Animal Behaviour* 65:531–41.

Schaefer, H. M., V. Schmidt, and H. Winkler. 2003. "Testing the defence trade-off hypothesis: how contents of nutrients and secondary compounds affect fruit removal." *Oikos* 102(2):318–28.

Scharff, C., and S. Haesler. 2005. "An evolutionary perspective on FoxP2: strictly for the birds?" *Current Opinion in Neurobiology* 15(6):694–703.

Scharff, C., and S. A. White. 2004. "Genetic components of vocal learning." In *Behavioral Neurobiology of Birdsong*, edited by H. Philip Zeigler and Peter Marler, 325–47. Annals of the New York Academy of Sciencs.

Schirtzinger, Erin E., Erika S. Tavares, Lauren A. Gonzales, Jessica R. Eberhard, Cristina Y. Miyaki, Juan J. Sanchez, Alexis Hernandez, et al. 2012. "Multiple independent origins of mitochondrial control region duplications in the order Psittaciformes." *Molecular Phylogenetics and Evolution* 64(2):342–56. doi:10.1016/j.ympev.2012.04.009.

Schnell, G. D., J. S. Weske, and J. J. Hellack. 1974. "Recent observations of Thick-billed Parrots in Jalisco." *Wilson Bulletin* 86(4):464–65.

Schodde, R. J., Jr., V. Remsen, E. E. Schirtzinger, L. Joseph, and T. F. Wright. 2013. "Higher classification of New World parrots (Psittaciformes; Arinae), with diagnoses of tribes." *Zootaxa* 3691(5):591–96.

Schubart, O., A. C. Aguirre, and H. Sick. 1965. "Contribuicao para o conhecimento da alimentacao das aves brasileiras." *Arquivos de Zoologia S. Paulo* 12:95–249.

Schuck-Paim, C., W. J. Alonso, and E. B. Ottoni. 2008 "Cognition in an ever-changing world: climatic variability is associated with brain size in Neotropical parrots." *Brain, Behavior and Evolution* 71(3):200–15.

Schuck-Paim, C., A. Borsari, and E. B. Ottoni. 2009. "Means to an end: Neotropical parrots manage to pull strings to meet their goals." *Animal Cognition* 12(2):287–301.

Schulte, P., L. Alegret, I. Arenillas, J. A. Arz, P. J. Barton, P. R. Bown, T. J. Bralower, et al. 2010. "The Chicxulub asteroid impact and mass extinction at the Cretaceous-Paleogene boundary." *Science* 327(5970):1214–18.

Schweizer, M., M. Guentert, and S. T. Hertwig. 2013. "Out of the Bassian province: historical biogeography of the Australasian platycercine parrots (Aves, Psittaciformes)." *Zoologica Scripta* 42(1):13–27. doi:10.1111/j.1463–6409.2012.00561.x.

Schweizer, M., S. T. Hertwig, and O. Seehausen. 2014. "Diversity versus disparity and the role of ecological opportunity in a continental bird radiation." *Journal of Biogeography* 41(7):1301–12. doi:10.1111/jbi.12293.

Schweizer, M., O. Seehausen, M. Guntart, and S. T. Hertwig. 2010. "The evolutionary diversification of parrots supports a taxon pulse model with multiple trans-oceanic dispersal events and other translocations." *Molecular Phylogenetics and Evolution* 54:984–94.

Schweizer, M., O. Seehausen, and S. T. Hertwig. 2011. "Macroevolutionary patterns in the diversification of parrots: effects of climate change, geological events and key innovations." *Journal of Biogeography* 38:2176–94. doi:10.1111/j.1365–2699.2011.02555.x.

Scott, John K., and Robert Black. 1981. "Selective predation by White-tailed Black Cockatoos on fruit of *Banksia attenuata* containing the seed-eating weevil *Alphitopis nives*." *Australian Wildlife Research* 8:421–30.

Searby, A., and P. Jouventin. 2004. "How to measure information carried by a modulated vocal signature?" *Journal of the Acoustical Society of America* 116(5):3192–98.

Searby, A., P. Jouventin, and T. Aubin. 2004. "Acoustic recognition in macaroni penguins: an original signature system." *Animal Behaviour* 67:615–25.

Seed, A., N. Emery, and N. Clayton. 2009. "Intelligence in corvids and apes: a case of convergent evolution?" *Ethology* 115(5):401–20.

Seibels, R. E., and W. E. McCullough. 1976. "Polygamy in the Sun Conure at Columbia Zoo." *International Zoological Yearbook* 18:105–06.

Selman, R. G., M. L. Hunter, and M. R. Perrin. 2000. "Ruppell's Parrot: status, ecology and conservation biology." *Ostrich* 71(1–2):347–48.

Selman, R. G., M. R. Perrin, and M. L. Hunter. 2002. "The feeding ecology of Ruppell's parrot, *Poicephalus rueppellii*, in the Waterberg, Namibia." *Ostrich* 73(3–4):127–34.

Serpell, J. A. 1982. "Factors influencing fighting and threat in the parrot genus *Trichoglossus*." *Animal Behaviour* 30:1244–51.

———. 1989. "Visual displays and taxonomic affinities in the parrot genus *Trichoglossus*." *Biological Journal of the Linnean Society* 36:193–211.

Serpell, James. 1981. "Duets, greetings and triumph ceremonies: analogous displays in the parrot genus *Trichoglossus*." *Zeitschrift für Tierpsychologie* 55:268–83.

Sewall, K. B. 2009. "Limited adult vocal learning maintains call dialects but permits pair-distinctive calls in red crossbills." *Animal Behaviour* 77(5):1303–11.

Shepherd, J. D., R. A. Ditgen, and J. Sanguinetti. 2008. "*Araucaria araucana* and the Austral parakeet: pre-dispersal seed predation on a masting species." *Revista Chilena De Historia Natural* 81(3):395–401.

Shieh, Bao-Sen, Ya-Hui Lin, Tsung-Wei Lee, Chia-Chieh Chang, and Kuan-Tzou Cheng. 2006. "Pet trade as sources of introduced bird species in Taiwan." *Taiwania* 51(2):81–86.

Shields, W. M., T. C. Grubb Jr., and A. Telis. 1974. "Use of native plants by Monk Parakeets in New Jersey." *Wilson Bulletin* 86:172–73.

Shirley, S. M., and S. Kark. 2006. "Amassing efforts against alien invasive species in Europe." *PLOS Biology* 4:1311–13.

———. 2009. "The role of species traits and taxonomic patterns in alien bird impacts." *Global Ecology and Biogeography* 18(4):450–59.

Shwartz, A., D. Strubbe, C. J. Butler, E. Matthysen, and S. Kark. 2009. "The effect of enemy-release and climate conditions on invasive birds: a regional test using the Rose-ringed Parakeet (*Psittacula krameri*) as a case study." *Diversity and Distributions* 15(2):310–18.

Shwartz, Assaf, Susan Shirley, and Salit Kark. 2008. "How do habitat variability and management regime shape the spatial heterogeneity of birds within a large Mediterranean urban park?" *Landscape and Urban Planning* 84:219–29.

Sibley, Charles G., and Jon E. Ahlquist. 1990. *Phylogeny and Classification of Birds*. New Haven, CT: Yale University Press.

Sick, H., and D. M. Teixeira. 1980. "Discovery of the home of the Indigo Macaw in Brazil." *American Birds* 34(2):118–19, 212.

Siegel, Rodney B., Wesley W. Weathers, and Steven R. Beissinger. 1999a. "Assessing parental effort in a Neotropical parrot: A comparison of methods." *Animal Behaviour* 57(1):73–79.

———. 1999b. "Hatching asynchrony reduces the duration, not the magnitude, of peak load in breeding green-rumped parrotlets (*Forpus passerinus*)." *Behavioral Ecology and Sociobiology* 45:444–50.

Silvius, K. M. 1995. "Avian consumers of cardon fruits (*Stenocereus griseus*, Cactaceae) on Margarita Island, Venezuela." *Biotropica* 27(1):96–105.

Simão, Isaac, Flavio Antonio Maës dos Santos, and Marco Aurelio Pizo. 1997. "Vertical stratification and diet of psittacids in a tropical lowland forest of Brazil." *Ararajuba* 5(2):169–74.

Simon, L. 1990. "New York's crusade for exotic birds." *Defenders* 65(6):26–31.

Simpson, M. B., Jr., and R. C. Ruiz. 1974. "Monk Parakeets breeding in Bumcombe County, North Carolina." *Wilson Bulletin* 86:171–72.

Simpson, K. G. 1972. "Feeding of the yellow-tailed black cockatoo on cossid moth larvae inhabiting *Acacia* species." *The Victorian Naturalist* 89:32–40.

Singer, M. S., E. A. Bernays, and Y. Carriere. 2002. "The interplay between nutrient balancing and toxin dilution in foraging by a generalist insect herbivore." *Animal Behaviour* 64:629–43.

Skead, C. J. 1964. "The overland flights and the feeding habits of the Cape Parrot *Poicephalusrobustus* (Gmelin) in the eastern Cape Province." *Ostrich* 35:202–23.

———. 1971. "The Cape Parrot in the Transkei and Natal." *Ostrich Supplement* 9:165–78.

Skeate, S. T. 1984. "Courtship and reproductive behaviour of captive White-fronted Amazon parrots *Amazona albifrons*." *Bird Behaviour* 5:103–09.

———. 1985. "Social play behaviour in captive White-fronted Amazon Parrots *Amazona albifrons*." *Bird Behaviour* 6:46–48.

Slack, K. E., F. Delsuc, P. A. McLenachan, U. Arnason, and D. Penny. 2007. "Resolving the root of the avian mitogenomic tree by breaking up long branches." *Molecular Phylogenetics and Evolution* 42(1):1–13.

Smales, I., P. Brown, P. Menkhorst, M. Holdsworth, and P. Holz. 2000. "Contribution of captive management of Orange-bellied Parrots *Neophema chrysogaster* to the recovery programme for the species in Australia." *International Zoo Yearbook* 37:171–78.

Smiet, F. 1985. "Notes on the field status and trade of Moluccan parrots." *Biological Conservation* 34:181–94.

Smith, G. A. 1975. "Systematics of Parrots." *Ibis* 117:18–117.

Smith, G. T. 1991. "Breeding ecology of the western long-billed corella, *Cacatua pastinator pastinator*." *Wildlife Research* 18:91–110.

Smith, G. T., and L. A. Moore. 1991. "Foods of corellas *Cacatua pastinator* in western Australia." *Emu* 91:87–92.

———. 1992. "Patterns of movement in the Western Long-billed Corella *Cacatua pastinator* in the south-west of Western Australia." *Emu* 92:19–27.

Smith, G. T., and I. C. R. Rowley. 1995. "Survival of adult and nestling western long-billed corellas, *Cacatua pastinator,* and Major Mitchell cockatoos, *C. leadbeateri,* in the wheatbelt of western Australia." *Wildlife Research* 22:155–62.

Smith, J., and A. Lill. 2008. "Importance of eucalypts in exploitation of urban parks by Rainbow and Musk Lorikeets." *Emu* 108(2):187–95.

Smith, Meredith J., and J. Le Gay Brereton. 1976. "Annual gonadal and adrenal cycles in the Eastern Rosella, *Platycercus eximius* (Psittaciformes: Platycercidae)." *Australian Journal of Zoology* 24:541–46.

Snyder, N. F. R. 2004. *The Carolina Parakeet: glimpses of a vanished bird.* Princeton, New Jersey: Princeton University Press.

Snyder, N. F. R., E. C. Enkerlin-Hoeflich, and M. A. Cruz-Nieto. 1999. "The Thick-billed Parrot (*Rhynchopsitta pachyrhyncha*)." In *The Birds of America Online,* edited by A. Poole. Ithaca: Cornell Lab of Ornithology.

Snyder, N. F. R., S. Koenig, and T. B. Johnson. 1995. "Ecological relationships of the Thick-billed Parrot with the pine-forests of southeastern Arizona." In *Biodiversity and Management of the Madrean Archipelago,* edited by F. DeBano, P. Foillott, A. Ortega-Rubio, G. J. Gottfried, R. H. Hamre, and C. B. Edminster, 288–93. Fort Collins, CO: U.S. Department of Agriculture, Forest Service, Rocky Mountain Forest and Range Experiment Station.

Snyder, N. F. R., S. E. Koenig, J. Koschmann, H. A. Snyder, and T. B. Johnson. 1994. "Thick-billed Parrot releases in Arizona." *Condor* 96(4):845–62.

Snyder, N. F. R., P. McGowan, J. Gilardi, and A. Grajal. 2000. *Parrots: status survey and conservation action plan 2000–2004.* Cambridge: IUCN.

Snyder, N. F. R., and K. Russell. 2002. "Carolina Parakeet (*Conuropsis carolinensis*)." In *The Birds of North America,* edited by A. Poole and F. Gill. Philadelphia, PA: Birds of North America.

Snyder, N. F. R., and J. D. Taapken. 1978. "Puerto Rican Parrots and nest predation by Pearly-eyed Thrashers." In *Endangered Birds: management techniques for preserving threatened species,* edited by S. A. Temple, 113–20. Madison: University of Wisconsin Press.

Snyder, N. F. R., J. W. Wiley, and C. B. Kepler. 1987. *The Parrots of Luquillo: natural history and conservation of the Puerto Rican Parrot.* Camarillo, CA: Western Foundation of Vertebrate Zoology.

Sodhi, Navjot S., and Kazuhiro Eguchi. 2004. "Invasive bird species: introduction." *Ornithological Science* 3(1):1–2.

Sol, D., D. M. Santos, E. Feria, and J. Clavell. 1997. "Habitat selection by the Monk Parakeet during colonization of a new area in Spain." *Condor* 99(1):39–46.

Soltis, J., K. Leong, and A. Savage. 2005. "African elephant vocal communication II: rumble variation reflects the individual identity and emotional state of callers." *Animal Behaviour* 70:589–99.

South, J. M., and S. Pruett-Jones. 2000. "Patterns of flock size, diet, and vigilance of naturalized Monk Parakeets in Hyde Park, Chicago." *Condor* 102(4):848–54.

Spano, S., and G. Truffi. 1986. "Records of free-living individuals of Rose-ringed Parakeet *Psittacula krameri* from Europe with particular reference to presences in Italy and first data on Monk Parakeet *Myiopsitta monachus*." *Rivista Italiana di Ornitologia* 56:231–39.

Spoon, T. R., J. R. Millam, and D. H. Owings. 2004. "Variation in the stability of cockatiel (*Nymphicus hollandicus*) pair relationships: the roles of males, females, and mate compatibility." *Behaviour* 141:1211–34.

———. 2006. "The importance of mate behavioural compatibility in parenting and reproductive success by cockatiels, *Nymphicus hollandicus*." *Animal Behaviour* 71:315–26.

———. 2007. "Behavioural compatibility, extrapair copulation and mate switching in a socially monogamous parrot." *Animal Behaviour* 73:815–24.

Stahala, C. 2008. "Seasonal movements of the Bahama Parrot (*Amazona leucocephala bahamensis*) between pine and hardwood forests: implications for habitat conservation." *Ornitologia Neotropical* 19:165–71.

Stamps, J. 1990. "When should avian parents differentially provision sons and daughters?" *American Naturalist* 135:671–85.

Stamps, Judy, Anne Clark, Pat Arrowood, and Barbara Kus. 1985. "Parent-offspring conflict in budgerigars." *Behaviour* 94:1–40.

———. 1989. "Begging behavior in budgerigars." *Ethology* 81:177–92.

Stamps, Judy, Anne Clark, Barbara Kus, and Pat Arrowood. 1987. "The effects of parent and offspring gender on food allocation in budgerigars." *Behaviour* 101:177–99.

Stamps, Judy, Barbara Kus, Anne Clark, and Patricia Arrowood. 1990. "Social relationships of fledgling budgerigars, *Melopsitticus undulatus*." *Animal Behaviour* 40:688–700.

Stattersfield, Alison J., and David R. Capper. 2000. *Threatened birds of the world: the official source for birds on the IUCN Red List.* Cambridge: BirdLife International.

Steadman, D. W. 1993. "Biogeography of Tongan birds before and after human impact." *Proc. Natl. Acad. Sci.* 90:818–22.

———. 2006a. "A new species of extinct parrot (Psittacidae: *Eclectus*) from Tonga and Vanuatu, South Pacific." *Pacific Science* 60(1):137–45.

———. 2006b. *Extinction and Biogeography of Tropical Pacific birds.* Chicago, IL: University of Chicago Press.Steadman, D. W., and P. V. Kirch. 1990. "Prehistoric extinction of birds on Mangaia, Cook Islands, Polynesia." *Proceedings of the National Academy of Sciences USA* 87:9605–09.

———. 1998. "Biogeography and prehistoric exploitation of birds in the Mussau Islands, Bismarck Archipelago, Papua New Guinea." *Emu* 98(1):13–22. doi:10.1071/MU98002.

Steadman, D. W., and M. C. Zarriello. 1987. "Two new species of parrots (Aves: Psittacidae) from archeological sites in the Marquesas Islands." *Proceedings of the Biological Society of Washington.* 100(3):518–28.

Stidham, T. A. 1999. "Did parrots exist in the Cretaceous period? Reply." *Nature* 399(6734):318.

———. 2009. "A lovebird (Psittaciformes: *Agapornis*) from the Plio-Pleistocene Kromdraai B locality, South Africa." *South African Journal of Science* 105(3–4):155–57.

Stock, M. J., and C. H. Wild. 2005. "Seasonal variation in Glossy Black-Cockatoo *Calyptorhynchus lathami* sightings on the Gold Coast, Queensland." *Corella* 29:88–89.

Stoleson, Scott H., and Steven R. Beissinger. 1997. "Hatching asynchrony, brood reduction, and food limitation in a Neotropical parrot." *Ecological Monographs* 67(2):131–54.

———. 1999. "Egg viability as a constraint on hatching synchrony at high ambient temperatures." *Journal of Animal Ecology* 68(5):951–62.

———. 2001. "Does risk of nest failure or adult predation influence hatching patterns of the Green-rumped Parrotlet?" *Condor* 103(1):85–97.

Stradi, R., E. Pini, and G. Celentano. 2001. "The chemical structure of the pigments in *Ara macao* plumage." *Comparative Biochemistry & Physiology. Part B, Biochemistry & Molecular Biology* 130B(1):57–63.

Strahl, S.D., P.A. Desenne., J.L. Jimenez, and I.R. Goldstein. 1991. "Behavior and biology of the Hawk-headed Parrot *Deroptyus accipitrinus* in southern Venezuela." *Condor* 93:177–80.

Striedter, G. F. 1994. "The vocal control pathways in Budgerigars differ from those in songbirds." *Journal of Comparative Neurology* 343(1):35–56.

Striedter, G. F., and C. J. Charvet. 2008. "Developmental origins of species differences in telencephalon and tectum size: morphometric comparisons between a parakeet (*Melopsittacus undulatus*) and a quail (*Colinus virgianus*)." *Journal of Comparative Neurology* 507(5):1663–75.

Striedter, G. F., L. Freibott, A. G. Hile, and N. T. Burley. 2003. "For whom the male calls: an effect of audience on contact call rate and repertoire in budgerigars, *Melopsittacus undulatus*." *Animal Behaviour* 65:875–82.

Strubbe, D., and E. Matthysen. 2009a. "Establishment success of invasive Ring-necked and Monk Parakeets in Europe." *Journal of Biogeography* 36(12):2264–78.

———. 2009b. "Experimental evidence for nest-site competition between invasive Ring-necked Parakeets (*Psittacula krameri*) and native Nuthatches (*Sitta europaea*)." *Biological Conservation* 142(8):1588–94.

———. 2009c. "Predicting the potential distribution of invasive Ring-necked Parakeets *Psittacula krameri* in northern Belgium using an ecological niche modelling approach." *Biological Invasions* 11(3):497–513.

Strubbe, D., E. Matthysen, and C. H. Graham. 2010. "Assessing the potential impact of invasive Ring-necked Parakeets *Psittacula krameri* on native Nuthatches *Sitta europeae* in Belgium." *Journal of Applied Ecology* 47(3):549–57.

Suh, Alexander, Martin Paus, Martin Kiefmann, Gennady Churakov, Franziska Anni Franke, Juergen Brosius, Jan Ole Kriegs, and Juergen Schmitz. 2011. "Mesozoic retroposons reveal parrots as the closest living relatives of passerine birds." *Nature Communications* 2:443. doi:10.1038/ncomms1448.

Summers, K., and B. J. Crespi. 2010. "Xmrks the spot: life history tradeoffs, sexual selection and the evolutionary ecology of oncogenesis." *Molecular Ecology* 19(15):3022–24.

Sutherland, W. J. 2002. "Conservation biology: science, sex and the Kakapo." 419(6904):265–66.

Swaisgood, R. R. 2007. "Current status and future directions of applied behavioral research for animal welfare and conservation." *Applied Animal Behaviour Science* 102(3–4):139–62.

Symes, C., M. Brown, L. Warburton, M. Perrin, and C. Downs. 2004. "Observations of Cape Parrot, *Poicephagus robustus*, nesting in the wild." *Ostrich* 75(3):106–109.

Symes, C. T., and S. J. Marsden. 2007. "Patterns of supra-canopy flight by pigeons and parrots at a hill-forest site in Papua New Guinea." *Emu* 107(2):115–125.

Symes, Craig T., and Michael R. Perrin. 2003a. "Seasonal occurrence and local movements of the grey-headed (brown-necked) parrot *Poicephalus fuscicollis suahelicus* in southern Africa." 41(4):299–305.

———. 2003b. "Feeding biology of the Greyheaded Parrot, *Poicephalus fuscicollis suahelicus* (Reichenow), in Northern Province, South Africa." *Emu* 103(1):49–58.

———. 2004a. "Behaviour and some vocalisations of the Grey-headed Parrot *Poicephalus fuscicollis suehelicus* (Psittaciformes: Psittacidae) in the wild." *Durban Museum Novitates* 29:5–13.

———. 2004b. "Breeding biology of the Greyheaded Parrot (*Poicephalus fuscicollis suahelicus*) in the wild." *Emu* 104(1):45–57.

Szamado, S., and E. Szathmary. 2006. "Selective scenarios for the emergence of natural language." *Trends in Ecology & Evolution* 21(10):555–61.

Tavares, E. S., A. J. Baker, S. L. Pereira, and C. Y. Miyaki. 2006. "Phylogenetic relationships and historical biogeography of Neotropical parrots (Psittaciformes : Psittacidae : Arini) inferred from mitochondrial and nuclear DNA sequences." *Systematic Biology* 55(3):454–70.

Tavares, E. S., C. Yamashita, and C. Y. Miyaki. 2004. "Phylogenetic relationships among some Neotropical parrot genera (Psittacidae) based on mitochondrial sequences." *Auk* 121(1):230–42.

Taylor, S., and M. R. Perrin. 2004. "Intraspecific associations of individual Brown-Headed Parrots (*Poicephalus cryptoxanthus*)." *African Zoology* 39(2):263–71.

———. 2005. "Vocalisations of the Brown-headed Parrot, *Poicephalus cryptoxanthus:* their general form and behavioural context." *Ostrich* 76(1–2):61–72.

———. 2006a. "Aspects of the breeding biology of the Brown-headed parrot *Poicephalus cryptoxanthus* in South Africa." *Ostrich* 77(3–4):225–28.

———. 2006b. "The diet of the Brown-headed Parrot (*Poicephalus cryptoxanthus*) in the wild in southern Africa." *Ostrich* 77(3–4):179–85.

———. 2008a. "Adaptive hatching hypotheses do not explain asynchronous hatching in Brown-headed Parrots *Poicephalus cryptoxanthus.*" *Ostrich* 79(2):205–09.

———. 2008b. "Parent-offspring recognition in the Brown-headed Parrot *Poicephalus cryptoxanthus.*" *Ostrich* 79(2):211–14.

Taylor, T. D., and D. T. Parkin. 2008. "Sex ratios observed in 80 species of parrots." *Journal of Zoology* 276(1):89–94.

———. 2009. "Preliminary evidence suggests extra-pair mating in the endangered echo parakeet, Psittacula eques." *African Zoology* 44(1):71–74.

Tella, J.L. 2001. "Sex-ratio theory in conservation biology." *Trends in Ecology and Evolution* 16:76–77.

Temby, I. D., and W. B. Emison. 1986. "Foods of the Long-billed Corella *Cacatua tenuirostris.*" *Australian Wildlife Research* 13(1):57–64.

Terborgh, J. 1992. "Maintenance of diversity in tropical forests." *Biotropica* 24(2):283–92.

Terborgh, J., S. K. Robinson, T. A. Parker III, C. A. Munn, and N. Pierpont. 1990. "Structure and organization of an Amazonian forest bird community." *Ecological Monographs* 60(2):213–38.

Terman, A., and U. T. Brunk. 2006. "Oxidative stress, accumulation of biological 'garbage', and aging." *Antioxidants & Redox Signalling* 8(1–2):197–204.

Theuerkauf, J., S. Rouys, J. M. Meriot, R. Gula, and R. Kuehn. 2009. "Cooperative breeding, mate guarding, and nest sharing in two parrot species of New Caledonia." *Journal of Ornithology* 150(4):791–97.

Thompson, J. J. 1990. "Inferences of breeding patterns from moult data of lovebirds *Agapornis* spp. at Lake Naivasha, Kenya." *Scopus* 14(1):1–5.

Thompson, J. J. , and W. K. Karanja. 1989. "Interspecific competition for nest cavities by introduced lovebirds *Agapornis* spp. at Lake Naivasha, Kenya." *Scopus* 12:73–78.

Tinbergen, N. 2005. "On aims and methods of Ethology (Reprinted from Zeitschrift für Tierpsychologie, vol 20, pg 410, 1963)." *Animal Biology* 55(4):297–321.

Toft, Catherine A., and Patrick J. Shea. 1983. "Detecting community-wide patterns: estimating power strengthens statistical inference." *American Naturalist* 122(5):618–25.

Tokita, M. 2003. "The skull development of parrots with special reference to the emergence of a morphologically unique cranio-facial hinge." *Zoological Science* 20(6):749–58.

Tokita, M., T. Kiyoshi, and K. N. Armstrong. 2007. "Evolution of craniofacial novelty in parrots through developmental modularity and heterochrony." *Evolution & Development* 9(6):590–601.

Toor, H. S., and M. Ramzan. 1974. "The extent of losses to sunflower due to Rose-ringed Parakeet, *Psittacula krameri,*(Scopoli) at Ludhiana (Punjab)." *Journal of Research: Punjab Agricultural University* 11(2):197–99.

Toral, G. M. 2008. "Multiple ways to become red: pigment identification in red feathers using spectrometry." *Comparative Biochemistry and Physiology B: Biochemistry & Molecular Biology* 150(2):147–52.

Tovar-Martinez, Adriana Elizabeth. 2009a. "Growth and plumage development of Azure-winged Parrot (*Hapalopsittaca fuertesi*) nestlings in the Central Andes of Colombia." *Ornitologia Colombiana* 8:5–18.

———. 2009b. "Reproductive parameters and nesting of Indigo-winged Parrot (*Hapalopsittaca fuertesi*) in artificial cavities." *Ornitholoia Neotropical* 20(3):357–68.

Towns, D. R., and M. Williams. 1993. "Single-species conservation in New Zealand: towards a redefined conceptual approach." *Journal of the Royal Society of New Zealand* 23(2):61–78.

Toyne, E. P., and J. N. M. Flanagan. 1997. "Observations on the breeding, diet and behaviour of the Red-faced Parrot *Hapalopsittaca pyrrhops* in southern Ecuador." *Bulletin of the British Ornithologists' Club* 117(4):257–63.

Toyne, E. P., J. N. M. Flanagan, and M.T. Jeffcote. 1995. "Vocalizations of the endangered red-faced parrot *Hapalopsittaca pyrrhops* in southern Ecuador." *Ornitologia Neotropical* 6:125–28.

Traveset, A., J. Rodriguez-Perez, and B. Pias. 2008. "Seed trait changes in dispersers' guts and consequences for germination and seedling growth." *Ecology* 89(1):95–106.

Trewick, S. A. 1997. "On the skewed sex ratio of the Kakapo *Strigops habroptilus*: sexual and natural selection in opposition?" *Ibis* 139(4):652–63.

Trewick, Steve. 1996. "The diet of Kakapo (*Strigops habroptilus*), Takahe (*Porphyrio mantelli*) and Pukeko (*P. porphyrio melanotus*) studied by faecal analysis." *Notornis* 43:79–84.

Triggs, S. J., R. G. Powlesland, and C. H. Daugherty. 1989. "Genetic variation and conservation of Kakapo (*Strigopshabroptilus*: Psittaciformes)." *Conservation Biology* 3(1):92–96.

Trillmich, Fritz. 1976a. "Learning experiments on individual recognition in budgerigars (*Melopsittacus undulatus*)." *Zeitschrift für Tierpsychologie* 41:372–95.

———. 1976b. "Spatial proximity and mate-specific behaviour in a flock of budgerigars (*Melopsittacus undulatus*; Aves Psittacidae)." *Zeitschrift für Tierpsychologie* 41:307 31.

Trivedi, M. R., F. H. Cornejo, and A. R. Watkinson. 2004. "Seed predation on Brazil nuts (*Bertholletia excelsa*) by macaws (Psittacidae) in Madre de Dios, Peru." 36(1):118–22.

Tsahar, E., Z. Ara, I. Izhaki, and C. M. Del Rio. 2006. "Do nectar- and fruit-eating birds have lower nitrogen requirements than omnivores? An allometric test." *Auk* 123(4):1004–12.

Tsahar, E., C. M. del Rio, I. Izhaki, and Z. Arad. 2005. "Can birds be ammonotelic? Nitrogen balance and excretion in two frugivores." *Journal of Experimental Biology* 208(6): 1025–34.

Tschudin, A., H. Rettmer, R. Watson, M. Clauss, and S. Hammer. 2010. "Evaluation of hand-rearing records for Spix's Macaw *Cyanopsitta spixii* at the Al Wabra Wildlife Preservation from 2005 to 2007." *International Zoo Yearbook* 44:201–11.

Tyack, PL. 2008. "Convergence of calls as animals form social bonds, active compensation for noisy communication channels, and the evolution of vocal learning in mammals." *Journal of Comparative Psychology* 122(3):319–31.

Valdes-Pena, R. A., S. G. Ortiz-Maciel, S. O. V. Juarez, E. C. E. Hoeflick, and N. F. R. Snyder. 2008. "Use of clay licks by Maroon-fronted Parrots (*Rhynchopsitta terrisi*) in northern Mexico." *Wilson Journal of Ornithology* 120(1):176–80.

Van Bael, S., and S. Pruett-Jones. 1996. "Exponential growth of Monk Parakeets in the United States." *Wilson Bulletin* 108:584–88.

Van Horik, J., B. Bell, and K. C. Burns. 2007. "Vocal ethology of the North Island kaka (*Nestor meridionalis septentrionalis*)." *New Zealand Journal of Zoology* 34(4):337–45.

van Tuinen, M., and S. B. Hedges. 2004. "The effect of external and internal fossil calibrations on the avian evolutionary timescale." *Journal of Paleontology* 78(1):45–50.

van Tuinen, M., T. Paton, O. Haddrath, and A. Baker. 2003. "'Big bang' for Tertiary birds? A reply." *Trends in Ecology & Evolution* 18(9):442–43.

van Tuinen, M., T. A. Stidham, and E. A. Hadley. 2006. "Tempo and mode of modern bird evolution observed with large-scale taxonomic sampling." *Historical Biology* 18(2): 205–21.

van Woerden, J. T., K. Isler, and C. P. van Schaik. 2009. "Seasonality and brain size: What's the link?" *American Journal of Physical Anthropology*:260–60.

Vaughan, C., N. Nemeth, and L. Marineros. 2006. "Scarlet macaw, *Ara macao*, (Psittaciformes : Psittacidae) diet in Central Pacific Costa Rica." *Revista De Biologia Tropical* 54(3):919–26.

Vehrencamp, Sandra L., A. F. Ritter, M. Keever, and J. W. Bradbury. 2003. "Responses to playback of local vs. distant contact calls in the orange-fronted conure, *Aratinga canicularis*." *Ethology* 109(1):37–54.

Venuto, V., V. Ferraiuolo, L. Bottoni, and R. Massa. 2001. "Distress call in six species of African *Poicephalus* parrots." *Ethology Ecology & Evolution* 13(1):49–68.

Veran, S., and S. R. Beissinger. 2009. "Demographic origins of skewed operational and adult sex ratios: perturbation analyses of two-sex models." *Ecology Letters* 12(2):129–43.

Vicentini, A., and E. A. Fischer. 1999. "Pollination of *Moronobea coccinea* (Clusiaceae) by the Golden-winged Parakeet in the Central Amazon." *Biotropica* 31(4):692–96.

Visalberghi, E., G. Sabbatini, N. Spagnoletti, F. R. D. Andrade, E. Ottoni, P. Izar, and D. Fragaszy. 2008. "Physical properties of palm fruits processed with tools by wild bearded capuchins (*Cebus libidinosus*)." *American Journal of Primatology* 70(9):884–91.

Vogel, R. M., J. R. M. Hosking, C. S. Elphick, D. L. Roberts, and J. M. Reed. 2009. "Goodness of fit of probability distributions for sightings as species approach extinction." *Bulletin of Mathematical Biology* 71(3):701–19.

Völker, O. 1936. "Ueber den gelben Federfarbstoff des Wellensittichs (*Melopsittacus undulatus* (Shaw))." *Journal of Ornithology* 84:618–30.

———. 1937. "Ueber flooreszierende, gelbe Federpigmente dei Papagein, eine neue Klasse von Federbarbstoffen." *Journal of Ornithology* 85:136–46.

———. 1942. "Die gelben und roten Federfarbstoffe der Papageien." *Biologisches Zentralblatt* 62:8–13.

Volodina, E. V., I. A. Volodin, I. V. Isaeva, and C. Unck. 2006. "Biphonation may function to enhance individual recognition in the dhole, *Cuon alpinus*." *Ethology* 112(8): 815–25.

Wada, K., H. Sakaguchi, E. D. Jarvis, and M. Hagiwara. 2004. "Differential expression of glutamate receptors in avian neural pathways for learned vocalization." *Journal of Comparative Neurology* 476(1):44–64.

Walker, J. S., A. J. Cahill, and S. J. Marsden. 2005. "Factors influencing nest-site occupancy and low reproductive output in the critically endangered Yellow-crested Cockatoo Cacatua sulphurea on Sumba, Indonesia." *Bird Conservation International* 15(4):347–59.

Wallace, A. R. 1869. *The Malay Archipelago: the Land of the Orang-Utan, and the Bird of Paradise. A narrative of travel with studies of man and nature.* New York: Harper and Brothers.

Walsh, Julie, Kerry-Jayne Wilson, and Graeme P. Elliott. 2006. "Seasonal changes in home range size and habitat selection by Kakapo (*Strigops habroptilus*) on Maud Island." *Notornis* 53:143–49.

Waltman, J. R., and S. R. Beissinger. 1992. "Breeding behavior of the Green-rumped Parrotlet." *Wilson Bulletin* 104(1):65–84.

Wang, Ning, Edward L. Braun, and Rebecca T. Kimball. 2012. "Testing hypotheses about the sister group of the Passeriformes using an independent 30-locus data set." *Molecular Biology and Evolution* 29(2):737–50. doi:10.1093/molbev/msr230.

Wanker, R. 2002. "Social system and acoustic communication of spectacled parrotlets (*Forpus conspicillatus*): research in captivity and in the wild." In *Bird Research and Breeding*, edited by Claudia Mettke-Hofmann and Udo Gansloßer (83–109). Fürth: Filander.

Wanker, R., J. Apcin, B. Jennerjahn, and B. Waibel. 1998. "Discrimination of different social companions in spectacled parrotlets (*Forpus conspicillatus*): evidence for individual vocal recognition." *Behavioral Ecology and Sociobiology* 43(3):197–202.

Wanker, R., L. C. Bernate, and D. Franck. 1996. "Socialization of spectacled parrotlets *Forpus conspicillatus*: the role of parents, creches and sibling groups in nature." *Journal Für Ornithologie* 137(4):447–61.

Wanker, R., and J. Fischer. 2001. "Intra- and interindividual variation in the contact calls of spectacled parrotlets (*Forpus conspicillatus*)." *Behaviour* 138:709–26.

Wanker, R., Y. Sugama, and S. Prinage. 2005. "Vocal labelling of family members in spectacled parrotlets, *Forpus conspicillatus*." *Animal Behaviour* 70:111–18.Warburton, L. S., and M. R. Perrin. 2005a. "Conservation implications of the drinking habits of Black-cheeked Lovebirds *Agapornis nigrigenis* in Zambia." *Bird Conservation International* 15(4):383–96.

———. 2005b. "Foraging behaviour and feeding ecology of the Black-cheeked Lovebird *Agapornis nigrigenis* in Zambia." *Ostrich* 76(3–4):118–29.

———. 2005c. "Nest-site characteristics and breeding biology of the Black-cheeked Lovebird *Agapornis nigrigenis* in Zambia." *Ostrich* 76(3–4):162–74.

Waring, George H. 1996. *Free-ranging parrot population of Haiku District, Maui, Hawaii*. Puunene, HI: Hawaiian Ecosystems at Risk. www.hear.org/alienspeciesinhawaii/waringreports/parrot.htm.

Warren, Denice K., Dianne K. Patterson, and Irene M. Pepperberg. 1996. "Mechanisms of American english vowel production in a grey parrot (Psittacus erithacus)." *Auk* 113(1):41–58

Wasser, D. E., and P. W. Sherman. 2010. "Avian longevities and their interpretation under evolutionary theories of senescence." *Journal of Zoology* 280(2):103–55.

Waterhouse, D. M. 2008. "Two new parrots (Psittaciformes) from the Lower Eocene Fur Formation of Denmark." *Palaeontology* 51:575–82.

Waterhouse, R. D. 1997. "Some observations on the ecology of the Rainbow Lorikeet *Trichoglossus haematodus* in Oatley, South Sydney." *Corella* 21:17–24.

Watling, Dick. 1995. "Notes on the status of Kuhl's Lorikeet *Vini kuhlii* in the northern Line Islands, Kiribati." *Bird Conservation International* 5:481–89.

Weathers, W.W., and D.F. Caccamise. 1978. "Seasonal acclimatization to temperature in Monk Parakeets." *Oecologia* 35:173–83.

Weaver, C.M. 1987. "A comparison of temperatures recorded in nest chambers excavated in termite mounds by the Golden-Shouldered Parrot." *Emu* 87:57–59.

Webb, D.M., and J. Zhang. 2005. "FoxP2 in song-learning birds and vocal-learning mammals." *Journal of Heredity* 96(3):212–16.

Webster, D. M. S., L. J. Rogers, J. D. Pettigrew, and J. D. Steeves. 1990. "Orgins of descending spinal pathways in prehensile birds: do parrots have a homologue to the corticospinal tract of mammals?" *Brain Behavior and Evolution* 36:216–26.

Weib, B.M., H. Symonds, P. Spong, and F. Ladich. 2007. "Intra- and intergroup vocal behavior in resident killer whales, *Orcinus orca*." *Journal of the Acoustical Society of America* 122(6):3710–16.

Weinert, B. T., and P. S. Timiras. 2003. "Theories of aging." *Journal of Applied Physiology* 95(4):1706–16.

Weiserbs, Anne. 2010. "Invasive species: the case of Belgian Psittacidae. Impacts, risks assessment and range of control measures." *Aves* 47(1):21–35.

Weisman, Ronald G., Milan G. Njegovan, Mitchel T. Williams, Jerome S. Cohen, and Christopher B. Sturdy. 2004. "A behavior analysis of absolute pitch: sex, experience, and species." *Behavioural Processes* 66(3):289–307.

Wendelken, P. W., and R. F. Martin. 1987. "Avian consumption of *Guaiacum sanctum* fruit in the arid interior of Guatemala." *Biotropica* 19(2):116–21.

Wermundsen, T. 1997. "Seasonal change in the diet of the Pacific Parakeet *Aratinga strenua* in Nicaragua." *Ibis* 139(3):566–68.

Westcott, David A., and Andrew Cockburn. 1988. "Flock size and vigilence in parrots." *Australian Journal of Zoology* 36:335–49.

Westfahl, C. 2008. "Estimation of protein requirement for maintenance in adult parrots (*Amazona* spp.) by determining inevitable N losses in excreta." *Journal of Animal Physiology and Animal Nutrition* 92(3):384–89.

Wetmore, A. 1935. "The Thick-billed Parrot in southern Arizona." *Condor* 37:18–21.

White, Nicole E., Matthew J. Phillips, M. Thomas P. Gilbert, Alonzo Alfaro-Nunez, Eske Willerslev, Peter R. Mawson, Peter B. S. Spencer, and Michael Bunce. 2011. "The evolutionary history of cockatoos (Aves: Psittaciformes: Cacatuidae)." *Molecular Phylogenetics and Evolution* 59(3):615–22. doi:10.1016/j.ympev.2011.03.011.

White, S. A., S. E. Fisher, D. H. Geschwind, C. Scharff, and T. E. Holy. 2006. "Singing mice, songbirds, and more: models for FOXP2 function and dysfunction in human speech and language." *Journal of Neuroscience* 26(41):10376–379.

White, T. H., W. Abreu-Gonzalez, M. Toledo-Gonzalez, and P. Torres-Baez. 2005. "From the field: artificial nest cavities for *Amazona* parrots." *Wildlife Society Bulletin* 33(2):756–60.

Whitehead, H., L. Rendell, R. W. Osborne, and B. Wursig. 2004. "Culture and conservation of non-humans with reference to whales and dolphins: review and new directions." *Biological Conservation* 120(3):427–37.

Wiedenfeld, David A., José Morales Molina, and Martín Lezama L. 1999. *Status, management, and trade of psittacines in Nicaragua, 1999*. Managua, Nicaragua: Oficina de CITES Nicargua y Ministerio de Recursos Naturales.

Wiersma, P., A. Munoz-Garcia, A. Walker, and J. B. Williams. 2007. "Tropical birds have a slow pace of life." *Proceedings of the National Academy of Sciences* 104(22):9340–45.

Wilcove, D. 1996. "Is there a cure for the blues? The fates of two rare bird species and nearby humans are subtly linked." *Wildlife Conservation* (March/April): 45–52.

Wild, J. M., H. Reinke, and S. M. Farabaugh. 1997. "Non-thalamic pathway contributes to a whole body map in the brain of the budgerigar." *Brain Research* 755(1):137–41.

Wiley, J. 1981. "The Puerto Rican Amazon (*Amazona vittata*): its decline and the program for its conservation." In *Conservation of New World Parrots*, edited by R. F. Pasquier, 104–33. Washington, DC: Smithsonian Institution Press.

———. 1985a. "Habitat loss and its role in the decline of the Puerto Rican Parrot (*Amazona vitttata*)." *Proceedings of the International Ornithological Congress* 18:1035.

———. 1985b. "The Puerto Rican Parrot and competition for its nest sites." In *Conservation of Island Birds*, edited by P. J. Moore (213–23). Technical Publication Number 3. International Council for Bird Preservation.

———. 1991. "Status and conservation of parrots and parakeets in the Greater Antilles, Bahama Islands, and Cayman Islands." *Bird Conservation International* 1:187–214.

Wilkie, S. E., P. R. Robinson, T. W. Cronin, S. Poopalasundaram, J. K. Bowmaker, and D. M. Hunt. 2000. "Spectral tuning of avian violet- and ultraviolet-sensitive visual pigments." *Biochemistry* 39(27):7895–7901.

Wilkinson, R., and T. R. Birkhead. 1995. "Copulation behaviour in the Vasa parrots *Coracopsis vasa* and *C. nigra*." *Ibis* 137(1):118–19.

Williams, G. R. 1956. "The Kakapo (*Strigops habroptilus*, Gray): a review and re-appraisal of a near-extinct species." *Notornis* 7:29–56.

Williams, Murray, and Don Merton. 2006. "Saving Kakapo: an illustrated history." *Notornis* 53(1):i–viii.

Wilmshurst, J. M., T. F. G. Higham, H. Allen, D. Johns, and C. Phillips. 2004. "Early Maori settlement impacts in northern coastal Taranaki, New Zealand." *New Zealand Journal of Ecology* 28(2):167–79.

Wilson, D. S. 2002. *Darwin's Cathedral: evolution, religion and the nature of society*. Chicago: University of Chicago Press.

Wilson, Deborah J., A. D. Grant, and N. Parker. 2006. "Diet of Kakapo in breeding and non-breeding years on Codfish Island (Whenua Hoa) and Stewart Island." *Notornis* 53:80–89.

Wilson, K. 1993. "Observations of the Kurämoó (*Vini peruviana*) on Aitutaki Island, Cook Islands." *Notornis* 40:71–75.

Wilson, Karen A., Rebecca Field, and Marcia H. Wilson. 1995. "Successful nesting behavior of Puerto Rican Parrots." *Wilson Bulletin* 107(3):518–29.

Wilson, Karen A., Marcia H. Wilson, and Rebecca Field. 1997. "Behavior of Puerto Rican parrots during failed nesting attempts." *Wilson Bulletin* 109(3):490–503.

Wilson, Marcia H., Cameron B. Kepler, Noel F. R. Snyder, Scott R. Derrickson, F. Josh Dein, James W. Wiley, Joseph M. Wunderle, Ariel E. Lugo, David L. Graham, and William D. Toone. 1994. "Puerto Rican Parrots and potential limitations of the metapopulation approach to species conservation." *Conservation Biology* 8(1):114–23.

Wirminghaus, J. O., C. T. Downs, M. R. Perrin, and C. T. Symes. 2001a. "Abundance and activity patterns of the Cape parrot (*Poicephalus robustus*) in two afromontane forests in South Africa." *African Zoology* 36(1):71–77.

———. 2001b. "Breeding biology of the Cape Parrot, *Poicephalus robustus*." *Ostrich* 72(3–4):159–64.

Wirminghaus, J. O., C. T. Downs, C. T. Symes, E. Dempster, and M. R. Perrin. 2000. "Vocalizations and behaviours of the Cape Parrot *Poicephalus robustus* (Psittaciformes: Psittacidae)." *Durban Museum Novitates* 25:12–17.

Wirminghaus, J. O., C. T. Downs, C. T. Symes, and M. R. Perrin. 1999. "Conservation of the Cape parrot *Poicephalus r. robustus* in southern Africa." *South African Journal of Wildlife Research* 29(4):118–29.

———. 2000. "Abundance of the Cape parrot in South Africa." *South African Journal of Wildlife Research* 30(1):43–52.

———. 2001. "Fruiting in two afromontane forests in KwaZulu-Natal, South Africa: the habitat type of the endangered Cape Parrot *Poicephalus robustus*." *South African Journal of Botany* 67(2):325–32.

———. 2002. "Diet of the Cape Parrot, *Poicephalus robustus*, in Afromontane forests in KwaZulu-Natal, South Africa." *Ostrich* 73(1–2):20–25.

Witmer, M. C., and P. J. Van Soest. 1998. "Contrasting digestive strategies of fruit-eating birds." *Functional Ecology* 12(5):728–41.

Wolf, P., A. C. Habich, M. Burkle, and J. Kamphues. 2007. "Basic data on food intake, nutrient digestibility and energy requirements of lorikeets." *Journal of Animal Physiology and Animal Nutrition* 91(5–6):282–88.

Wood, Jamie R. 2006. "Subfossil Kakapo (*Strigops habroptilus*) remains from near Gibraltar Rock, Cromwell Gorge, Central Otago, New Zealand." *Nortornis* 53:191–97.

Wright, T. F. 1996. "Regional dialects in the contact call of a parrot." *Proceedings of the Royal Society of London Series B: Biological Sciences* 263(1372):867–72.

———. 1997. *Vocal communication in the Yellow-naped Amazon (Amazona auropalliata)*. University of California, San Diego.

Wright, T. F., Kathryn A. Cortopassi, Jack W. Bradbury, and Robert J. Dooling. 2003. "Hearing and vocalizations in the orange-fronted conure (*Aratinga canicularis*)." *Journal of Comparative Psychology* 117(1):87–95.

Wright, T. F., and C. R. Dahlin. 2007. "Pair duets in the yellow-naped amazon (*Amazona auropalliata*): phonology and syntax." *Behaviour* 144:207–28.

Wright, T. F., C. R. Dahlin, and A. Salinas-Melgoza. 2008. "Stability and change in vocal dialects of the yellow-naped amazon." *Animal Behaviour* 76:1017–27.

Wright, T. F., J. R. Eberhard, E. A. Hobson, M. L. Avery, and M. A. Russello. 2010. "Behavioral flexibility and species invasions: the adaptive flexibility hypothesis." *Ethology, Ecology & Evolution* 22(4):393–404.

Wright, T. F., A. M. Rodriguez, and R. C. Fleischer. 2005. "Vocal dialects, sex-biased dispersal, and microsatellite population structure in the parrot *Amazona auropalliata*." *Molecular Ecology* 14(4):1197–1205.

Wright, T. F., E. E. Schirtzinger, T. Matsumoto, J. R. Eberhard, G. R. Graves, J. J. Sanchez, S. Capelli, H. Mueller, J. Scharpegge, G. K. Chambers, and R. C. Fleischer. 2008. "A multilocus molecular phylogeny of the parrots (Psittaciformes): support for a Gondwanan origin during the Cretaceous." *Molecular Biology and Evolution* 25(10):2141–56.

Wright, T. F., C. A. Toft, E. Enkerlin-Hoeflich, J. Gonzalez-Elizondo, M. Albornoz, A. Rodriguez-Ferraro, F. Rojas-Suarez, et al. 2001. "Nest poaching in Neotropical parrots." *Conservation Biology* 15(3):710–20.

Wright, T. F., and G. S. Wilkinson. 2001. "Population genetic structure and vocal dialects in an amazon parrot." *Proceedings of the Royal Society Biological Sciences Series B* 268(1467):609–16.

Wyndham, E. 1980a. "Breeding and mortality of budgerigars *Melopsittacus undulatus*." *Emu* 81:240–43.

———. 1980b. "Environment and food of the budgerigar *Melopsittacus undulatus*." *Australian Journal of Ecology* 5:47–61.

———. 1983. "Movements and breeding seasons of the budgerigar." *Emu* 82:276–82.

Yamashita, C. 1987. "Field observations and comments on the Indigo Macaw (*Anodorhynchus leari*), a highly endangered species from northeastern Brazil." *Wilson Bulletin* 99(2):280–82.

———. 1992. "Comportamento de ararazna (*Anodorhynchus hyacinthinus*) Psittacidae, Aves." *Anais de Etologia* 10:158–62.

———. 1997. "*Anodorhynchus* macaws as followers of extinct megafauna: an hypothesis." *Ararajuba* 5:176–82.

Yamashita, C., and Y. Machado de Barros. 1997. "The blue-throated macaw *Ara glaucogularis*: characterization of its distinctive habitats in savannahs of the Beni, Bolvia." *Ararajuba* 5:141–50.

Yamashita, C., and M. de Paula Valle. 1993. "On the linkage between *Anodorhyncus* macaws and palm nuts, and the extinction of the Glaucous Macaw." *Bulletin of the British Ornithologists' Club* 113(1):53–60.

Yamazaki, S., S. Ohi, and R. Satoh. 2000. "Functional auditory pathways in a budgerigar (*Melopsittacus undulatus*)." *International Journal of Psychology* 35(3–4):235.

Young, A.M., E.A. Hobson, L.B. Lackey, and T.F. Wright. 2012. "Survival on the ark: lifespan records for captive parrots." *Animal Conservation* 15:28–53. doi:DOI:10.1111/j.1469–1795.2011.00477.x.

Zampiga, E., H. Hoi, and A. Pilastro. 2004. "Preening, plumage reflectance and female choice in budgerigars." *Ethology Ecology & Evolution* 16(4):339–49.

Zhang, J.X., W. Wei, J.H. Zhang, and W.H. Yang. 2010. "Uropygial gland-secreted alkanols contribute to olfactory sex signals in Budgerigars." *Chemical Senses* 35(5): 375–82.

Zhang, J.Z., D.M. Webb, and O. Podlaha. 2002. "Accelerated protein evolution and origins of human-specific features: FOXP2 as an example." *Genetics* 162(4):1825–35.

Zhou, Z.H. 2004. "The origin and early evolution of birds: discoveries, disputes, and perspectives from fossil evidence." *Naturwissenschaften* 91(10):455–71.

Ziembicki, Mark. 2005. Jewels lost in an ocean: the plight of the *Vini* lorikeets of the South Pacific Islands. http://www.lorikeets.com/vini.htm.

INDEX

Note: Page numbers followed by (f) indicate figures; page numbers followed by (t) indicate tables.

ability *(as in cognition)*, 131 (t), 132
abnormal repetitive behaviors (ARB), 139–143
absence vs. presence *(as in cognition)*, 132–133
adaptive response, 186, 242
adaptive strategy, 200, 203, 214–215
adaptive syndrome, 175, 261
additive mortality, 222, 257, 259
affiliative behaviors, 161
Agapornis canus, 184
Agapornis personatus, 184, 251
Agapornis pullarius, 184, 188 (f)
Agapornis roseicollis, 184–185
Agapornis taranta, 184
Alexandrine Parakeet, 257
alkaloid, 61
Allee Effect on the Carolina Parakeet, 219
altricial, 155, 186, 196, 199, 261
Amazona aestiva, 5 (f), 35, 83, 93 (f), 143 (f),
    187 (f), 194 (f), 250–251, 257
Amazona albifrons, 40 (f), 46, 48, 250
Amazona amazonica, 59, 141, 251
Amazona auropalliata, 96 (t), 109–111, 113, 115,
    123 (f)
Amazona autumnalis, 34 (f), 48
Amazona brasiliensi, 257
Amazona finschi, 187, 250 (f)
Amazona leucocephala bahamensis, 178
Amazona ochrocephala, 48, 135, 187, 251
Amazona rhodocorytha, 148 (f), 257
Amazona ventralis, 251

Amazona versicolor, 100
Amazona vinacea, 257
Amazona violacea, 225 (t), 236
Amazona vittata, 150–151, 158, 159 (f), 187,
    208–209, 222, 237
ammonotely, 73
ancestral parrot, 13, 14, 16, 35, 50, 238
ancestral state reconstruction, 13
androstenedione, 202–203
Anodorhynchus glaucus, 48, 223, 225 (t),
    230–233
Anodorhynchus hyacinthinus, 35, 48, 52 (f)
Anodorhynchus leari, 48, 178, 209–210, 230,
    232 (f)
anthropomorphism, 140, 262–264
Antipodes Island Parakeet, 76, 78–79
Aprosmictus erythopterus, 90 (f)
Ara ararauna, 35, 43, 251
Ara atwoodi, 225 (t), 236
Ara chloroptera, xiv, 43
Ara glaucogularis, 53, 214 (f), 233, 234 (f), 257
Ara guadeloupensis, 225 (t), 236
Ara macao, 4 (f), 43, 48, 181 (f), 196, 210, 257
Ara rubrogenys, 209, 210 (f)
Ara severa, 251
Aratinga acuticaudata, 21, 249
Aratinga aurea, 21
Aratinga auricapillus, 20 (f)
Aratinga finschi, 21, 187
Aratinga holochlora, 21

Aratinga jandaya, 20 (f)
Aratinga labati, 225 (t), 236
Aratinga nana, 21
Aratinga solstitialis, 20 (f), 21, 175
Aratinga weddellii, 211, 250
assortative mating, 159
Austral Parakeet, 54, 67
Australian Ringneck, 43 (f), 111, 123 (f)

Bahamian Parrot, 178
Barnardius zonarius, 43 (f), 111, 123 (f)
basal clade, 15–17
beak morphology, 49, 50–52, 74, 77
biodiversity *(as in population)*, 217, 223
biogeographical patterns, 21
Black-winged Lovebird, 184
Black-winged Parrot, 235
Blue Lorikeet, 32 (f), 184, 237
Blue-and-yellow Macaw, 35, 43, 251
Blue-bellied Parrot, 43
Blue-crowned Parakeet (Conure), 21, 249
Blue-fronted Amazon, 5 (f), 35, 83, 93 (f),
      143 (f), 187 (f), 194 (f), 252, 257
Blue-headed Parrot, 35, 60 (f), 262 (f)
Blue-throated Macaw, 53, 214 (f), 233,
      234 (f), 257
Blue-winged Macaw, 149 (f), 157 (f)
Blue-winged Parrot, 63 (f)
Broad-billed Parrot, 225 (t), 236
brood reduction, 186, 190–196
Brotogeris chiriri, 45, 249–250
Brotogeris chrysopterus, 66
Brotogeris jugularis, 45–46, 48, 150
Brotogeris versicolurus, 45, 150, 249–250
Brown-headed Parrot, 46, 96, 99–100, 196
Brown-necked Parrot, 46, 96 (t)
Brown-throated Conure, 21
Budgerigar, 22, 29, 30, 31 (f), 32, 49–50,
      62–64, 72, 83–84, 86, 90, 91 (f), 92, 97 (f),
      98 (f), 99, 100 (f), 101–122, 123 (t), 143, 147,
      150, 157, 189, 191–194
Burrowing Parrot, 43, 59, 151–152 (f), 153 (f),
      156, 157 (f), 158 (f), 159, 160, 178–184, 187,
      188, 193–200, 251

Cacatua alba, 248
Cacatua galerita, 64, 94, 95 (f), 124 (f), 248
Cacatua goffini, 248

Cacatua leadbeateri, 150–151, 156, 188
Cacatua moluccensis, 137 (f), 138–139, 141,
      212, 257
Cacatua pastinator, 64, 151, 206
Cacatua sanguinea, 197, 206, 208
Cacatua sulphurea, 136 (f), 248
Cacatua tenuirostris, 64, 188, 206, 208
cage stereotypies, 139, 141, 143
call convergence, 107–111
Callocephalon fimbriatum, 24–25, 51
Calyptorhynchus banksii, 51, 54
Calyptorhynchus funereus, 150 (f)
Calyptorhynchus lathami, 53–55 (f), 56–57,
      187
Calyptorhynchus latirostris, 96 (t), 126 (f), 187,
      195–198, 219
camouflage, 235, 239
Canary-winged Parakeet, 45, 150, 249–250
Cape Parrot, 46, 54, 57, 100, 236, 257
captivity (effect on parrots), 138–143
Carolina Parakeet, 19, 20 (f), 21, 218 (f), 219,
      225 (f)
Carolina Parakeet, *history and extinction of*, 18,
      219–223, 225 (t)
category concept *(as in cognition)*, 131–133
cerebellum, 125
cerebrotypes, 127
characters (measurements of traits), 6–8, 15,
      30, 161
Chestnut-fronted Macaw, 251
chronic stress, 140, 143
chronograms, 28–29
clay licks *(colpas)*, 58, 61
Cockatiel, 23, 24 (f), 25, 33, 49, 51, 63–64,
      143, 160 (f), 161, 188, 196, 202, 219,
      247–252
cognition, cognitive learning, 117, 125–140,
      205
cognitive ethology, 128, 130, 132–135, 205
cognitive functions, 119, 139
colonization, 22, 28–29, 35, 151, 153, 157, 182,
      184–185, 191, 252–253
communal nesting, 79, 109, 138, 175, 184
compensatory mortality, 222, 242
conjunctive task *(as in cognition)*, 134
Conuropsis carolinensis, 19, 20 (f), 21, 218 (f),
      219, 225 (f)
convergent evolution, 9, 32

Coracopsis nigra, 44, 166
Coracopsis vasa, 22, 44, 166, 167 (f), 168–174, 187
correlates of longevity, 203, 213–215
Crimson Rosella, 166, 187, 189–190 (f), 191–196
Cyanoliseus patagonus, 43, 59, 151–152 (f), 153 (f), 156, 157 (f), 158 (f), 159, 178–184, 187, 188, 193–200, 252
Cyanopsitta spixii, 35, 43–44, 222–223, 224 (f), 225 (t), 230–231, 236, 265
Cyanoramphus novaezelandiae, 76, 166
Cyanoramphus saisseti, 166, 174
Cyanoramphus unicolor, 76, 78–79

density *(as in population)*, 226, 235, 248
Derbyan Parakeet, 12 (f)
Deroptyus accipitrinus, 35, 43
development, growth stages, 178, 200–203, 208, 210
dietary requirements, 39—79
dimorphic, 156, 165
divergence, 26–29, 49 (f), 77
diversification *(as in evolution)*, 9, 11, 13–14, 25, 27–28, 35, 185, 236
divorce *(as in mating systems)*, 150–151, 153, 158–162
DNA (deoxyribonucleic acid), 7–9, 12 (f), 19, 20 (f), 23–25, 27, 30, 35, 156, 159, 172, 212, 215
Dominican Green-and-yellow Macaw, 225 (t), 236
Dusky-billed Parrotlet, 210
Dusky-headed Parakeet, 211, 250

Eastern Rosella, 95, 96 (t), 203
Eclectus infectus, 225 (t), 236
Eclectus Parrot, 22, 156, 165–166, 170, 171 (f), 172–174, 187–188, 225 (t), 226, 236
Eclectus roratus, 22, 156, 165–166, 170, 171 (f), 172–174, 187–188, 225 (t), 226, 236
egg and chick rearing *(general)*, 186–211
    in Budgerigars, 191–192
    in Burrowing Parrots, 193–195
    in Crimson Rosellas, 189–190
    in Green-rumped Parrotlet, 201, 202
    in Short-billed Black Cockatoos, 195–196, 197–199

Enicognathus ferrugineus, 54, 67
Eolophus roseicapillus, 24–25, 36, 64, 111, 123 (t), 151, 185, 188, 197, 206, 207 (f), 208, 212
Eos bornea, 72
Eos histrio, 32 (f)
Eunymphicus cornutus, 166, 174–175, 187
Eunymphicus uvaeensis, 187, 259
Eupsittula canicularis, 46, 48, 96 (t), 100, 110 (f), 123 (t), 150, 188, 250
Eupsittula pertinax, 21, 257
evolution, 3–36
explosive radiation *(as in evolution)*, 10, 11, 26
extinction, *risk and causes*, 219, 220–223–225 (t), 226–227 (t), 228 (t), 229 (t), 230 (t), 231, 235–237, 240–241, 243, 246, 253
extinction threshold, 222
extinction vortex, 220–222
extrapair copulation (EPC), 149, 152–153, 161

feather picking, plucking, 139, 141, 143
female-only incubation, 188
feral (established species), 246–250
Finsch's Conure, 21, 187
fitness, 34, 124, 135, 148, 165, 177, 192, 199, 205, 211, 213, 215, 240, 254
folivore, 40, 71, 74–75, 163, 238
foraging ecology, 39–80
Forpus conspicillatus, 114–115, 123
Forpus passerinus, 71, 107, 108 (f), 150–151, 153–154 (f), 155, 187, 200–203, 253
Forpus sclateri, 211
fossils, parrot, 5–6, 13–15, 25–29, 33
frugivore, 40, 45, 69–74

Gang-gang Cockatoo, 24–25, 51
genetic markers, 9, 10, 35
Geoffroyus geoffroyi, 34 (f)
geophagy (dirt–eating), 57–62
Glaucous Macaw, 48, 223, 225 (t), 230–231, 233
Glossopsitta concinna, 69
Glossopsitta porphyrocephala, 67
Glossy Black Cockatoo, 53–55 (f), 56–57, 187
Goffin's Cockatoo, 248. *See also:* Tanimbar Corella
Golden Parakeet (Conure), 35, 175
Golden-capped Conure, 20 (f)
Golden-winged Parakeet, 66

granivore, 40–44, 46–49, 53, 57, 62, 64, 67–68, 71–74
Graydidascalus brachyurus, 44
Great-billed Parrot, 23 (f)
Greater Vasa Parrot, 22, 44, 166, 167 (f), 168–174, 187
Green Conure, 21
Green Rosella, 69
Green-rumped Parrotlet, 71, 107, 108 (f), 150–151, 153–154 (f), 155, 187, 200–203, 253
Grey Parrot, 18, 36, 44, 46, 73, 107, 109, 118 (f), 120, 123 (t), 128, 130, 131 (t), 135, 257, 258 (f), 264
Grey-headed Lovebird, 184
Grey-Headed Parrot, 46–47, 96 (t)
Ground Parrot, 62, 64, 156, 178, 235
Guadeloupe Parakeet, 225 (t), 236
Guadeloupe Parrot, 225 (t), 236
Guaruba guarouba, 35, 175, 257

Hapalopsittaca amazonina,
Hapalopsittaca fuertesi, 233
Hapalopsittaca melanotis, 235
Hapalopsittaca pyrrhops, 233
hatching asynchrony, 186, 191, 193, 196, 201
Hawk-headed Parrot, 35, 43
herbivore, 40–42, 46, 48, 52, 54, 58–59, 61–62, 64, 69, 71–72, 74–75, 77, 79, 214, 239
Hispaniolan Parrot, 251
homologous *(as trait)*, 33–35
Hooded Parrot, 62
Horned Parakeet, 166, 175, 187
Hyacinth Macaw, 35, 48, 52 (f)

inbreeding *(as in Kakapo)*, 166, 243
Indigo-winged Parrot, 235
invasion biology, 246–252
Invasive species *(parrots as)*, 246, 250–254

Jandaya Conure, 20 (f)

Kaka, 16, 50, 77, 78 (f), 79, 156, 205
Kakapo, 16, 33, 41, 50, 62, 64, 74–76, 156, 163, 164 (f), 165–167, 178, 200, 205, 206, 208, 213 (f), 214, 222, 235–239 (f), 240–242 (f), 243–246

Kea, 16, 18 (f), 36, 50, 77–79, 111, 112 (f), 113, 123 (t), 156, 200, 204 (f), 205–206 (f), 210
Kuhl's Lorikeet, 244, 245 (f), 246. *See also:* Rimatara Lorikeet

Lathamus discolor, 30, 66, 68–69, 77
Leadbeater's Cockatoo, 150–151, 156, 188. *See also:* Major Mitchell's Cockatoo
Lear's Macaw, 48, 178, 209, 231, 232 (f), 233
Lek Mating System *(as in Kakapo)*, 163–166
Lesser Antillean Macaw, 225 (t), 236
Lesser Vasa Parrot, 44, 166
Lilac-crowned Amazon, 187, 250 (f)
Little Corella, 197, 206, 208
Long-billed Corella, 64, 188, 206, 208
Lophopsittacus bensoni, 225 (t), 236
Lophopsittacus mauritianus, 225 (t), 236

Major Mitchell's Cockatoo, 150–151, 156, 188. *See also:* Leadbeater's Cockatoo
Maroon-bellied Parakeet, 46
Maroon-bellied Parrot, 54
marriage *(as in mating systems)*, 147, 149, 162, 211, 238
mass recession, 194, 197–200
mating systems, 16, 147–149, 155, 162–175, 239
Mauritius Grey Parrot, 225 (t), 236
Melopsittacus undulatus, 22, 29, 30, 31 (f), 32, 49–50, 62–64, 72, 83–84, 86, 90, 91 (f), 92, 97 (f), 98 (f), 100 (f), 101–122, 123 (t), 143, 147, 150, 157, 189, 191–194
mesoscale foragers *(parrots as)*, 79
Meyer's Parrot, 46, 47 (f)
Mitred Parakeet (Conure), 21, 249 (f), 250
molecular phylogenies, 15–16, 23–29
Moluccan Cockatoo, 137 (f), 138–139, 141, 212, 257. *See also:* Salmon–crested Cockatoo
Monk Parakeet, 35, 100, 120, 175, 178, 180, 182, 183 (f), 184–185, 209, 250–254
monogamy in parrots, 16, 148,–149, 151, 153–154, 156 –163, 165–166, 169–170, 174–175, 203, 223
monogamy: genetic, 151, 153, 154, 175
monogamy: social, 16, 147–149, 151
monomorphic, 93, 156–157, 185

morphology, morphological traits, 7–10, 14, 16, 30, 49–50, 56, 74, 77, 125–126, 156, 170

Musk Lorikeet, 69

Myiopsitta monachus, 35, 98, 120, 175, 178, 180, 182, 183 (f), 184–185, 209, 250–254

Nanday Conure, 19–20, 44 (f), 251

Nandayus nenday, 19–20, 44 (f), 251

natural selection, 4, 14, 33–35, 49–50, 92, 122, 124–126, 136, 159, 162, 199, 203, 205, 211, 215, 262, 263

naturalized species, 246, 250

Necropsittacus rodericanus, 225 (t), 236

nectarivore, 32, 40, 62, 64–65, 68–69, 71–73

Neophema chrysogaster, 235, 247

Neophema chrysostoma, 63 (f)

nest construction, 180–185 (f)

Nestor meridionalis, 16, 50, 76, 78 (f), 79, 156, 205

Nestor notabilis, 16, 18 (f), 36, 50, 77–79, 111, 112 (f), 113, 123 (t), 156, 200, 204 (f), 205–206 (f), 210

neuroarchitecture, 121–122, 125–126, 132

New Caledonian Parakeet, 166, 174

Night Parrot, 22, 64, 235

nitrogen, 54, 69, 71–73, 87

nocturnal, 16, 174, 235

nomadic species, 30, 107, 191, 219, 235, 247–248

nullius filius, 152

Nymphicus hollandicus, 23, 24 (f), 25, 33, 49, 51, 63–64, 143, 160 (f), 161, 188, 196, 202, 219, 247–252

object permanence *(as in cognition)*, 133

Oceanic Eclectus Parrot, 223 (t), 234

Ognorhynchus icterotis, 43, 233

Olive-throated Conure, 21

omnivore, 16, 40–41, 75, 77–79

Orange-bellied Parrot, 235, 247

Orange-chinned Parakeet, 45–46, 48, 150

Orange-fronted Parakeet, 46, 48, 96 (t), 100, 110 (f), 123 (t), 150, 188, 250

Orange-winged Parrot, 59, 141, 251

Ouvéa Parakeet, 187, 259

Painted Conure, 35

pair bond, 107, 149, 150, 151, 155–156, 161, 203

Palm Cockatoo, 23–25, 48, 51, 52, 53 (f), 59, 127 (f), 129 (f), 156, 186

Paradise Parrot, 225 (t), 230, 231, 233

Peach-fronted Conure, 21

Pesquet's Parrot, 22, 69, 70 (f), 72–74, 77. *See also:* Vulturine Parrot

Pezoporus occidentalis, 22, 64, 235

Pezoporus wallicus, 62, 64, 156, 178, 235

Pezoporus wallicus flaviventris, 31 (f), 65 (f)

Phylogenetic Tree, 7–8, 19–20, 29–30. *See also:* Tree Of Life 11 (f), 17 (f)

phylogeny (evolutionary history), 6, 7, 9, 11, 13, 15–23, 25, 26–28, 30, 25–26, 50 –51, 62, 64, 66, 74, 88, 122, 184–185, 235

Pionus menstruus, 35, 60 (f), 262 (f)

Platycercus caledonicus, 69

Platycercus elegans, 166, 187, 189–190 (f), 191–196

Platycercus eximius, 95, 96 (t), 203

Poicephalus cryptoxanthus, 46, 96, 99–100, 196

Poicephalus fuscicollis fuscicollis, 46, 96 (t)

Poicephalus fuscicollis suahelicus, 46–47, 96 (t)

Poicephalus meyeri, 46, 47 (f)

Poicephalus robustus, 46, 54, 57, 100, 236, 257

Poicephalus rueppellii, 46

Poicephalus senegalus, 251

polyandry mating system, 166–170, 174

polygynandry, 147, 149, 155–156, 163–170, 174–175, 239

polygyny, 147, 155, 163, 169, 239

population ecology, 217–260

predators, parrots as, 41, 76–78

Primolius maracana, 149 (f), 157 (f)

Probosciger aterrimus, 23–25, 48, 51, 52, 53 (f), 59, 127 (f), 129 (f), 156, 186

prostaglandins, 59, 61

Psephotus dissimilis, 62

Psephotus pulcherrimus, 225 (t), 230, 231, 233

Psittacara erythrogenys, 248, 250, 251

Psittacara leucophthalmus, 211

Psittacara mitratus, 21, 249 (f), 251

Psittacula derbiana, 12 (f)

Psittacula eupatria, 257

Psittacula krameri, 22, 250, 251 (f), 252–254

Psittacus erithacus, 18, 36, 44, 46, 73, 107, 109, 118 (f), 120, 123 (t), 128, 130, 131 (t), 135, 257, 258 (f), 264

Psittrichas fulgidus, 22, 69, 70 (f), 72–74, 77
Puerto Rican Parrot, 150–151, 158, 159 (f), 160, 187, 208–209, 222, 237
Purple-crowned Lorikeet, 67
Purpureicephalus spurius, 62, 85 (f)
Pyrrhura frontalis, 46
Pyrrhura picta, 35

Rainbow Lorikeet, 67 (f), 68–69, 82 (f), 178
rapid radiation *(as in evolution)*, 26, 185
recursive task *(as in cognition)*, 134
Red Lory, 72
Red-and-blue Lorikeet, 32 (f)
Red-and-green Macaw, 43
Red-browed Parrot (Amazon), 148 (f), 257
Red-capped Parrot, 62, 85 (f)
Red-cheeked Parrot, 34 (f)
Red-faced Lovebird, 184, 188 (f). *See also:* Red–headed Lovebird
Red-faced Parrot, 233
Red-fronted Macaw, 209, 210 (f)
Red-fronted Parakeet, 76, 166
Red-headed Lovebird, 184, 188 (f). *See also:* Red–faced Lovebird
Red-lored Parrot (Amazon), 34 (f), 48
Red-masked Parakeet, 248, 250, 252
Red-tailed Black Cockatoo, 51, 53
Red-tailed Parrot, 257
Red-winged Parrot, 90 (f)
Rhynchopsitta pachyrhyncha, 20 (f), 21, 54, 56–58 (f), 100, 187, 222
Rhynchopsitta terrisi, 54
Richard Henry, conservationist, 238, 241, 243
Richard Henry, Kakapo, 213 (f)
Rimatara Lorikeet, 244, 245 (f), 246. *See also:* Kuhl's Lorikeet
Rodrigues Parrot, 225 (t), 236
Rose-ringed Parakeet, 22, 250, 251 (f), 252–254
Rosy-faced Lovebird, 184–185
Rüppell's Parrot, 46
Rusty-faced Parrot, 233

Salmon-crested Cockatoo, 137 (f), 138–139, 141, 212, 257. *See also:* Moluccan Cockatoo
Scaly-breasted Lorikeet, 69

Scarlet Macaw, 4 (f), 43, 48, 181 (f), 196, 210, 257
Senegal Parrot, 251
senescence, 179, 202, 211, 215
sex role reversal mating system, 167, 170–171
sexual size dimorphism *(as in Kakapo)*, 164
shared incubation, 187–188
Short-billed Black Cockatoo, 96 (t), 126 (f), 187, 195–198, 219
Short-tailed Parrot, 44
Sinoto's Lorikeet, 225 (t), 236
social learning, 113, 118, 135, 137, 175
social structure, 106, 110, 111, 254
Spectacled Parrotlet, 114–115, 123
Spix's Macaw, 35, 43–44, 222–223, 224 (f), 225 (t), 230–231, 236, 265
St. Lucia Parrot, 100
stress indicators, 140
Strigops habroptilus, 16, 33, 41, 50, 62, 64, 74–76, 156, 163, 164 (f), 165–167, 178, 200, 205, 206, 208, 212, 213 (f), 214, 222, 235–239 (f), 240–242 (f), 243–246
Sulphur–crested Cockatoo, 64, 94, 95 (f), 124 (f), 248
Sun Parakeet (Conure), 20 (f), 21, 175
Swift Parrot, 30, 66, 68–69, 77
symbolic labels *(as in cognition)*, 130

Tanimbar Corella, 248. *See also:* Goffin's Cockatoo
Tanygnathus megalorynchos, 23 (f)
Thick–billed Parrot, 20 (f), 21, 54, 56–58 (f), 100, 187, 222
trade in wild–caught birds, 138, 178, 217, 247, 255–257, 265
Tree of Life, 11 (f), 17 (f). *See also:* Phylogenetic Tree
Trichoglossus chlorolepidotus, 69
Trichoglossus haematodus, 67 (f), 68–69, 82 (f), 178
Triclaria malachitacea, 43

vicariance, 27–28
Vinaceous Parrot, 257
Vini kuhlii, 244, 245 (f), 246
Vini peruviana, 237
Vini sinotoi, 225 (t), 236
Vini vidivici, 225 (t), 236

Vulturine Parrot, 22, 69, 70 (f), 72–74, 77. *See also:* Pesquet's Parrot

Western Corella, 64, 151, 206
Western Ground Parrot, 31 (f), 65 (f)
White Cockatoo, 248
White-eyed Parakeet, 211
White-fronted Parrot (Amazon), 40 (f), 46, 48
widowed *(as in mating systems)*, 151, 159, 162

Yellow-chevroned Parakeet, 45, 249–250
Yellow-collared Lovebird, 184, 250
Yellow-crested Cockatoo, 136 (f), 248
Yellow-crowned Parrot, 48, 135, 187, 251
Yellow-eared Parrot, 43, 233
Yellow-naped Amazon, 96 (t), 109–111, 113, 115, 123 (t)
Yellow-tailed Black Cockatoo, 150 (f)